GAS ADSORPTION EQUILIBRIA

Experimental Methods and Adsorptive Isotherms

This page intentionally left blank

GAS ADSORPTION EQUILIBRIA

Experimental Methods and Adsorptive Isotherms

Jürgen U. Keller
Reiner Staudt
Universität Siegen
Germany

Springer

ISBN: 978-1-4419-3640-0
eBook ISBN: 978-0-387-23598-1
eBook ISBN: 0-387-23598-1

Created in the United States of America

Visit Springer's eBookstore at: http://ebooks.kluweronline.com
and the Springer Global Website Online at: http://www.springeronline.com

CONTENTS

This page intentionally left blank

PREFACE

This book is intended to present for the first time experimental methods to measure equilibria states of pure and mixed gases being adsorbed on the surface of solid materials. It has been written for engineers and scientists from industry and academia who are interested in adsorption based gas separation processes and/or in using gas adsorption for characterization of the porosity of solid materials.

This book is the result of a fruitful collaboration of a theoretician (JUK) and an experimentalist (RS) over more than twelve years in the field of gas adsorption systems at the Institute of Fluid- and Thermodynamics (IFT) at the University of Siegen, Siegen, Germany. This collaboration resulted in the development of several new methods to measure not only pure gas adsorption, but gas mixture or coadsorption equilibria on inert porous solids. Also several new theoretical results could be achieved leading to new types of so-called adsorption isotherms based on the concepts of molecular association and – phenomenologically speaking – on that of thermodynamic phases of fractal dimension. Naturally, results of international collaboration of the authors over the years (1980-2000) also are included.

Both, traditional and new measurement methods for gas adsorption equilibria are presented in Chaps. 2-6 and elucidated by quite a number of experimental data sets, most of them having been measured in our laboratories. Special emphasis is given to uncertainties of data and pros and cons of all measurement methods are given to the best of our knowledge. Also the basic concepts underlying interpretation of measurements and calculations of adsorbed masses from measurement signals, are discussed in Chap. 1.

In publishing this book the authors hope to contribute to

- the development of effective and reliable methods to measure pure gas and gas mixture adsorption equilibria;
- preventing young (and old) experimenters from doing all the mistakes we have done during our laboratory work*);
- making experimental gas adsorption data measured today in many laboratories all over the world more easily comparable to each other, as methods and procedures should be come more and more similar and possibly also will be standardized (IUPAC) in the years to come.

In view of the complexity of interaction of molecules from fluid, i. e. gaseous or liquid phases with the atoms of the surface of a solid material the authors have put their emphasis on experimental measurement methods approaching especially mixture adsorption phenomena. Of course we are well aware that simulation of adsorption systems based on molecular models is making considerable progress. This especially is promoted by still growing computer capacities and new and powerful software and simulation programs. However, reality is in experiment, not in computer's silica. There only our present knowledge and model of physical-chemical reality can be reflected. Nevertheless, we expect in future a combination of highly selective chosen key experiments and computer simulations to be the most effective way to make progress in the complex field of gas mixture adsorption equilibria and probably also in some neighboring fields like adsorption kinetics. However, all these interesting fields of adsorption science including applications of adsorption phenomena to chemical engineering are not considered here but left to other authors.

In view of space limitations neither all of the experimental details and tricks of the various measurement methods nor all of the analytic arguments of the underlying theories could be presented. If readers do have questions they are cordially invited to approach the authors, namely for the former RS**) for the later JUK**).

*) A true experimenter pursues his goal till everything in the lab is ruined. Often only then he becomes aware that nobody has taken notes of what was done and what has really happened (W. Sibbertsen, 1990).

**) keller@ift.maschinenbau.uni-siegen.de
Staudt@inc.uni-leipzig.de

As we are well aware of the fact that not many readers do have time to read a book like this cover to cover, we always have tried to present the material in nearly self-contained separate chapters. For this reason we also have provided the literature separately for each chapter being aware of the fact that some books and papers on gas adsorption may have been cited more than once.

Acknowledgements

It is now our pleasure to express our grateful thanks to all of our undergraduate and graduate students who have been engaged in project work at IFT in the field of gas adsorption during the years (1984-2004). Among them especially the contributions of F. Dreisbach, N. Iossifova, H. Rave, M. Seelbach and M. Tomalla are highly appreciated.

Thanks for cooperation and discussions at international conferences (FOA, COPS, PBCAST) and at private meetings are due to our colleagues

W. Arlt
G. V. Baron
W. Bongartz
St. Brandani
J. Cyprian
D. D. Do
J. Fritzsche
L. Fuller
U. von Gemmingen
A. Guillot
Ch. Haynes
R. He
K.-D. Henning
U. Hoffmann
M. Jaroniec
J. Kärger
K. Kaneko
H. von Kienle
K. S. Knaebel
K.-F. Krebs
E. Krumm
M. D. Le Van
H.-W. Lösch
F. Metz
P. Monson
A. L. Myers
A. W. Neimark
B. Roehl-Kuhn
J. Rouquerol
W. Rudzinski
D. M. Ruthven
A. Sakoda
M. Sakuth
Sh. Sircar
F. Stoeckli
D. Sunderer
M. Suzuki
O. Talu
M. Thommes
K. Unger
R. T. Yang
H. Yoshida
Li Zhou
W. Zimmermann

Special thanks are given to W. A. Steele, College Park, for reading the manuscript of the book and helping to improve the English wording. Thanks are also due to J. M. Prausnitz, Berkeley and J. A. Clark, Ann Arbor, USA, for reading chapter 3, Gravimetry, of the manuscript and contributing valuable hints and remarks to its contents.

Special tribute is also paid to K. S. W. Sing, Exeter, UK for several stimulating lectures given at IFT during (1992-1998) and also for discussions on fundamental aspects of gas adsorption systems.

Cordial thanks are also given to our colleague and friend Prof. h. c. Erich Robens, Friedrichsdorf and Mainz, for fruitful and interesting discussions over many years on the porosity of solids and also for valuable hints to experimental measurement procedures.

Several people have contributed to realize this monography by processing the manuscript: Mrs U. Schilk did the excellent typing and formatting of the text with never ending patience and Mr M. U. Göbel did the art work, contributing also many ideas to Figures and Diagrams. Both of them are given our sincere thanks for devotion and dedication to this work.

Last not least we would like to express our gratitude to the Publishers, especially to Mrs C. Day and Mrs D. Doherty for providing useful information in layout and styling of the manuscript and for several encouraging e-mails and notes.

Siegen - Weidenau | Leipzig

J. U. Keller | R. Staudt

INTRODUCTION

Abstract This introductory chapter provides some background information of the material to be presented: experimental methods to measure adsorption equilibria of pure and mixed gases on inert porous solids. Applications of gas adsorption processes in science and technology are outlined. An overview of the contents of the book is given. Remarks on subjects, measurement methods and other fields of adsorption science which could not be considered within this monography are mentioned. Hints to respective literature and references are given.

1. INTRODUCTION

Physisorption processes of pure and mixed gases on porous solids are of growing importance in both science and engineering [0.1-0.3]. This is reflected – for example – in a growing number of chemical, petrochemical and biochemical processes including adsorption based separation processes. As most of these processes today still are driven by the respective adsorption equilibria, for design of new or up-scaling of laboratory sized processes, adsorption equilibria data in a broad range of pressure and temperature must be known. These data are decisive for selection of type, size and number of adsorption reactors at given gas feed, product specifications and environmental conditions. As gas adsorption equilibria data up to now cannot be calculated accurately by theoretical or analytical simulation based models, it is necessary to measure them, i. e. to determine them by reliably and accurately performed experiments.

The purpose of this book is to present

a) classical and new experimental methods to measure adsorption equilibria of
pure gases and
gas mixtures
on inert rigid or deformable porous solids, and

b) adsorption isotherms for data correlation allowing to calculate gas-adsorption-equilibria data at other gas concentrations, pressures and temperatures.

These data and correlation functions are needed in simulation programs to develop and check new or better, i. e. smaller, faster and more energy-efficient adsorption based processes for a large variety of engineering, health and environmental purposes, cp. Sect. 2.

In Sect. 3 the measurement methods for gas adsorption equilibria which are presented in this book are outlined. Several other phenomena in gas adsorption systems like the kinetics of the mass exchange process, which could not be considered here are mentioned in brief in Section 4. There also some general information on gas adsorption systems will be given and references for the various fields mentioned will be provided.

2. GAS ADSORPTION PROCESSES IN SEPARATION TECHNOLOGY

The sticking of molecules of gases or liquids to the surface of a solid material is called a**d**sorption. It should not be mixed up with the phenomenon of a**b**sorption where molecules of gases or liquids are dissolved in another liquid or solid material. Adsorption is a surface phenomenon which in principle occurs at any pressure and temperature. Absorption is a bulk or volume phenomenon which may or may not occur at given pressure and temperature. The difference between both effects simply can be demonstrated by the sketch shown below. Here the cake symbolizes the molecule of the gas or liquid. The person represents the solid material. Absorption means eating the cake. Adsorption occurs if the cake is splashed on the persons face.

The interactions of a gas – normally a mixture – with the surface of a solid material can be fairly complex. This is due to the fact that the gas molecules can vary considerably in size, structure and electric properties (dipole and quadrupole moments), and also the surface of the solid may offer different types of sites for adsorption, reflected in both the pore spectrum and the enthalpies of adsorption, cp. Chap. 1, [0.4-0.6]. Hence one has to expect that interactions between adsorbed molecules of different type will be different from their possible interactions in a bulk gas or liquid phase.

Figure 0.1. The difference between absorption and adsorption.
Symbols: Person: solid material, cake: molecule from gas or liquid phase.
Absorption: The cake is eaten by the person.
Adsorption: The cake is splashed on the persons face.

Consequently, concentrations of gas mixtures and mixture adsorbed phases – so-called adsorbates or coadsorbates – will be different. This surface effect of the solid material can be used for several technical processes, the most important of which are:

1. Gas separation processes [0.7-0.10]
2. Drying processes of gases and solid materials [0.11]
3. Cleaning processes of air, water, soil [0.12, 0.13]
4. Adsorption based energetic processes, air conditioning refrigerating processes [0.14, 015]
5. Gas storage processes [0.16]
6. Characterization of porous solid materials [0.6].

As there are many presentations of the above mentioned fields in adsorption science and technology available in literature [0.17-0.19], we here restrict the discussion to mentions of only a few of the most important separation processes, cp. Table 0.1. In it the most important adsorption-based gas separation processes are mentioned and the feed and the products are designated. Possible sorbent materials are not given here but can be found in Chap. 1, Tab. 1.3. Also we have chosen not to provide more information on the processes themselves, for example regeneration procedures of the sorbent materials used, proposals for flowsheets, typical data of pressures, temperatures and energy demands, as those can be found in the respective literature [0.7-0.13] and expert programs for gas separation process design [0.20], [0.21].

The gas separation process itself can be based on one or more of the following physical effects:

a) mixture adsorption equilibria, i. e. one component is much more adsorbed than all the others [0.1, 0.19];
b) adsorption kinetic effects, i. e. one component is diffusing much faster within the adsorption material than all the others [0.20];
c) molecular sieve effects, i. e. steric effects of bulky molecules preventing them from entering a pore (system) [0.21];
d) quantum sieve effects in so-called nanopores. This effect is only of importance for separating hydrogen or deuterium from other gases cp. Chap. 1.

Table 0.1. Adsorption based gas separation processes

Process	**Feed**	**Product(s)**
Air separation	Air, dry, clean	Nitrogen (N_2) Oxygen – Argon mixture
Pressure swing adsorption (PSA) [0.8-0.10, 0.22, 0.25]		Molar concentrations N_2 : 0.1 % < y (O_2) < 20 % O_2 : 6 % < y (N_2) < 10 % Oxygen enriched air to burn inflammable gas and for automotive, medical and leisure purposes. (20 % < y (O_2) < 50 %)
Air conditioning	Air loaded with exhaust gases from industry, traffic, residential heating etc. N_2, O_2, H_2O, CO_2, H_2S, aromatics etc.	Clean air N_2, O_2, (H_2O)

Process	**Feed**	**Product(s)**
Air purification Solvent recovery	Air loaded with a) VOCs (BTX) Benzene Toluol Xylol b) Smells, odors Heavy metal atoms	Clean air (N_2, O_2, Ar) VOCs (Volatile Organic Compounds)
Biogenic gas conditioning	Methane containing gases from coal mines, refuse dumps, sewage plants, land fills, etc.	Town gas with high content of (CH_4, H_2, CO)
Carbon dioxide removal from Source: Direct reduction of iron or (COREX) $FeO + H_2 \rightarrow Fe + H_2O$ $FeO + CO \rightarrow Fe + CO_2$	Blast-furnace gas CO_2, CO, H_2, H_2O	Syngas (H_2, CO)
Drying of air prior to pressurization	Humid air	Dry air for pneumatic systems
Flue-gas purification	Exhaust gases of power stations etc. N_2, O_2, CO_2, SO_2,NO_x, etc. Hg from crematories, Isotopes from nuclear power plants	"Clean flue gases" N_2, O_2, H_2O, CO_2
Hydrogen separation from primary gas mixture	Reforming gas Blast-furnace gas H_2, CO, CO_2, CH_4, H_2O $\leqq$200.000Nm^3/h	Hydrogen rich product gas (Syngas: H_2, CO)
Natural gas Upgrading/enrichment of methane content	Raw gas from well CH_4, CO_2, N_2, H_2S, CO, etc.	Town gas CH_4, H_2, CO
Sewage gas purification	Sewage gas or other biogenic gases CH_4, H_2S, NH_3, CO_2, H_2O, N_2, O_2 silanes etc.	Methane enriched gas to be fed to gas supply system

As new materials with tailored micro- and nanostructures are developed and synthesized in an increasing number of laboratories around the world, it is to be expected that adsorption as a separation technology for gaseous (and liquid) mixtures will be of growing importance and impact to chemical, biochemical and environmental technology as well as to other fields of sciences (medicine, pharmacy) and engineering.

3. EXPERIMENTAL METHODS

Gas adsorption equilibria can be measured by several basically different methods. In this section we are going to outline the classical ones, namely volumetry/manometry and gravimetry as well as some newer ones, oscillometry and impedance spectroscopy. Emphasis is given to the underlying physical principles. Complementary remarks deal with possibilities to measure binary coadsorption equilibria with and without gas phase analysis. Technical details of all the measurement methods are given in the subsequent chapters, Chaps. (2-6). Prior to considering the measurement methods some general remarks on experimental work with gas adsorption systems are in order.

Most important in all kinds of experiments is monitoring of the procedure and of all data. A notebook, either paper based or electronic can be very helpful in this respect. The record of the experiment should include

a) Name and place of the laboratory, the experimenter, date and environmental data like temperature, pressure and humidity of ambient air.

b) Detailed description of the solid material (sorbent) used for adsorption including manufacturer, chemical analysis, purity, form, information on particle size, bulk density, helium atmosphere density etc.

c) Activation or preparation procedure of the sorbent material prior to adsorption of gases on it, i. e. degasification procedure, vacuum treatment, heating and cooling procedure, sampling and storage conditions, All sorbent materials may change their adsorption properties over the years due to internal physico-chemical processes, but also due to uptake of gases and vapors (humidity) from the ambient air. This especially for carbon based sorbent materials should be taken into account.

d) Information concerning the gas to be adsorbed including concentrations of components and possible impurities (humidity).

e) Measurement method including information about calibration of instruments, carrier gases used, duration of experiment, reproducibility of data etc.

f) Data evaluation and correlation, consistency tests, uncertainties of data, discussion of possible systematic uncertainties of measurements [0.26]. Example: gravimetric measurements using microbalances may be influenced by drifts of the base line, i. e. changes in the zero position of the data recording system of about $(\Delta m / m^s \Delta t) \geq 5.10^{-7}\, g/(gh)$. Here (Δm) is the fictitious change of sorbent mass (m^s) over the time of observation (Δt) corresponding to the drift of the balance.

Useful information on measurement methods of standard thermodynamic parameters like temperature, pressure, density of gases etc. can be found in the literature [0.27, 0.28].

Thermal equations of state (EOS) of pure gases and gas mixtures are represented in [0.29-0.30].

The standard method to measure pure gas adsorption equilibria most often used today is the volumetric or manometric method, Chap. 2. Basically it is the mass balance of a certain amount of gas partly adsorbed on the sorbent material. This method can be realized in either open or closed systems, the former ones often using a carrier gas, the adsorption of which normally being neglected. Complemented by a gas analyzer (chromatograph, mass spectrometer) this method also can be used to measure multicomponent or coadsorption equilibria.

Volumetric measurements also can be combined with caloric measurements. A special instrument allowing measurements of this type is presented in Chap. 2, Sect. 5. It does not use thermocouples for temperature measurements but instead a sensor gas, the temperature caused pressure changes of which leading to time dependent signals allowing one finally to determine the (integral and differential) heat of adsorption of the system.

The volumetric method has specific disadvantages discussed in Chap. 2. More accurate and reliable measurements can be performed by weighing the sorbent mass exerted to the gas atmosphere using a very sensitive microbalance, preferently a magnetic suspension balance. This so-called gravimetric method is presented in Chap. 3.

In Chap. 4 a combination of both the volumetric/manometric and the gravimetric method is discussed. For pure gas adsorption systems it does not lead to new information but only is resulting in a consistency relation of the volumetric and the gravimetric data. However, for binary mixtures with non-isomeric components it does allow one to determine coadsorption equilibria without analyzing the sorptive gas phase, i. e. without using either a gas chromatograph or a mass spectrometer. Similar measurement methods result in combining direct gas density measurements using buoyancy effects of sample masses, with either volumetric or gravimetric (or calorimetric) measurements. These methods, namely the densimetric-volumetric or the densimetric-gravimetric method, are discussed in brief in Chap. 4, Sect. 3.5.

In Chap. 5 measurements of gas adsorption by slow rotational oscillations of the sorbent material are discussed. This method uses the inertia of mass to detect changes caused by gas adsorption. Combined with gravimetric or volumetric measurements it allows the measurement of gas solubilities in non-rigid, i. e. swelling sorbent materials as for example polymers.

The dielectric properties of a sorbent material are changed upon gas adsorption. This effect can be used to indirectly determine masses adsorbed by monitoring the (frequency dependent) dielectric permittivity of the sorbent material. After combining these data with either volumetrically or gravimetrically determined calibration data, the mass of the adsorbed gas can be measured at other pressures and temperatures of the gas by dielectric measurements only. Measurements of this type are very useful in industrial applications. For example an increasing content of carbon monoxide in an activated carbon adsorption reactor indicating local heating effects, can be detected immediately and thus help to avoid overheating and even burning.

Adsorption isotherms, i. e. the thermal equations of state of the masses adsorbed are discussed in Chap. 7 for pure and mixture gas adsorption systems as well. This information should allow the reader to choose the isotherm for his data correlation problem properly and also to extend the range of adsorption data known of the system by cautious extrapolation.

As mentioned above multicomponent gas adsorption equilibria can be measured by

a) a method allowing one to measure the total mass adsorbed like volumetry or gravimetry, and to analyze the gas phase to determine the masses of components adsorbed via the mass balance related to this component, or

b) combining two or more of the measurement methods mentioned above for single gas component systems.

Indeed, procedures (b) open various interesting possibilities to measure binary coadsorption equilibria and to design respective instruments for fully automated measurements, cp. for example Chap. 4, Fig. 4.11b. To get an overview, the various possibilities of coadsorption measurements by combining single component methods are sketched in Table 0.2. The numbers in the upper right portion of this matrix scheme indicate the number of components in the gas mixture which can be determined by the respective method. The numbers in the lower left portion of the matrix give the Chapter and Section where more information on this method can be found. Empty fields indicate that we did not do respective measurements and also are not aware of any institution where such measurements might have been realized.

Table 0.2. Measurement methods for single and multicomponent gas adsorption equilibria. Figures in the upper right portion of the matrix indicate number of gas components to which the respective method could be applied. Empty fields imply lack of experimental experience with this method. More information on the various methods can be found in Chaps./Sections indicated in the lower left portion of the matrix.

No.	Method	1	2	3	4	5	6	7
1	Volumetry	–	2	2	1*)	1**)	1+)	N ≥ 2
2	Gravimetry	4	–	2	1*)		1+)	N ≥ 2
3	Densimetry	4	4	–				
4	Oscillometry	5	5		–		1+)	
5	Calorimetry	2				–		
6	Spectroscopy	6	6				–	
7	Chromatography		2	3				–

*) Swelling polymeric sorbent materials.

**) Sensor gas calorimeter.

+) Calibration measurements for adsorption based sensor systems.

In practice combined volumetric-gravimetric measurements have been fairly successful [0.31]. Also densimetric-volumetric and densimetric gravimetric measurements using magnetic suspension balances (2 positions and 3 positions types respectively) can be recommended. If swelling sorbent materials are considered (slow) oscillometric measurements are recommended, Chap. 5. In case of multicomponent sorption systems (N > 2) a gas analyzing system has to be used in any case.

4. WHAT IS NOT CONSIDERED

In view of limitations in time and number of printed pages not all of the experimental methods to measure gas adsorption equilibria, which are discussed in today's literature, could be taken into account. To give reason for this the following remarks should be helpful.

1. Dynamic methods using sorbent material filled columns with open gas flows are not considered. Their main advantages are that apparatus and measurements are fairly simple, cp. Tab. 2.1, [0.32, 0.33] and pressure (p) and Temperature (T) of the sorptive gas can be measured directly. However, the amount of gas adsorbed cannot be determined directly from measured data but models of both equilibria and kinetics of the adsorption column have to be introduced. Naturally, results will depend on the respective models which makes it difficult to compare them to other experimental data. However, this method does have the advantage that by a single, fairly simple experiment information not only on adsorption equilibria but also on the kinetics of the adsorption process may be gained.

2. Spring balances for gravimetric and/or oscillometric measurements are not considered. Uncertainties of measurements often are to large and, in case of oscillations, the flow field of the surrounding gas becomes turbulent, i. e. the friction forces exerted by the gas on the sorbent sample cannot reliably be calculated from the Navier-Stockes-equations, cp. Chap. 5.

3. High frequency oscillating disks or rods using sometimes Piezo-effects are not considered [0.34, 0.35]. Here again the geometry of the oscillating elements is too complicated to allow calculation of the gas flow field surrounding it. Hence, the friction force exerted by the gas on the instrument including the sorbent sample cannot be exactly calculated and hence masses adsorbed cannot be determined.

4. The zero length column (ZLC) method is not considered here [0.36]. This is a fairly new and interesting measurement method allowing in principle to get information of adsorption equilibria as well as of adsorption kinetics, i. e. diffusion coefficients by fairly simple experiments. However, there are still open questions about the actual state of the sorbent material filled with adsorbed gas in the surrounding gas flow. This state often will be kind of transient non-equilibrium state, i. e. a corresponding equilibrium state is not directly observed but data are gained by extrapolation which sometimes may be misleading. Also thermal polarization of the sorbent sample in the gas flow may occur, i. e. small temperature differences between its front

and rear portions, this leading to inhomogeneous sorbate distributions within the sample.

5. Calorimetry as a method to determine gas adsorption equilibria is not considered here in view of excellent presentations of this field in the literature [0.6, 0.37].

Summarizing we want to emphasize that measurement methods (1-4) mentioned above do have certain advantages and hence potential for further development, especially if one is interested in the kinetics of the gas adsorption process. Hence, only future developments in experimental techniques and theory will show which method can best serve the needs of adsorption science and technology.

As a final remark we would like to draw reader's attention to some neighboring fields of gas adsorption on solid surfaces which for obvious reasons could not be encountered here, namely

- adsorption from liquid phases on solid surfaces,
- adsorption from gases or liquids on the surface of another liquid,
- adsorption of biomolecules, especially proteins on synthetic or organic membranes and
- ion exchange phenomena between fluid and solid phases.

Introductory and review articles of all these areas can be found in the literature [0.13, 0.38] which is recommended especially to the young reader's attention.

REFERENCES

[0.1] **Le Van D. M.**
Adsorption Processes and Modeling:
Present and Future, article in Fundamentals of Adsorption 6 (FOA6),
F. Meunier, Ed., p. 19-29, Elsevier, Paris etc., 1998.

[0.2] **Rodriguez-Reinoso F., Mc Enaney B., Rouquerol J., Unger K.**
Characterization of Porous Solids VI, Proceedings of the 6th Int. Symposium on the Characterization of Porous Solids, (COPS-VI), Alicante, May 2002, Studies in Surface Science and Catalysis, Vol. 144 Elsevier, Amsterdam etc., 2002.

[0.3] **Robens E., Krebs K.-F., Meyer K., Unger K. K.**
Standardization of Sorption Measurements and Reference Materials for Dispersed and Porous Solids, Chap. 3 of A. Dabrowski (ed.):"Adsorption and its Application in Industry and Environmental Protection, Vol. 1: Application in Industry,
Studies in surface science and catalysis, Vol. 120, p. 14,
Elsevier, Amsterdam 1999, p. 95-116, ISBN 0-444-50165-7.

[0.4] **Steele W.**
The Interaction of Gases with Solid Surfaces,
Pergamon, New York, 1974.

[0.5] **Schüth F., Sing K. S. W., Weitkamp J.**
Handbook of Porous Solids, Vols 1-5, p. 3141,
Wiley-VCH, Weinheim etc., 2002, ISBN 3-527-30246-8.

[0.6] **Rouquerol F., Rouquerol J., Sing K.S.W.**
Adsorption by Powders and Porous Solids,
Academic Press, San Diego, USA, 1999.

[0.7] **Ruthven D. M.**
Principles of Adsorption and Adsorption Processes,
J. Wiley & Sons, New York etc., 1984.

[0.8] **Yang R. T.**
Gas Separation by Adsorption Processes,
Imperial College Press, London, 1997.

[0.9] **Bathen D., Breitbach M.**
Adsorptionstechnik, VDI-Buch, Springer, Berlin, New York etc., 2001.

[0.10] **Notaro F., Ackley M. W., Smolarek J.**
Recover Industrial Gases Via Adsorption. Here is a look at modern design for gas recovery and air prepurification alike.
Chem. Eng., 106 (1999), p. 104-108.

[0.11] **Sircar S.**
Drying Processes
Article in [0.2], Vol. 4, p. 2533-67.

[0.12] **Meunier F.**
Adsorption for Environment,
article in [0.16], p. 1-12.

[0.13] **Ullmann's**
Encyclopaedia of Industrial Chemistry,
Wiley – VHC, Weinheim, 6th Edition, 2001.

[0.14] **Meunier F.**
Solid Sorption Heat Powered Cycles for Cooling and Heat Pumping Applications,
Applied Thermal Engineering, 18 (1998), p. 715-729.

[0.15] **Vasiliev L. L. et al.**
Solar-Gas Solid Sorption Refrigerator,
Adsorption, 7 (2001), p. 149-161.

[0.16] **Crittenden B., Thomas W. J.**
Adsorption Technology and Design,
Butterworth – Heinemann, Oxford, 1998.

[0.17] **Ray, M. S.**
Adsorptive and Membrane-type Separation:

A Bibliographical Update (1998),
Adsorption Science & Technology, 17 (1999), 205.

A Bibliographical Update (1999),
Adsorption Science & Technology, 18 (2000), 439.

A Bibliographical Update (2000),
Adsorption Science & Technology, 19 (2001),821,

[0.18] **Sircar S.**
Publications on Adsorption Science and Technology,
Adsorption, 6 (2000), p. 359-365.

[0.19] **Kaneko K. (Editor)**
Fundamentals of Adsorption 7,
Proceedings of 7th FoA Conference, Nagasaki, May 2001, IK International, 2002.

[0.20] **Röhm H.-J., Mielke B.**
GPEX: Gas Purification Expert System,
Inst. Thermo- und Fluiddynamik, Verfahrens- und Umwelttechnik,
Ruhr Universität Bochum, Bochum, Germany, Fax +49(0)234-32-14164

[0.21] **gPROMS Process Systems Enterprises Ltd.**
Imperial College, London, UK
http://www.psenterprise.com/gPROMS

[0.22] **Basmadjian D.**
The Little Adsorption Book
CRC Press, Boca Raton, 1996.

[0.23] **Do D. D.**
Adsorption Analysis: Equilibria and Kinetics,
Imperial College Press, London, 1998.

[0.24] **Yang R. T.**
Adsorbents, Fundamentals and Applications,
Wiley-Interscience, Hoboken, New Jersey, 2003.

[0.25] **Ruthven D. M., Farooq D., Knaebel K. S.**
Pressure Swing Adsorption,
VCH-Wiley, Weinheim, New York, 1994.

[0.26] **International Organization for Standardization, ISO**
Guide to the expression of uncertainty in measurement,
2nd Printing, 1995, CH – 1211 Geneva

[0.27] **Goodwin A., Marsh K. N., Wakeham W. A.**
Measurement of the Thermodynamic Properties of Single Phases,
Series Experimental Thermodynamics, 6, Elsevier Science,
Amsterdam etc., 2003, ISBN 0444 50 9313

[0.28] **Eder F. X.**
Arbeitsmethoden der Thermodynamik, Bd. 1: Temperaturmessung,
Bd. 2: Thermische und kalorische Stoffeigenschaften,
Springer, Berlin etc., 1983.

[0.29] **Sengers J. V., Kayser R. F., Peters C. J., White Jr. H. F., Eds.**
Equations of State for Fluids and Fluid Mixtures, Part I
IUPAC Series in Experimental Thermodynamics, Vol. V,
Elsevier, New York etc., 2000.

[0.30] **Reid R. C., Prausnitz J. M., Poling B. E.**
The Properties of Gases and Liquids,
Mc Graw Hill, New York etc., 4th Ed.., 1986.

[0.31] **Schein E., He R., Keller J. U.**
Measurement of adsorption of gas mixtures including inert components on zeolite,
Report of IFT– USI, Thermodynamics Series, 2003.

[0.32] **Keller J. U., Robens E., du Fresne von Hohenesche, C.**
Thermogravimetric and Sorption Measurement Techniques / Instruments,
Proceedings of 6th Int. Symposium on the Characterization of Porous Solids (COPS VI), Alicante, Spain, May 8-11, 2002, Studies in Surface Sciences and Catalysis,
Vol. 144, F. Rodriguez-Reinoso et al., Eds., p. 387-394, Elsevier,
New York etc., 2003.

[0.33] **Kast W.**
Adsorption aus der Gasphase, Ingenieurwissenschaftliche Grundlagen und
technische Verfahren, Verlag Chemie, Weinheim, Germany, 1988.

[0.34] **Ward M. D., Buttry D. A.**
In Situ Interfacial Mass Detection with Piezoelectronic Transducers,
Science, 249 (1990), p. 1000.

[0.35] **Van Y., Bein T.**
Molecular Recognition on Acoustic Wave Devices: Sorption in Chemically
Anchored Zeolite Monolayers,
J. Phys. Chem., 96 (1992), p. 9387.

[0.36] **Brandani St., Ruthven D. M.,**
Ind. Eng. Chem. Research, 42 (2003), p. 1462-1469.

[0.37] **Buelow M., Shen D.**
Comparison of Experimental Techniques for Measuring Isosteric Heats of Adsorption,
Kluwer and University of Pennsylvania, Dordrecht, 2000.

[0.38] **Rehm H.-J., Reed G., Pühler A., Stadler P., Eds.**
Biotechnology, Vols. 1-12, 2nd Completely Revised Edition,
VCH Weinheim etc., 1993.

Chapter 1

BASIC CONCEPTS

Abstract The basic concepts of adsorption phenomena of gases on the surface of solid materials are presented and discussed in brief. Different types of adsorption processes are characterized by their molecular mechanism and energy of adsorption or desorption respectively. Technically important classes of sorbent materials are mentioned and characterized. The concepts of mass and volume of an adsorbed phase are illustrated with regard to the experimental techniques available today for investigation.
List of Symbols. References.

1. INTRODUCTION

In this chapter we will discuss some of the basic concepts which are used to describe adsorption phenomena of pure and mixed gases on the surface of solids. We here prefer a physical point of view, restricted to physisorption phenomena where adsorbed molecules (admolecules) always are preserved and are not subject to chemical reactions or catalysis. Also, we always have industrial applications of physisorption processes in mind, i. e. we prefer simple and phenomenological concepts based on macroscopic experiments often being embedded within the framework of thermodynamics. That is, we prefer to take only those aspects of the molecular situation of an adsorption system into account which have been or at least can be proved experimentally and are not subject to mere speculation.

Adsorption phenomena can be due to several different molecular mechanisms. These are described and characterized in brief by their respective enthalpies in Sect. 2. Several classes of sorbent materials used for different industrial purposes such as separation of gas mixtures, recovery of volatile solvents or energetic purposes like adsorption-based air conditioning systems, are presented in Sect. 3. This section is complemented by an overview of most often used methods to characterize porous sorbent materials given in Sect. 4. The basic concepts of mass and – to a lesser extent – volume of a sorbed phase are discussed in Sect. 5. In Sect. 6 a short overview of the experimental methods used to measure adsorption equilibria and enthalpies of pure and mixed gases in rigid and swelling sorbent materials is given. In view of space limitations we have to restrict the discussion to classical methods like volumetry/manometry and gravimetry. However, introduction to on some new measurement methods like oscillometry and impedance spectroscopy emerging in today's literature also will be given and their pros and cons will be discussed in brief.

2. ADSORPTION PHENOMENA

Molecules of fluid phases (f), i. e. gases, vapors, and liquids, can stick to the surface of solids (s) or other liquid phases (l). This phenomenon is called adsorption. It occurs in principle at any temperature and pressure and for all chemical species known so far [1.1-1.3]. The adsorbed molecules may have their place on the surface of the solid and return to the gaseous phase. This phenomenon is called desorption. Often one can observe dynamic equilibrium between the number of molecules adsorbed and those desorbed in a certain time interval. Such a situation is called adsorption equilibrium. If these molecular flows to and from the surface do not match, we have either an adsorption process or a desorption process [1.4-1.6].

Additionally, in highly porous solids like zeolites and activated carbons there may be internal diffusion processes of the adsorbed molecules (admolecules). These can occur without external exchange of mass, i. e. at constant mass adsorbed, cp. Sects. 4, 5. An example for such a phenomenon is presented in Chap. 6, Fig. 6.29, [1.4, 1.7-1.9].

In Figure 1.1 a schematics of the molecular situation of an adsorption system is presented [1.3, 1.10].

Nomenclature:

Adsorptive: Gas or liquid whose molecules are interacting with the surface atoms of a solid phase.

Adsorbent: Solid phase with external and internal surfaces exposed to the molecules of a gas or liquid phase.

Adsorbate: Set of molecules being adsorbed on the surface of an (often porous) solid material and forming a separate phase in the sense of thermodynamics, cp. Sect. 5.

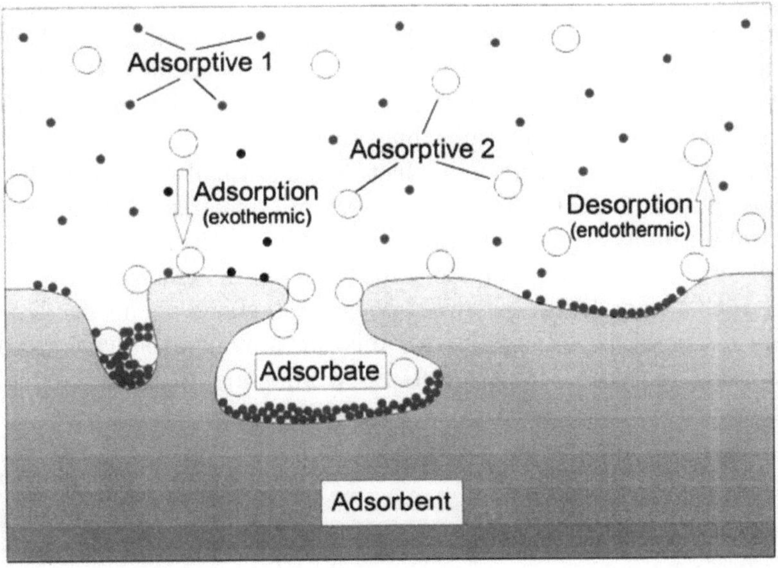

Figure 1.1. Adsorption system consisting of a two component sorptive gas (Adsorptive 1, Adsorptive 2), a sorbed phase or adsorbate (Adsorbate) including also the 2 components (1, 2) showing however different concentrations than in the gas phase due to their different interaction with the sorbent atoms, and a solid sorbent phase (Adsorbent).
Adsorption is the transfer of molecules from the gas or liquid phase to the surface of the solid phase, normally an exothermic process.
Desorption is the transfer of molecules sticking to the surface of the solid back to the gas or liquid phase, normally an endothermic process.

Due to the complexity by which adsorbed molecules (admolecules) can interact with the atoms and molecules of the sorbent and with each other, a variety of phenomena can be expected to occur during an adsorption process. Depending on the strength or interaction energy by which admolecules are bound to sorbent's surface, one can distinguish physisorption, physico-chemical adsorption and chemisorption phenomena [1.11, 1.12].

In physisorption systems admolecules are weakly bound, often by van der Waals- and/or dispersion forces due to induced dipole-dipole interactions. They also can be desorbed reversibly by lowering the sorptive gas pressure or increasing the temperature. Admolecules are basically preserved and not subject to chemical reactions i. e. changes in the character of their electron shells due to interactions with the atoms and/or molecules of the sorbent.

Physico-chemical adsorption phenomena are characterized by weak interactions of admolecules and sorbent atoms or molecules. However, due to catalytic properties of the sorbent surface either dissociations or fairly strong associations between admolecules may occur.

In chemisorption systems admolecules are normally strongly bound to the surface atoms or molecules of the sorbent material and are subject to chemical reactions. They also cannot reversibly be desorbed from the sorbent, but only irreversibly by which the sorbent material is changed.

Though a clear decision between physisorption and chemisorption states is not always possible – examples for this are ammonia or water sorption on hydrophilic zeolites – it seems to be worthwhile to illustrate their basic differences qualitatively in Table 1.1 as follows [1.2, 1.3, 1.11]:

Table 1.1. Basic properties of physisorption and chemisorption systems and phenomena.

Property	Physisorption	Chemisorption
Selectivity of sorptive gases	low	high
Intensity of adsorption increased for sorptive gas pressure (p) and temperature (T)	$p \to \infty$ $T \to 0$	--- T
Kinetics	rapid	slow
Desorption	reversible	irreversible
Ratio (δ) of enthalpy of desorption and evaporation	$\delta = 2\text{-}3$	$\delta > 5$
Structure of sorbate	monolayer multilayer pore fluid	monolayer

We here restrict in what follows to physisorption phenomena. However, some of the examples presented in the subsequent Chapters refer to physico-chemical adsorption systems, cp. Fig. 6.29, and chemisorption systems, cp. Fig. 3.24.

Physisorbates, i. e. adsorbed phases caused by physisorption phenomena can exhibit many different structures reflecting the underlying molecular mechanism. The most often types of these can be described as follows:

1. Monolayer adsorbates
   Sorbent offers many energetically nearly homogenous adsorption sites.
   Sorptive gas pressure (p) is well below saturation pressure at system temperature: $p_s(T) > p$.

2. Multilayer adsorbates
   Sorbent offers many adsorption sites which energetically can be homogenous or inhomogeneous.
   Sorptive gas pressure (p) may approach the saturation pressure at system temperature.*)

3. Pore fluids
   Sorbent has cylindrical or slit like pores which gradually can be filled with admolecules depending on sorptive gas pressure.
   In sorptive gas mixtures molecular sieve or size exclusion effects may occur as only the smaller molecules can enter the pores whereas the bigger ones are prevented from doing this due to their size.

4. Steric sorbates
   Sorbent offers specially formed adsorption sites provided for example by organic molecules impregnating an activated carbon. These anchor molecules only accept (biochemical) admolecules having an appropriate complementary atomic group (key-lock-mechanism).

5. Ionic sorbates
   Sorbent has ions on its surface which can be replaced by other ions diffusing freely in a sorptive liquid or are part of molecules being dissolved in the liquid (ion exchange).

6. Quantum sorbates
   Sorbent has very narrow submicropores – so-called nanotubes – the diameter of which is about 1 nm = 10^{-9} m, [1.17], in which light molecules like hydrogen (H_2), deuterium (D_2) tritium (T_2), or helium (He^3, He^4) can penetrate. As the pore diameter at low temperatures becomes comparable with the de Broglie wave length of the admolecules, quantum sieve effects may occur allowing separation of the different types of admolecules due to different diffusion velocities (quantum resonance effect), [1.13]. **)

Technical adsorbents often are heterogeneous, i. e. include pores of very different size, shape, and connectivity. Hence, the above mentioned types of adsorbates may occur simultaneously or in a mixed way, one of the other

*) For supercritical temperatures ($T > T_c$) the so-called Riedel pressure $p_r(T) = (\partial p / \partial T)_c (T - T_c)$ should be considered, the index "c" indicating the critical state of the sorptive gas.

**) The de Broglie wavelength of H_2 is at T_s (1 atm) = 20 K about 0.5 nm and at 300 K about 0.1 nm.

dominating for different types of sorptive gases and ranges of temperature and pressure.

Industrial adsorption processes normally are cyclic processes in which adsorption and desorption steps of the sorbent material alterate periodically. Often the desorption or regeneration step is crucial and essentially determine the period and the energetic efficiency of the cycle [1.2, 1.14-1.16]. An important quantity to characterize the desorption process is the (molar) enthalpy (ΔH_m) needed to desorb the leading component either of product or waste – of a gas mixture from the sorbent. In Table 1.2 some examples of desorption processes and their industrial applications together with typical values of the molar desorption enthalpy are given. Summarizing it can be stated that in reversible physidesorption processes molar enthalpies of about (10-50) kJ/mol are needed whereas in irreversible chemisorption processes (70-200) kJ/mol are necessary for desorption.*)

Table 1.2. Desorption enthalpies of cyclic industrial adsorption processes for gas/vapor separation and air conditioning systems.

Desorption Process	Industrial Application	ΔH_m/(kJ/mol)
Reduction of sorptive gas pressure, ($p \rightarrow 0$) vacuum technology. Sorbent: Activated carbon, zeolites. Cycle period: 1'-30'	Air separation (N_2, O_2) Recovery of solvents (VOCs) Natural gas cleaning (CH_4, H_2S)	10-30
Reversible thermal regeneration using a) hot air (200 °C < T < 500 °C) or b) water vapor (T < 900 °C). Cycle period: 1 h – 8 h Sorbent: Activated carbon, zeolites	Flue-gas purification Recovery of volatile solvents	20-40 30-60
Thermal regeneration (Ohm heat, Solar)	Air conditioning systems zeolite – water	$40 < r_{H2O} < 60$ r : Heat of condensation
Thermal regeneration including catalytic chemical reactions. Sorbent: Activated carbon Cycle periods: 1d - ...	Recovery of complex organic compounds from the flue gas or waste water	≥ 70

*) It can be shown that adsorption or desorption energies at room temperature (293 K) always have to be larger than 4 kJ/mol which is about 20 % of the energy of hydrogen bonds (20 kJ/mol). This is a consequence of Heisenberg's uncertainly relation ($\Delta E \cdot \Delta t \geq h$) and the fact that atoms of solids at room temperature are oscillating with frequencies about (10^{13} - 10^{14}) Hz.

Adsorbates, i. e. sets of adsorbed molecules, can assume, from a macroscopic point of view, many different types of thermodynamic phases. The structure of these phases mainly depends on the number of molecules adsorbed, on the strength of interactions between admolecules and sorbent atoms and on the geometry of the surface of the sorbent material or, equivalently, on its pore size distribution.

At low adsorption loads admolecules often form type of lattice gas on the surface of the sorbent, i. e. admolecules mainly are isolated from each other but jump around from one adsorption site to another. An example for such a situation is given in Figure 1.2 showing an electron microscope picture of Cu-atoms being adsorbed at 7 K on the (1, 1, 1)-surface of an Ag-crystal as either monomers or dimers. Monomers are performing a random walk type of motion whereas dimers are rotating at their local places.

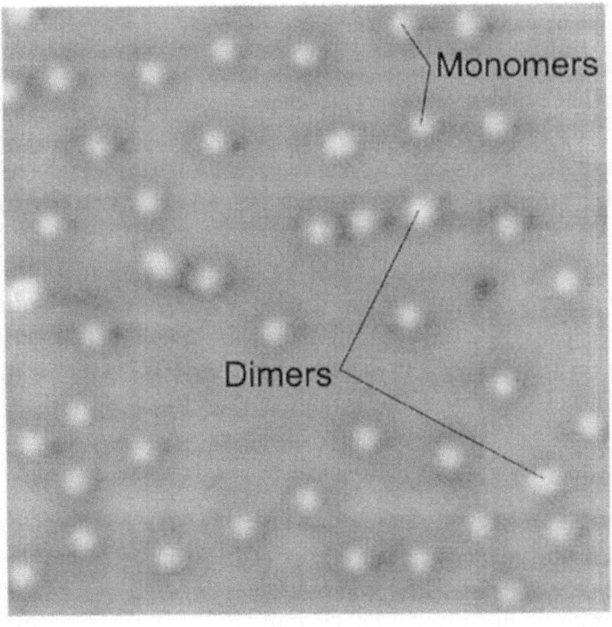

Figure 1.2. Copper atoms adsorbed at T = 7 K on the open (1, 1, 1)-surface of an Ag-crystal forming monomers (small dots) and dimers (larger dots). The electron microscopic photo shows an area of 14 nm x 14 nm. Courtesy is due to K. Morgenstern and K. H. Rieder, Phys. Rev. Lett. 93 (2004) 056102, Free University of Berlin, Berlin, 1998.

Gas like structures of the adsorbed phase also can occur in very narrow pores, i. e. submicropores with diameters about 1 nm [1.17], in which for example He-atoms can enter one by one and form quasi one dimensional string gas structures. Similar arrangements of admolecules also may occur in micropores (diameter: 1 nm < d ≤ 2 nm) [1.18, 1.19], depending of course on the size of the molecules adsorbed [1.20, 1.21].

At increasing amounts of molecules adsorbed, i. e. if the sorptive gas pressure (p) is approaching the saturation pressure ($p_s(T)$), adsorbates often form liquid like structures. These may occur as monolayer patches, liquid films or pore fluids, especially in mesoporous systems, i. e. pores with diameters 2 nm < d < 50 nm, [1.4, 1.11]. The density of these liquid like phases (adliquids) may be higher than that of the bulk liquid phase in saturation state at the same temperature. Also these adliquids can occur at subtriple temperatures and high pressures where in bulk only solid phases exist (surface melting [1.22]). An example for this is given by water which even at temperatures of 77 K seems to form near the surface of mesoporous solids a few, i. e. 2-4 molecular layers which are in a liquid like state, the frozen solid state only starting above these [1.23].

Admolecules diffuse within the pore system of a solid sorbent. This process can last many hours, days, and even weeks, as has been observed for adsorption of helium in activated carbon (NORIT R1), [1.23]. As a consequence it can take the same time till thermodynamic equilibrium between the sorptive gas phase and the adsorbate is realized. In view of practical and industrial needs it is therefore necessary to introduce the concept of "technical equilibrium" defined as a state in which the relative uptake ($\Delta m/m$) of mass at total mass (m) due to adsorption is less than a given value (ε) typically $\varepsilon = 10^{-5}$, within a certain time interval (Δt), typically $\Delta t = 30'$. These data will allow, together with cycle periods (t_c) of an industrial process, one to define characteristic Deborah numbers

$$De = \frac{\Delta m}{\Delta t}\left(\frac{t_c}{m}\right) \qquad 1.1$$

for this process. These allow one to approximately describe the "distance" of an actual state of an adsorption system from its thermodynamic equilibrium state at given temperatures and pressure [1.2, 1.4, 1.6, 1.15, 1.16]. For $De \to 0$, the process is near equilibrium, whereas for $De \geq 100$ essentially non-equilibrium phenomena in heat and mass transfer should be taken into account [1.24]. For more information about the kinetics of sorbate phases the reader is referred to the (still growing) literature [1.4, 1.6, 1.7, 1.8, 1.25].

A few examples of gravimetrically investigated gas adsorption processes will be graphically presented in the next Chapters 3, 4, 6.

In conclusion it can be said that due to the complexity of the interactions of

a) admolecule – surface atoms of sorbent,
b) admolecule – admolecule

and also due to the energetic and geometric heterogeneity of most sorbent surfaces, adsorbates can have many different structures. Hence a unifying model that describes all the different structures probably does not exist. Needless to say that the situation becomes even more complicated if mixture adsorption phenomena or kinetic processes within adsorbed phases have to be considered. Consequently, all models for equilibria and non-equilibria states of single or multicomponent adsorbates presently discussed in the literature have practical limitations which should be taken into account [1.1-1.5]. To investigate the structure and properties of adsorbed phases joint efforts of classical and new experimental measurement methods, refined thermo-analytic models and molecular simulation models are needed. Exchange of results, gained by the various approaches, undoubtedly will lead to progress in understanding the phenomenon of porosity of solids and the design of industrial adsorption processes.

3. SORBENT MATERIALS

Today there are many different types of materials available designed for adsorption of molecules from gases and liquids, i. e. having considerable internal surfaces which are – based on the BET surface*) - mostly in the range $500\ m^2 / g \ldots 2500\ m^2 / g$ [1.2-1.3, 1.26].

For industrial purposes the most important sorbents are activated carbons and zeolites which are available in a great variety of different forms (powder, pellets, fibers, membranes etc.) having different properties [1.27, 1.28]. Besides many other sorbents are investigated and synthesized today being based on either natural materials like peat or coal or natural gas and crude oil leading – for example – finally to porous polymeric materials etc. [1.26].

In view of the abundance of porous materials already available and actual space limitations, here the purpose of this section only can be to provide the

*) This surface is usually determined by the amount of N_2 adsorbed on the surface of the pores at the boiling temperature of N_2 at 1 atm, i. e. 77 K. Experimental data are correlated by use of a special adsorption isotherm due to Brunauer, Emmett and Teller, cp. Chap 7, Sect. 3.4. From this curve the BET surface is determined assuming the N_2 molecules form a monolayer, each molecule occupying an area of $\sigma(N_2) = 0.162\ (nm)^2$. Sometimes instead of N_2, Ar at $T = 87$ K is measured. In this case the respective area $\sigma(Ar) = 0.138\ (nm)^2$ should be used [1.3].

reader with a certain overview of classes of sorbent materials and their most important applications in industrial adsorption processes. For more information special literature of the field should be consulted [1.3, 1.5, 1.8, 1.14-1.16, 1.26-1.28, 1.29-1.31]. For sake of brevity information on materials is given in a table in alphabetical order indicating chemical properties, characteristics of porosity and technical adsorption processes in which the material can be used.

Table 1.3. A selection of sorbent materials used in gas adsorption processes and others [1.2, 1.5, 1.14, 1.26, 1.30, 1.34].

Sorbent material **Chemical elements** **Surface area (BET, m^2/g)** **Porosity*), Pore Volume (V_p / cm^3 / g)**	**Main use** **Technical Processes** **Desorption Enthalpies (kJ/mol)**
Alumina, activated Aluminates $Al(OH)_x$, $3 > x \rightarrow 0$ BET: ca. 300-400 m^2 / g i : e : a = 50 : 30 : 20 $V_p = 0.25$ cm^3 / g	Air separation (N_2, O_2) Gas purification (COS, H_2S, NH_3, PH_3, SO_2 etc.) Dehydration agent (20 – 60) kJ / mol [1.32]
Aluminum Oxide γ-Al_2O_3, porous Aluminum-Oxide-Hydroxide $(AlOOH)_n$, $n \rightarrow \infty$ BET: 100 – 500 m^2 / g	Drying of gases and liquid petrochemicals (40 – 60) kJ / mol [1.33]
Biopolymers Raw materials: starch, cellulose, food products etc. BET: 300 – 800 m^2 / g i : e : a = 10 : 60 : 30	Sorption of water, organic vapors and odorants from ambient air Drying of food products (30 – 60) kJ / mol

*) IUPAC-Classification [1.18, 1.19]:
Micropores: 0 < d < 2 nm … (i) Macropores: 50 nm < d … (a)
Mesopores: 2 nm < d < 50 nm ... (e) d … pore diameter

Sorbent material **Chemical elements** **Surface area (BET, m^2/g)** **Porosity, Pore Volume (V_p / cm^3 / g)**	**Main use** **Technical Processes** **Desorption Enthalpies (kJ/mol)**
Carbon, activated (AC) CO_xH_y ..., $x \rightarrow 0$, $y \rightarrow 0$ BET : 2300 ± 1400 m^2 / g i : e : a ≅ 70 : 20 : 10 $V_p = 0.3 - 1.5$ cm^3 / g	Removal of organic substances (VOCs) from air and technical gases Air separation (N_2, O_2) Purification of natural gas, biogas, syngas and others (H_2S, CO_2,...) Adsorption of biochemicals, herbicides etc.
Carbon molecular sieves Carbon Nanotubes (CNT) Carbon Fibres (ACF) Superactivated carbons BET : 2800 – 3500 m^2 / g $V_p = 1.4 - 2.0$ cm^3 / g	Deodorisation of air Gas masks Waste water purification (organic chemicals, metal ions) Storage material for CH_4, H_2 at 300 K Catalytic properties of ACs should be observed. (10 – 40) kJ/mol [1.3], [1.35]
Most ACs are hydrophobic. Activated carbons with organic adsorbates are flammable.	
Carbon aerogels i : e : a = 30 : 10 : 60 Bulk density: 100 – 300 kg / m^3	Storage of carbon dioxide (CO_2)
Ceramic Sorbents Perovskite (Al_2O_3, SiC) i : e : a = 0 : 30 : 70	Air separation (N_2, O_2) at high temperature (700 K)
Compound / Hybrid materials Example 1: Ordered mesoporous silica (OMS) with block copolymers Example 2: Mixtures of zeolite and activated carbon (Engelhard)	Design materials with specified adsorption isotherm for certain gases

Sorbent material **Chemical elements** **Surface area (BET, m^2/g)** **Porosity, Pore Volume (V_p / cm^3 / g)**	**Main use** **Technical Processes** **Desorption Enthalpies (kJ/mol)**
Impregnated Activated Carbons (IAC) Activated carbons impregnated with organic molecules (often from supercritical CO_2 atmosphere) to provide anchor sites for (big) biochemical molecules	Substrates for retarded release of organic or pharmaceutical substances Removal of heavy metal ions from solutions [1.36]
Impregnated Zeolites Impregnation: $CuCl_2$, LiCl	Removal of CO from syngas Storage of hydrogen (H_2)
Metal Hydrates $NaAlH_4$, $LiAlH_4$, $Mg(AlH_4)_2$, $FeTiH_2$, TiV_2H_4 CeO_2, $Cu_xCe_{1-x}O_4$, $x < 1$	Storage of hydrogen (H_2) by chemisorption Dehydrogenation often slow and only possible at $T > 500$ K
Metal Oxide Powders (MOP) Examples; ZnO, MgO, Al_2O_3 Zinc ferrite $ZnFe_2O_4$	Removal of NO_x from exhaust gas Hot exhaust gas desulfurization [1.41]
Metallo-Organic-Complexes Metallo-Organic-Molecular Sieves (MOMS)	Storage of natural gas Gas separation (paraffins, olefins)
Peat (sphagnum) cellulose, water $\frac{m_{water}}{m_{cellulose}} = 0...10$ BET: $500 - 800\ m^2 / g$ $i : e : a = 5 : 5 : 90$ pH: $5 - 6$ Pore volume variable with water content Adsorption isotherm of water often shows hysteresis (IUPAC Type IV, cp. Chap. 7)	Drying processes Air conditioning Sorption of complex molecules and bacteria from air and other technical gases [1.37, 1.38]
Polymeric sorbents Polymeric resins BET: $300 - 1100\ m^2 / g$ $i : e : a = 10 : 10 : 80$	Waste water treatment (ion exchange) Chromatography Hydrophobicity Regeneration often easily

Sorbent material **Chemical elements** **Surface area (BET, m^2/g)** **Porosity, Pore Volume(V_p / cm^3 / g)**	**Main use** **Technical Processes** **Desorption Enthalpies (kJ/mol)**
Polydimethylsiloxane (PDMS) (porous elastomere)	Removal of VOCs from air
Protein Crystals Biological zeolites	Biocatalytic processes Chromatography
Silica Gel $(SiO_4)^{4-}$, tetrahedral, amorphous BET: 100 – 800 m^2 / g i : e : a = 20 : 70 : 10 V_p = 0.4 – 1.2 cm^3 / g	Drying of air and technical gases and liquids Removal of volatile organic compounds (VOCs) from air Air separation (N_2, O_2) Gas storage (H_2, CH_4) (20 – 40) kJ / mol [1.39]
Mesoporous silica, hexagonally ordered (MCM41)*) BET: 660 m^2 / g V_p = 0.6 cm^3 / g Adsorption isotherms (N_2, 77 K) show hysteresis (IUPAC Type IV, cp. Chap. 7)	
Dealuminated Y-type zeolite (DAY, Degussa)	Hydrophobic zeolite Storage of pharmaceutical substances
Silica aerogels V_p = 2 cm^3 / g	
Silica xerogels V_p = 0.4 cm^3 / g	

As mentioned before detailed information on all of the sorbent materials mentioned above is available in the printed literature and partly also online (www...). The reader is advised to especially consult the encyclopaedic "Handbook of Porous Solids", 5 Vols., edited by F. Schüth et al. (2002) [1.26]. For industrial applications also the monograph by R. Yang [1.27] as well as the representations of the field included in Rouquerol, Rouquerol & Sing [1.3] as well as in Dabrowski [1.5] wholeheartedly can be recommended.

*) Mobil Catalytic Material 41.

Sorbent material **Chemical elements** **Surface area (BET, m^2/g)** **Porosity, Pore Volume (V_p / cm^3 / g)**	**Main use** **Technical Processes** **Desorption Enthalpies (kJ/mol)**
Zeolites Zeolite Molecular Sieves (ZMS) of type A, X, Y $X_i\,(AL_2O_3)_j(SiO_2)_k(H_2O)_l$ X_i = Na, K, Mg, Ca … i, j, k, l = 1, 2, 3, … BET (273K): 500 m^2 / g i : e : a ≅ 80 : 10 : 10 Vp = 0.3 – 0.5 cm^3 / g Technical sorbents: zeolite crystals (0.1 - 10 nm) and binder formed into pellets and beads.	Air separation (N_2, O_2) Gas purification (SO_2, NO_x, CO_2) [1.40]
Zeolite A 8 sodalite cages forming a cube. $[Na_{12}(Al_{12}Si_{12}O_{48})27\,H_2O]_8$ (Si / Al) ≅1 i : e : a = 100 : 0 : 0 V_p ≅ 0.3 cm^3 / g	Drying of natural gas (CH_4, H_2O) (30 – 90) kJ / mol
Zeolite X and Zeolite Y Faujasite, 10 sodalite cages forming a tetrahedral diamond-like structure $(Na_2CaMg)_{29}(Al_{58}Si_{134}O_{384}) \bullet 240\,H_2O$ X: (Si/Al) = 1 – 1.5 Y: (Si/Al) = 1.5 – 3.0 i : e : a = 30 : 70 : 0 V_p = 0.4 cm^3 / g Most zeolites are hydrophobic	Removal of CO_2 from (CH_4, N_2, CO_2)- gas mixtures (40 – 60) kJ / mol Paraffine / olefine separation (C_2H_6 / C_2H_4) Removal of chlorinated hydrocarcons from industrial gases (40 – 80) kJ / mol

4. CHARACTERIZATION OF POROUS SOLIDS

The porosity or pore system of a porous sorbent material can be characterized in many different ways. The most important physical methods and their ranges of application, i. e. the range of diameter of cylindrical pores or width of slit like pores, are listed in Table 1.4, [1.42]. The basic physical principle of all of these methods is the same: one chooses a physical agent, namely a probe or yardstick molecule or atom, or elementary particles like neutrons, protons or photons of different wave length, i. e. energy and momentum. Contacting the sorbent material to this agent, part of it will interact with the material, i. e. either being absorbed or adsorbed or, as in case of radiation, scattered or reflected in various directions and at various energies. This provides a signal, imaging in a certain way the porous material and allowing it to be characterized in comparison to other porous materials.

Table 1.4. Physical methods for characterization of porous sorbent materials [1.42], [1.1 - 1.3], [1.26].

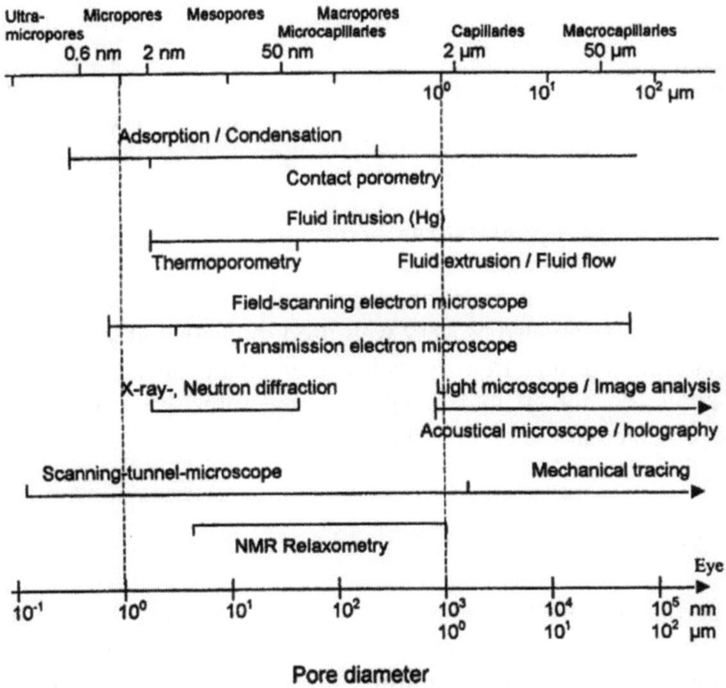

As extensive literature is available for all of the methods mentioned in Table 1.4 [1.1-1.3, 1.26, 1.29, 1.42] we do not consider them here in more detail but restrict the discussion only to those which are the most important for characterization of sorbent materials for industrial purposes:

1. Mercury intrusion porometry
2. Helium measurements
3. Gas adsorption (N_2, Ar, CO_2)

A variety of fairly accurate and reliable instruments for all of these methods is available commercially today, some of which are listed in Table 1.5 below.

Table 1.5. **Instruments for standardized characterization of porous materials by He-density-, mercury intrusion- and N_2-adsorption measurements.**

	Micromeritics	Porotec	Quantachrome
He-Density Measurements (Pycnometers)	Accu Pyc 1330	PYCNOMATIC TPD/R/O 1100 (20 °C – 750 °C)	Pentapycnometer ® He, N_2, SF_6
Mercury Porosimetry	Auto Pore IV $p < 200$ MPa Reference materials available	PASCAL 140/240/440 $p < 400$ MPa $d = 1.8$ nm – 0.3 mm	Pore Master ® $p < 200$ MPa $d = 3.6$ nm – 0.9 mm
Gas Adsorption Instruments (N_2, 77 K)	ASAP 2020 N_2, Ar, CH_4, SF_6	Sorptomatic 1990 fully automated N_2, Ar, SF_6	Autosorb-1Series N_2, Ar, SF_6
www.	micromeritics.com	Porotec.de	quantachrome.com

Additionally it should be mentioned that calorimetric measurements in gas adsorption systems can also be very well used to characterize the porosity of a sorbent material. As this field is well presented in the literature [1.3, 1.29, 1.61, 1.62] we do not go into details here but refer the reader to the (few) examples presented at the end of Chap. 2.

4.1 Mercury Intrusion Porosimetry

The porous material is immersed in a non-wetting liquid, preferably mercury (Hg). Increasing the pressure in the liquid will cause it to penetrate into the pores of the solid until equilibrium against the surface tension (σ) in the smaller and smaller pores is attained. The respective mechanical equilibrium condition leads to the so-called Washburn equation for the limiting pore radius (r) into which mercury at pressure (p) can penetrate [1.1, 1.2, 1.43, 1.44]:

$$r = -2\sigma(p,T)\cos\theta/p \qquad 1.2$$

Here $\sigma = \sigma$ (p, T) is the surface tension of mercury which on principle is a pressure and temperature dependent quantity*), $\theta = 140°$ is the contact angle between the mercury meniscus and the pore wall. Though the exact value of this parameter normally is unknown, a practical value of 140° turned out to lead to physically reasonable results in many cases and hence is recommended for practical use [1.44].

The total volume of mercury $V_{Hg}(p)$ penetrating the pores of the material at pressure p leads via equation (1.2) to the integral volume $V_p(r)$ of all pores with radii (ρ) larger than $r < \rho < \infty$, i. e. $V_p(r) = V_{Hg}(p)$. By differentiation to the pore radius r this yields the differential pore size distribution of the material. This method is valuable to investigate macro- and mesopores (IUPAC, cp. Sect. 3), but not for micropores, i. e. it is limited to pore radii r > 1 nm.

An example for the integral and differential pore distribution of an activated carbon (Norit R1 Extra) determined at BAM**), Berlin with a commercial mercury porosimeter is shown in Figures 1.3, 1.4.

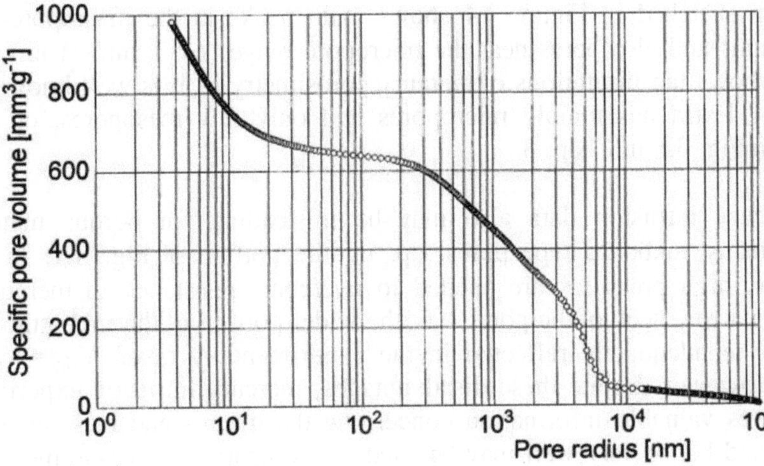

Figure 1.3. Cumulative or integral volume of pores $v_p(r)$ per unit mass of sorbent as function of the pore radius (r) for activated carbon Norit R1 Extra at 298 K (Hg-intrusion) [1.36].

*) The dependence on pressure normally can be neglected up to $p \leq 100$ MPa. At T = 298 K we have $\sigma = 0.48$ N/m.

**) BAM = Bundesanstalt für Materialforschung und –prüfung (Federal Institute for Materials Research and Testing), D-12200 Berlin, Germany.

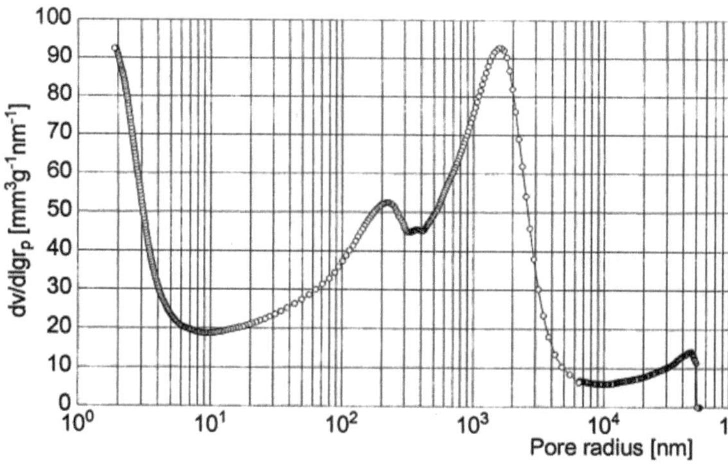

Figure 1.4. Differential pore volume distribution ($dv_p(r)/dr$) per unit mass of activated carbon Norit R1 Extra at 298 K (Hg-intrusion) [1.36].

The curve in Figure 1.3 represents the intrusion process of the mercury measured at increasing pressures. As often can be observed, this process is irreversible, i. e. at decreasing pressures some of the mercury is kept in the smaller pores, this leading to permanent changes, namely mercury inclusions in the sample investigated. This is a disadvantage of this method. The pore spectrum sketched in Figure 1.4 shows main peaks in the macropore-range ($r > 25$ nm) and also pores near the micropore range ($r < 1$ nm) at left. This clearly shows the limitations of mercury porosimetry as it is well known that Norit R1 Extra has mainly micropores and only few mesopores, cp. pore spectrum Fig. 3.8 in Chap. 3.

Mercury intrusion data also may be misleading for porous materials having many inkbottle type pores, cp. middle portion of Fig. 1.1. In such situations high pressures are needed to overcome resistance of mercury to pass the narrow neck of the pore, i. e. the wider portion of the inkbottle pores will not be adequately reflected in the experimentally taken $V_{Hg} = V_{Hg}(p)$ curve. However, despite these disadvantages, mercury intrusion experiments often gives valuable information concerning the macro- and mesopores of a sorbent and hence very well may be used for comparative measurements and quality tests of sorbent samples.

4.2 Helium Measurements

The material is, after preparation and possible activation, i. e. degassing at elevated temperatures, put in a vessel of known volume (V_{AC}) which after evacuation is filled with a known mass (m_{He}) of Helium gas. Assuming helium neither to be adsorbed nor absorbed, the volume of the sorbent material (mass: m^s) which is impenetrable to the helium molecules, i. e. the so-called helium volume (V^s_{He}) of the material, can be calculated from a mass balance of this gas expansion experiment.

$$m_{He} = \rho_{He}\left(V_{AC} - V^s_{He}\right), \qquad 1.3$$

viz.

$$V^s_{He} = V_{AC} - m_{He}/\rho_{He} \qquad 1.4$$

Here $\rho_{He} = \rho_{He}(p,T)$ is the density of the helium gas which can be determined from measured data of pressure (p) and temperature (T) normally by using the ideal gas equation of state (EOS) or a standardized real gas EOS [1.45].

This procedure is simple and effective and several types of He-gas pycnometers are commercially available, cp. Tab. 1.5.

However, it does have certain disadvantages which may lead to serious experimental errors in gas adsorption measurements [1.46-1.49]. This will be demonstrated by a set of helium gas expansion experiments performed at the Institute of Fluid- and Thermodynamics (IFT) at the University of Siegen, Siegen during 1994 – 2002 as follows.

a) Gas expansion experiments were performed in a commercial gas pycnometer to determine the known volume of a standard calibration cylinder (stainless steel, electro-polished surface). Gases used were He (5.0), N_2 (5.0), CH_4 (5.5)*) for pressures up to 0.14 MPa at T = 298 K. Results are depicted in Figure 1.5. As can be seen, the volume (51.10 cm^3) of the standardization sample was best reproduced by the N_2-measurement at p = 110 kPa. Neither He-measurements nor CH_4-measurements lead to accurate results but show deviations of 1 ‰ and more, uncertainties of data being about the size of the graphic symbols. The overshooting of the true value of the sample volume by He-measurements may well be due to ab- and adsorption of small amounts of helium in/on the walls of the adsorption chamber (stainless steel) and the surface of the sample. The undershooting of the sample volume by CH_4-measurements could not simply be explained. Real gas effects can be excluded. However, sensitivity of the pressure transducer system to CH_4 may have caused a systematic uncertainty of data [1.48].

*) The numbers indicate the purity / quality of the gas. Example: 5.5 = 99.9995 %Vol.

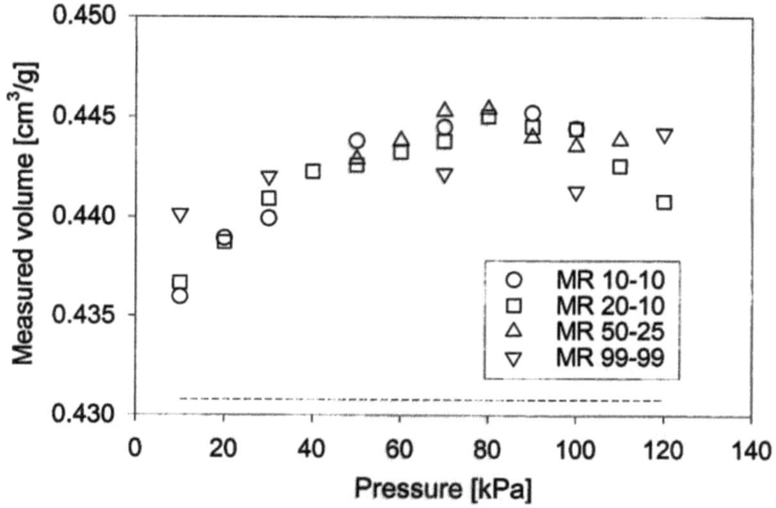

Figure 1.5. Determination of the (known) volume (51.10 cm^3) of a calibration cylinder by gas expansion experiments using gases He (5.0), N_2 (5.0) and CH_4 (5.5) at 298 K in a commercial gas pycnometer [1.48].

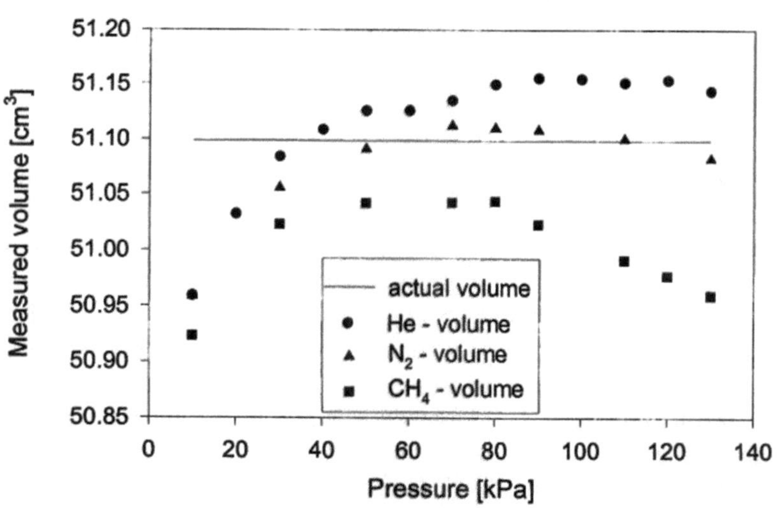

Figure 1.6. Determination of the volume of activated carbon Norit R1 Extra by helium expansion measurements at 298 K in a commercial gas pycnometer (Micromerites, Accu Pyc 1330).

b) The volume of a sample of (highly microporous) activated carbon (AC) Norit R1 Extra has been determined in a commercial helium pycnometer (Micromeritics, Accu Pyc 1330) at 293 K for various pressures up to 0.12 MPa. Equilibration times were 10 – 60 s. Results are sketched in

Figure 1.6. As can be seen, resulting values of the specific volume of the AC as "seen" by the He-molecules depend on the gas pressure. Differences make up to 2 % in the pressure range considered. Several measurement procedures have been used. Data presented in the figure are statistical averages of 10, 10, 25, 99 separate measurements [1.48].

c) The volume of activated carbon fibers (ACF) impenetrable to helium gas has been measured at 293 K with a commercial gas pycnometer for a series of increasing and then again decreasing gas pressures. Data are shown in Figure 1.7. As can be seen data for the specific volume of the ACF corresponding to the increasing pressure- and hence possibly adsorption-branch are always higher than those for the decreasing pressure- or desorption-branch, this indicating hysteresis behavior of the helium possibly adsorbed. However, equilibration times for single gas pressure measurements were only about 15 minutes. Hence we assume that data shown actually do not correspond to thermodynamic equilibria states, cp. also Figure 1.8, but to transient non-equilibrium states. Also it may be that the increase in volume during the upper adsorption related data branch is due to the desorption of preadsorbed gases in the ACF due to helium intrusion.

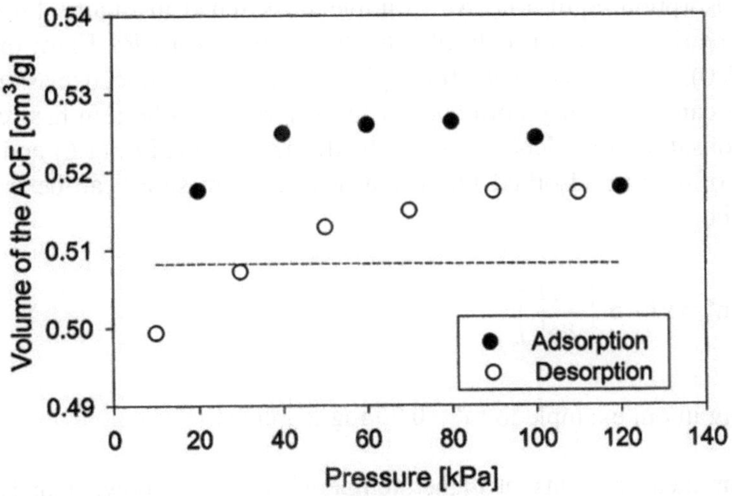

Figure 1.7. Measurements of the (specific) He-volume of activated carbon fibres (ACF) at 298 K for a set of pressure step up experiments (upper data •) and respective step down pressures (lower data ○). Measurements were performed in a commercial gas pycnometer using He (5.0) [1.47, 1.48].

d) The He-volume of a sorbent material also can be measured gravimetrically using either a two-armed beam balance (Cahn, C. I. Electronics, Setaram) or a single-armed suspension balance (Rubotherm GmbH, Bochum), cp. Chap. 3. The sorbent material is placed in a bowl at the balance. Then the balance- or adsorption-vessel is evacuated and after this filled with gas at certain pressure (p) and temperature (T). It can be shown, Chap. 3, that after certain calibration measurements finally the so-called reduced mass

$$\Omega = m^a - \rho^f V^{as} \quad 1.5$$

can be determined from the various balance readings. Here (m^a) is the mass of gas adsorbed on the sorbent of mass (m^s), $\rho^f = \rho^f(p, T)$ is the density of the sorptive gas and V^{as} is the volume of the sorbent material (s) filled with sorbate (a). If Ω for $p \rightarrow \infty$ is a linear function of the sorptive gas density (ρ^f), V^{as} can be determined from (1.5) by differentiation

$$p \rightarrow \infty \qquad V^{as} = -\left(\frac{\partial \Omega}{\partial \rho^f}\right)_{T,\infty} = -\left(\frac{\partial \Omega}{\partial p}\right)_{T,\infty} R_{He} T \quad 1.6$$

Once V^{as} is known, $m^a = m^a(p, T, m^s)$ can be calculated from Eq. (1.5). Such a situation is very common for gravimetric measurements of helium gas adsorption equilibria. An example is sketched in Figure 1.8 showing gravimetric adsorption data of activated carbon Norit R1 Extra exerted to He (5.0). As can be seen, for high pressures the reduced mass data (•) easily can be linearly correlated, i. e. adsorption of helium has reached a state of saturation. Hence, V_{He} can be determined via Eq. (1.6) and also the mass of helium adsorbed initially at low gas pressures can be calculated from Eq. (1.5) as:

$$m^a = \Omega - \rho^f \left(\frac{\partial \Omega}{\partial \rho^f}\right)_{T,\infty} \quad 1.7$$

leading in our example to $m^a = 0.25$ mg/g = const.

Similar measurements at higher temperatures have shown that helium is adsorbed in ACs to a certain extent even at temperatures about 500 K. The amount of helium adsorbed increases considerably at low temperatures ($T \leq 4$ K). This phenomenon has been investigated thoroughly by K. Kaneko and co-workers, Chiba University, Chiba, Japan, who also have shown that helium (diameter: 2 Å = 0.2 nm) even at very low pressures is adsorbed in so-called supramicropores (diameter $2r < 1$ nm). These authors also could demonstrate that helium in the adsorbed phase in submicropores has a density $\rho^a_{He} = 0.20\,g/cm^3$ which is about twice its value in the corresponding bulk liquid state at the same temperature [1.17, 1.20].

Gravimetric measurements of helium adsorption are very sensitive to the actual state of the sorbent material, i. e. to whether it has been activated or not. Also presorbed gases can change the helium adsorption isotherm considerably. An example for this is given in Figure 1.9 showing data of two samples of molecular sieves (MS5A) one of which having been prepared by evacuation of the adsorption chamber at ambient temperature (293 K), the other having been activated in vacuum at elevated temperature (573 K) for several hours [1.48]. In both cases the experimental data (♦, ■) of the reduced masses (Ω), cp. Eq. (1.6), for higher pressures ($p \rightarrow \infty$) can be linearly correlated as functions of either the gas pressure (p) or the density ($\rho = p/R_{He}T$), and hence indicate constant mass of helium adsorbed (m^a) in the initial period of the experiment, i. e. at very low gas pressures, cp. Eq. (1.7). The respective data are shown as symbols (◊, □) indicating for $p \rightarrow \infty$ a saturated state of adsorption ($m^a_\infty = \text{const}$).

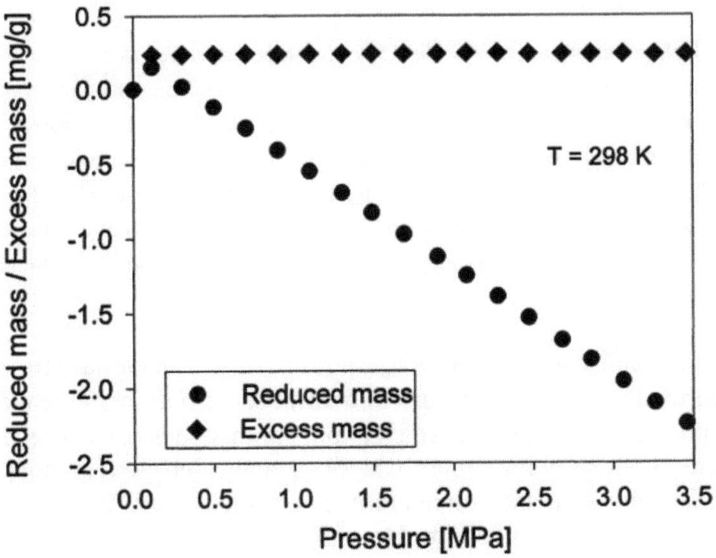

Figure 1.8. Adsorption isotherm of helium on activated carbon (AC) Norit R1 Extra at 293 K. Full circles present data of the reduced mass (Ω) defined by Eq. (1.5). As these can be correlated linearly as function of the helium gas density $\rho^f = p/R_{He}T$, the volume of the AC impenetrable to the helium molecules can be calculated from Eq. (1.6). From Eq. (1.5) the mass of helium initially adsorbed at low pressures can be calculated (via Eq. (1.7) leading to $m^a = 0.25$ mg/g = const.

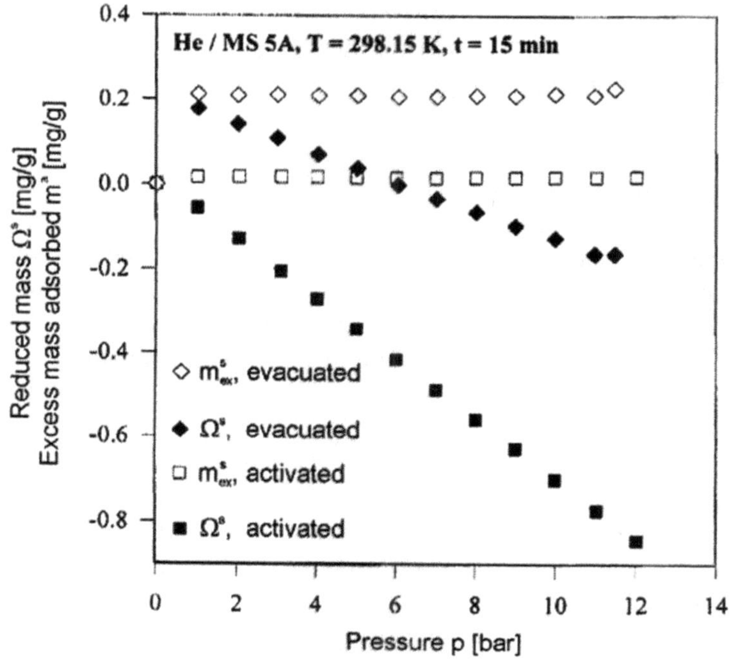

Figure 1.9. Helium adsorption at T = 298.15 K on molecular sieves (MS 5A) with different activation procedures. Full symbols (♦, ■) present the measured values of the reduced masses (**Ω**), cp. Eq. (1.6). Open symbols (◊, □) are masses of helium adsorbed on the samples calculated by Eq. (1.7). Data indicate saturated states of adsorption of helium at increasing pressures for both samples. However, the (specific) volumes (V^{as}) of both samples seen by the helium molecules and calculated by Eq. (1.6) from steepnesses of the linearly correlated **Ω-data** are quite different [1.48].

e) Helium molecules seem to penetrate porous sorbent materials for ever and ever, thermodynamic equilibrium in a strict sense only being realized after month, possibly years [1.48]. An example for this phenomenon is presented in Figure 1.10 showing microbalance signals detected at constant gas density, i. e. pressure and temperature for a sample of AC Norit R1 Extra of $m^s = 55.06$ g exposed to helium (He (5.0)) at T = 298.17 K, p = 0.109 MPa for nearly 60 hours. As can be seen, the reduced mass of the sample increases at nearly constant rate which, as the gas density was exactly kept constant only can be explained by an uptake of helium gas by the sorbent sample. Similar experiments have been performed at IFT for nearly two months showing basically the same result, i. e. not reaching equilibrium against exchange of mass within this period [1.46, 1.50].

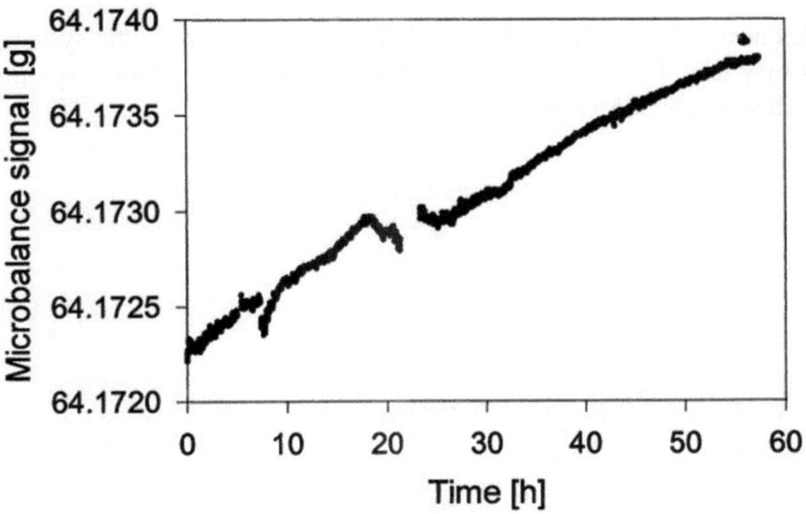

Figure 1.10. Adsorption process of helium (He(5.0)) on AC Norit R1 Extra at T = 298.17 K during 58 hours measured gravimetrically (magnetic suspension balance, Rubotherm). The interruption of measurement data at about 24 h is due to limitations in data storage capacity, i. e. an overflow of data, which made a change of the data storage device necessary.

f) Gravimetric helium adsorption experiments also have been performed at gas adsorption systems in equilibrium states at elevated sorptive gas pressures, i. e. helium at low pressures ($p < 0.2$ MPa) was isothermally added to the sorptive gas atmosphere [1.48]. An example for measurements of this type is given in Figure 1.11. Activated carbon Norit R1 Extra is exposed at T = 298.15K to nitrogen (N_2, 5.0) as the sorptive gas at different pressures $p^f = 0, 5, 7, 10, 13$ MPa. Equilibration times chosen were always 15 minutes. After this helium was added isothermally and the reduced mass (Ω), Eq. (1.6) referring to the helium atmosphere only was measured. The reduced mass (Ω_{N2}) of the sorbent sample in the N_2-atmosphere always was chosen as reference state, i. e. the scaling of the Ω-data refers to $\Omega_{N2} = 0$. The data at $p^f = 7, 10, 13$ MPa indicate that helium at these pressures is not adsorbed additionally to N_2, but that the volumes of the sorbent sample ($V^{as} \simeq -(\partial\Omega/\partial p)_{T,\infty}$) seen by the helium molecules increases with increasing sorptive gas pressure (p^f). The Ω-data referring to $p^f = 5$ MPa are quite unusual as they hint at considerable gas adsorption. Indeed this may occur immediately after helium has been applied to the system but not been distributed in the system to equalize gas concentrations. In such a situation helium acts like a stamp increasing simply the nitrogen pressure at the location of the sorbent which will lead to an increase in the amount of nitrogen adsorbed. Of course this effect fades away after a while when the helium is spread out in the system, but this may take 1-2 hours, i. e. exceed observation times of this experiments which were about 15 minutes [1.48].

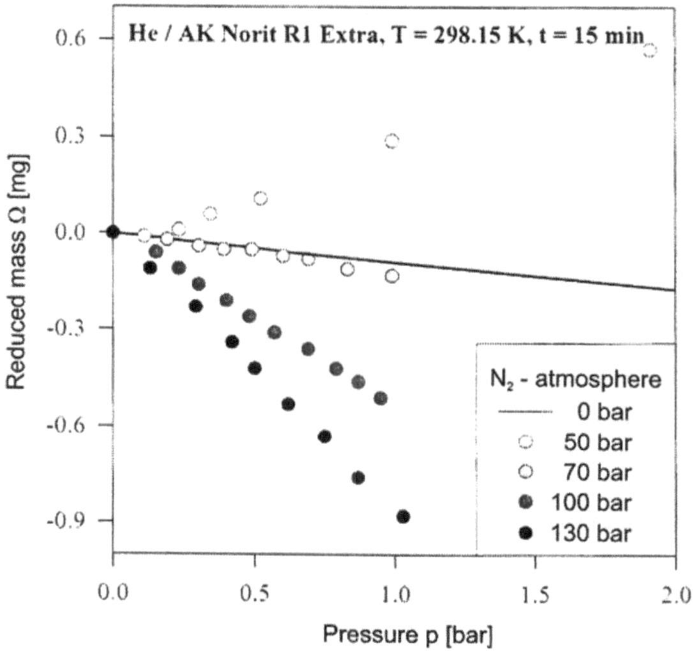

Figure 1.11. Adsorption of helium added to activated carbon Norit R1 Extra in a nitrogen atmosphere of $p^f = 0, 5, 7, 10, 13$ MPa at T = 298.15 K. The reduced mass (Ω), Eq. (1.6), referring to helium gas pressure. Data for p = 7, 10, 13 MPa indicate that helium is not adsorbed additionally to nitrogen, but that the volume (V^{as}) of the sorbent sample (s) loaded with increasing amounts of nitrogen (a) as seen by the helium molecules is also increasing monotonously. The p = 5 MPa-data show adsorption. But this simply may be adsorption of N_2 due to poor mixing of the He- and the N_2-gas immediately after adding helium to the system [1.48].

Summarizing our experiences with helium adsorption experiments it is save to state:

1. Helium is adsorbed or sorbed in all porous materials even at low pressures and high temperatures. The adsorption process is very quick, i. e. a preliminary type of saturation is reached within few seconds, sometimes minutes.

2. If preliminary saturation is reached, increase of the helium gas pressure does not lead to gravimetrically measurable increase of adsorption for several hours.

3. However, long time adsorption experiments lasting days and months show a slow but steady increase of helium being sorbed in the material. This also was observed for nonporous sorbent materials like dense polymers.

4. Helium adsorption experiments at gas adsorption systems in equilibria states show that the volume (V^{as}) of the combined sorbent/sorbate phase is not constant but depends, i. e. increases with the amount of gas adsorbed, cp. next Section 5 of this chapter and also Chaps. 2-4.

Consequently, the helium volume of a sorbent material can be considered as one of its characteristic quantities describing that volume of the material impenetrable for helium molecules on a short time scale of minutes, may be hours. It seems to be kind of lower boundary to the sorbent volumes impenetrable to other, bigger molecules.

Helium volume measurements are also very sensitive to the state of activation of the sorbent material and to preadsorption of – for example – other gases or water vapor. Hence measurements of this type may be used on an industrial level for quality control.

4.3 Gas Adsorption (N_2, Ar, CO_2)

Similarly to the helium adsorption experiments described in Sect. 4.2, adsorption experiments with other gases, preferably nitrogen, argon or carbon dioxide at high purity (> 4.0), can be used to characterize the pore system of a porous sorbent material. Early experiments have been performed by I. Langmuir, W. Ostwald and others prior to World War I. Meanwhile gas adsorption either by volumetric/manometric or by the gravimetric measurement method, has become a standard technology, though discussion of the physio-chemical interpretation of experimental data gained is still going on [5.51]. The reasons for this are threefold.

First, measurements are fairly sensitive to experimental circumstances, namely the preparation of the sorbent sample, i. e. activation, outgassing or still remnant preadsorbed substances such as water in zeolitic materials. Also gas pressure measurements, i. e. calibration of pressure gauges, leaks in the installation and insufficient approach to true equilibrium may considerably influence measurement results.

Second, as already mentioned, volumetric as well as gravimetric measurements in principle do not allow one to determine the mass (m^a) of gas adsorbed on the surface of a porous solid (mass m^s) but only the so-called reduced mass (Ω) defined by Eq. (1.5). Hence, in order to calculate m^a from (Ω, ρ^f)-data, a model assumption for the volume of the sorbent impenetrable to sorptive gas molecules (V^s) and that for the sorbed phase (V^a), or equivalently for their sum

$$V^{as} = V^a + V^s \qquad 1.8$$

has to be introduced. Obviously, resulting m^a-data will depend on this assumption. Unfortunately, this basic problem is not often addressed in today's literature, (for exceptions see [1.49, 1.52-1.53]), but most authors choose to use for (V^s) the helium volume approximation, i. e. assume $V^s = V^s_{He}$. Here V^s_{He} is gained from Eq. (1.5) applied to helium experiments and assuming helium not to be adsorbed or absorbed in the sorbent material during time of observation, i. e. by assuming $m^a_{He} = 0$. Additionally it often is assumed that $V^s = 0$, which leads to the concept of the Gibbs excess mass adsorbed (m^a_{GE}), cp. Chaps. 2, 3 [1.1-1.3, 1.18, 1.19, 1.42]. This will be discussed in the next Sect. 5 of this chapter. As for low gas pressures $m^a_{GE} \cong m^a$, the above mentioned model assumptions seem to be reasonable from a pragmatic point of view. However, as for high pressures considerable differences between m^a and m^a_{GE} normally occur, cp. examples given in Chap. 3, these assumptions should be abandoned and measurement methods should be used – at least for calibration data – allowing one to determine the absolute mass adsorbed (m^a) without introducing additional assumptions for the volume (V^{as}). Indeed, a method which in principle would allow to do this has been developed and preliminary measurements have been performed. The method consists on the combined measurements of two extensive quantities of the adsorbed phase, namely its (integral) enthalpy and its dielectric polarization. By combining the respective caloric and dielectric equations of state of the sorbate phase, an algebraic equation can be derived including the mass (m^a) as single unknown quantity. Hence, up to now (m^a) can be calculated by standard algebraic procedures. Unfortunately, measurements are laborious and even cumbersome. Hence, there are only a few data of measured absolute masses of adsorbates available in literature [1.54].

A third obstacle for determining pore spectra of a sorbent material from gas adsorption data is the fact that today there are several different methods in use to calculate the pore spectrum of the sorbent from a measured gas adsorption isotherm. Naturally, they often will lead to different results. Examples for these methods are the t-plot method, the α_s-method, the Horvath-Kavazoe procedure, the density functional theory (DFT) procedure and also its non-local generalization (NLDFT). All of these are presented and discussed in today's literature [1.2, 1.42]. Actually, the above mentioned difficulties with gas adsorption measurements are condensed in two important draft documents of the International Standardization Organization (ISO), Geneva [1.55, 1.56] providing recommendations for both the experimental procedure of measuring gas adsorption isotherms and to calculate from these data the macro-, meso-, and micropore spectrum of the solid material considered, However, it must be emphasized that these documents today (2004) are still drafts and do not provide international standards. Even though, it can be expected that they sometime will be promoted, probably in an only slightly different form.

A characteristic parameter of a porous material is its so-called BET-surface. This is the surface of a monolayer adsorbate of N_2-molecules at boiling temperature of N_2 at $p_0 = 1$ atm, namely 77.3 K [1.3]. It is determined by nitrogen adsorption experiments at this temperature for pressures $(p/p_0) \leq 1$. The mass of the monolayer load (m_l) is determined by fitting data (m^a) to the adsorption isotherm equation developed by Brunauer, Emmett and Teller in 1938, originally designed for multilayer adsorbates [1.1-1.3, 1.57].

$$\frac{p}{(p_0 - p)m^a} = \frac{C-1}{Cm_l}\left(\frac{p}{p_0}\right) + \frac{1}{Cm_l} \qquad 1.9$$

Here (C, m_l) are parameters to be determined by a data fitting procedure, preferably restricted to the region $0 < p/p_0 < 0.3$. A theoretical analysis of Eq. (1.9) is presented in brief in Sect. 3.4 of Chap. 7.

Assuming a cross section of $\sigma_{N2} = 0.162\ (nm)^2$ per N_2-molecule [1.3], the BET-surface A_{BET} of the monolayer simply can be calculated as

$$A_{BET} = \sigma_{N2} m_l / M_{N2}, \qquad 1.10$$

$M_{N2} = 28$ g/mol being the molar mass of nitrogen. An example for such measurements is given in Figure 1.12, [1.36]. It presents Gibbs excess masses of nitrogen adsorbed on reference material CRM BAM-PM-104 at 77 K ($p_0 = 1$ atm) determined by the volumetric/manometric method in a closed system and without carrier gas. Data show hysteresis between the adsorption-

and the desorption branch indicating existence of mesopores and pore condensation [1.3]. The resulting BET surface for this material is $A_{BET} = 78.2\ m^2/g$ which is in reasonable agreement with its standardized value of $79.8 \pm 2\ m^2/g$, [1.36].

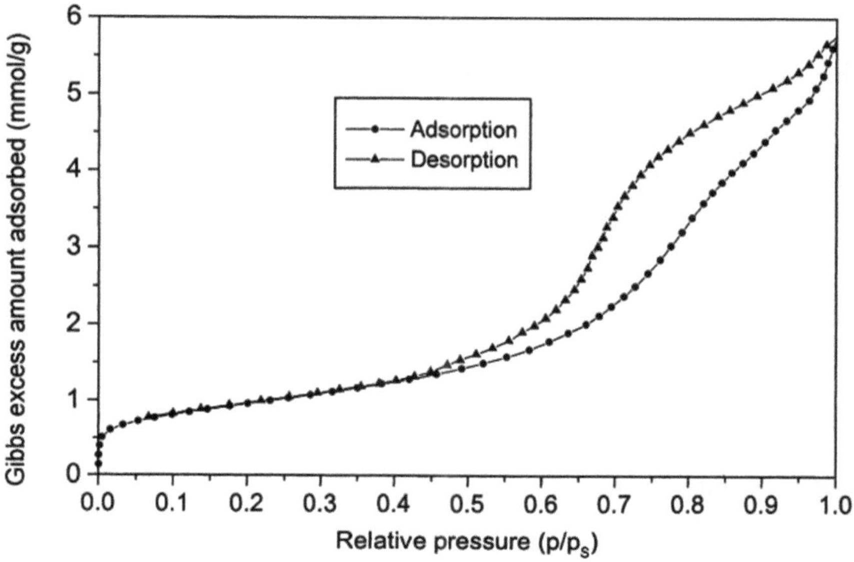

Figure 1.12. Adsorption- and desorption isotherm of N_2 (5.0) on standardized material CRMBAM-PM-104 at 77.3 K, $p_0 = 1$ atm, [1.36].

The BET-surface concept only should be used for nonporous or meso- and macropores including materials but not for microporous substances. The reason for this restriction is that in micropores and especially submicropores (diameter $d < 1$ nm) pore filling may occur prior to adsorption on the more "open" surfaces of meso- and macropores. Hence, part of the nitrogen adsorbed may form linear, string like aggregates in the micropores for which the concept of surfaces becomes obsolete [1.3, 1.56]. Nevertheless BET-data are still used today also for microporous materials in the technical literature.

Another example showing the usefulness of gas adsorption measurements for characterizing the porosity of solids is given in Figure 1.13. It refers to the adsorption of nitrogen N_2 (5.0) at 77.3 K on activated carbon ACTW 52 and certain of its modifications [1.36]. The activated carbon is based on charcoal and has a BET surface of $765\ m^2/g$. The modifications were produced by impregnating the activated carbon with benzoic acid dissolved in a supercritical CO_2-atmosphere and afterwards being pyrolized at 550 °C for 30 minutes and 90 minutes respectively. The steep increases of the adsorption data at low pressures $p/p_0 < 0.05$ indicate micropore filling which however is reduced by the impregnation procedure but partly restored by pyrolyzation afterwards.

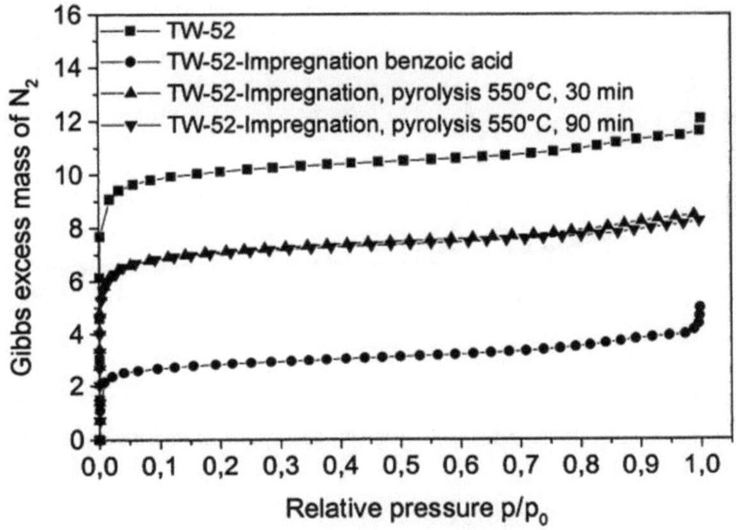

Figure 1.13. Adsorption of nitrogen N_2 (5.0) at 77.3 K on activated carbon ACTW 52 and certain of its modifications produced by impregnation with benzoic acid and pyrolyzation at 550 °C for 30 minutes and 90 minutes respectively [1.36].

From the adsorption isotherms in Figure 1.13 the mesopore spectra (pore radii $r > 1$ nm) of the carbons have been calculated applying the method of Barrett, Joyner, and Halenda [1.3, 1.55]. Results are sketched in Figure 1.14. They clearly show that the mesopore volume in the carbon is reduced considerably by the impregnation but regained partly by the pyrolyzation process.

In addition to the nitrogen measurements on the ACTW 52 presented in Figure 1.13, adsorption experiments were performed at IFT with carbon dioxide CO_2 (4.5) at $T = 273.15$ K, i. e. saturation pressure of $p_0 = 3.485$ MPa, cp. also [1.58]. Results are depicted in Figure 1.15. Data again show that the adsorption capacity of the activated carbon is reduced by the impregnation with benzoic acid, but regained partly by pyrolysis at 550 °C, the duration of which however does not seem to be of considerable influence [1.36].

From the adsorption data in Figure 1.15, the micropore spectra (pore radii $r < 1$ nm) of the carbons have been calculated using the method of Dubinin and Radushkevich [1.3, 1.56, 1.59]. Results are presented in Figure 1.16.

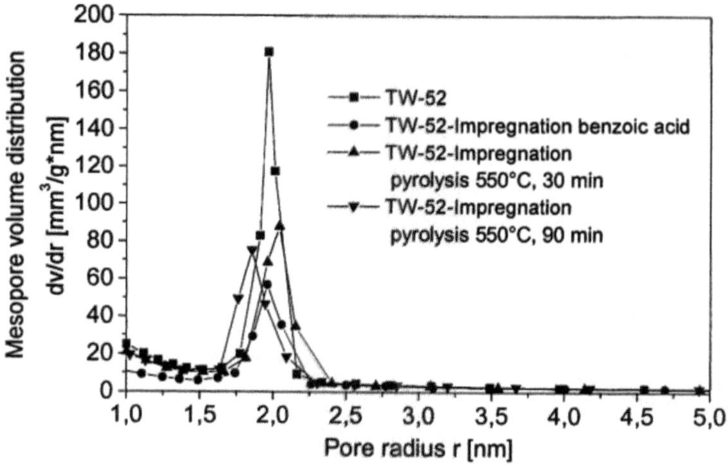

Figure 1.14. Distribution of the mesopore volume in activated carbon ACTW 52 and certain of its modifications calculated from N_2-adsorption isotherms by the BJH-method [1.3].

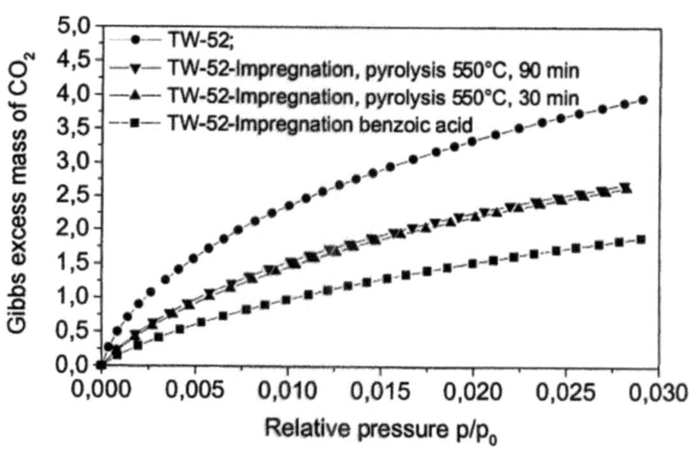

Figure 1.15. Adsorption of carbon dioxide CO_2 (4.5) at T = 273.15 K, p_0 = 3.485 MPa on activated carbon ACTW 52 and certain of its modifications gained by impregnation with benzoic acid and pyrolyzation at 550 °C for 30 minutes and 90 minutes respectively [1.36].

As can be recognized, the micropore volume of the carbon is reduced by the impregnation, but again increased by the pyrolysis procedure. Note also that the average pore radius is somewhat shifted to a lower value by the impregnation but again increases after pyrolysation [1.36].

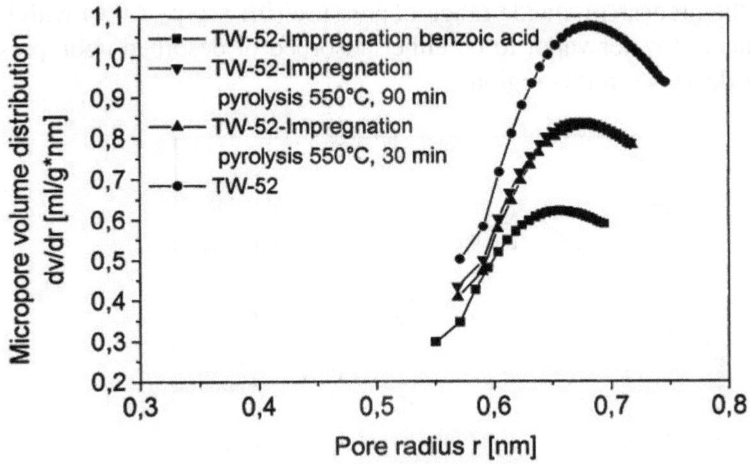

Figure 1.16. Micropore volume distribution function of activated carbon ACTW 52 and certain of its modifications calculated from CO_2 adsorption data by the DR-method [1.3, 1.59].

Data of the CO_2-based total micropore volume and of the N_2-based BET surface of the four activated carbons are given in Table 1.6.

Table 1.6. BET-surface areas (N_2) and micropore volume (CO_2) of activated carbon ACTW 52 and 3 of its modifications by benzoic acid impregnation and pyrolysis at 550 °C.

AC	Impregnation	Pyrolyzation Time	Micropore volume cm^3 / g	BET-Surface m^2 / g
TW 52	---	---	0.331	765
TW 52	Benzoic Acid	---	0.187	220
TW 52	Benzoic Acid	30	0.259	570
TW 52	Benzoic Acid	90	0.259	570

In many applications of porous sorbent materials to gas adsorption problems, especially in drying and air conditioning processes, the behavior of the sorbent in the presence water vapor is of importance. Figure 1.17 shows a set of adsorption isotherms describing the amount of water vapor adsorbed per unit mass of sorbent at given pressure and temperature of the vapor. Measurements were taken at $t = 298{,}15$ K = 25 °C (saturation pressure: $p_0 = 3.1$ kPa) for relative pressures $0 < p/p_0 < 0.8$, [1.39, p. 2537]. These curves are characteristic for the hydrophilic behavior of the zeolites reflected in the steep increase of the respective isotherms at very low (relative) vapor pressures, this indicating existence of micropores. The curves also show for activated carbon hydrophobicity and for activated alumina, amphoteric behavior. For adsorption based air conditioning systems the isotherms of silica gel and activated alumina H-151 are most advantageous, allowing one to operate the process in a wide range of pressure ($0.1 < p/p_0 < 0.6$) with fairly large amounts of water vapor to be either adsorbed or desorbed upon pressure increase or decrease in this region.

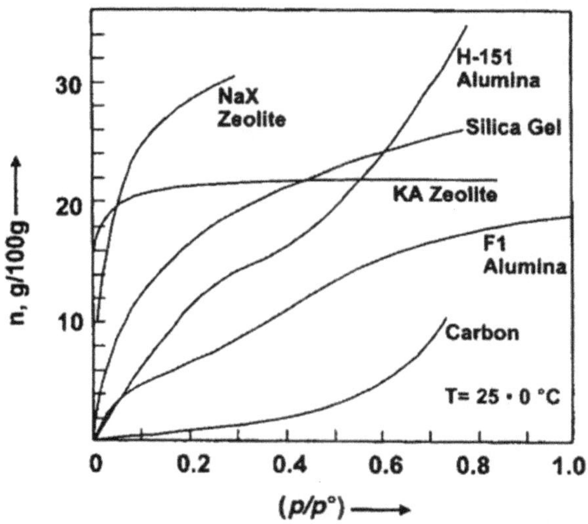

Figure 1.17. Adsorption isotherms of water vapor on different porous materials taken at T = 25 °C, the saturation pressure being $p_0 = 3.1$ kPa. Reprint taken from [1.39] with permission granted.

Gas adsorption systems normally do not only differ in their adsorption capacity, i. e. thermodynamic equilibria states, but also in their kinetic behavior, i. e. spontaneity of uptake or release of gas upon increase or decrease of sorptive gas pressure. Examples for this are given in Figures 1.18, 1.19, showing the uptake process of carbon dioxide (CO_2 (4.5)) on dry Fig. (1.18), and prewetted, Fig. (1.19) zeolite molecular sieve MS Na 13X (Linde, UOP) at T = 323 K due to an increase of the sorptive gas pressure from 0.5 MPa to 6 MPa, [1.60].

In both Figures the uptake of CO_2 (upper curves) and the gas temperature (lower curve), measured immediately below the microbalance connected sorbent sample are shown as functions of time, approaching equilibrium after ca. 5 hours. For the dry zeolite experiment an overshooting of the uptake of

CO_2 and a temporarily decrease in the gas temperature can be observed (Fig. 1.18). This is due to the adiabatic expansion of the high pressure CO_2-gas, released to the adsorption vessel including the sorbent sample and a low pressure CO_2-atmosphere. In the experiment shown in Figure 1.19 the gas valves were opened more slowly (and carefully) so that no adiabatic cooling of the CO_2 occured.

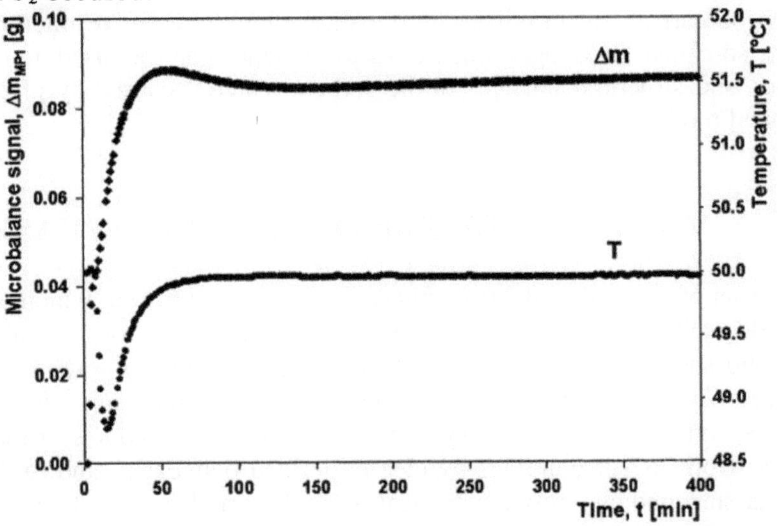

Figure 1.18. Adsorption process of CO_2 (4.5) on dry zeolite molecular sieve MS Na 13X at 323 K due to an increase in gas pressure from 0.5 MPa to 6 MPa [1.60].

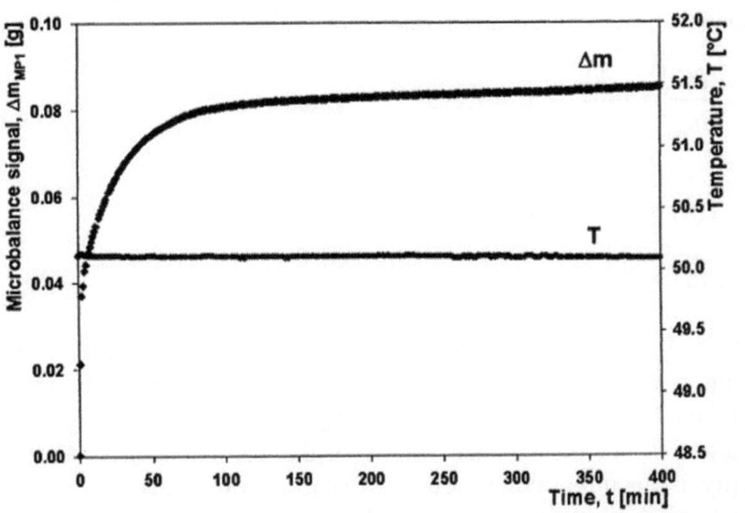

Figure 1.19. Uptake process of CO_2 (4.5) on zeolite molecular sieve MS Na 13X prewetted with 3.2 mmol/g H_2O at 323 K due to an increase in gas pressure from 0.5 MPa to 6 MPa [1.60].

Consequently, the temperature of the gas remained nearly constant and a monotonous increase of the mass of CO_2 adsorbed could be observed. More examples of gas adsorption processes are given in Chaps. 3, 4 and 6.

In concluding this section we want to emphasize that gas adsorption experiments also can be done at elevated temperatures. Preferently they should be carried out gravimetrically using a highly sensitive microbalance. These so-called thermogravimetric measurements can be used to monitor surface and structural changes of the sorbent material caused by physical or chemical modifications [1.3].

5. MASS AND VOLUME OF ADSORBED PHASES

The set of molecules of a fluid phase, i. e. a gas or a liquid, which is within the field of forces of the atoms or molecules of the external or internal surface of a solid sorbent is called the adsorbate of the fluid on this sorbent material. This set may be considered as a thermodynamic system or phase in the sense of W. Schottky [1.63]. Such a system is defined as set of bodies or molecules divided from its surroundings by clearly defined boundaries and exchanging with its surroundings only mechanical work, heat and mass.*) Obviously, any adsorbate is an inhomogeneous phase as – by definition – its molecules are subject to the surface forces of the sorbent atoms. Hence the conditions for local ($\mathbf{x} = (x_1, x_2, x_3)$) thermodynamic equilibrium conditions, derived from the Second Law of thermodynamics, are [1.64]

$$\frac{\partial p}{\partial x_\alpha} = \rho \frac{\partial \Phi}{\partial x_\alpha}, \quad \alpha = 1,2,3 \qquad 1.11$$

$$T(\mathbf{x}) = T^f = \text{const} \qquad 1.12$$

$$\mu(\mathbf{x}) + \Phi(\mathbf{x}) = \text{const} \qquad 1.13$$

Here p, ρ, T, μ indicate the pressure, mass density, temperature and chemical potential of the adsorbate at location $\mathbf{x} = (x_1, x_2, x_3)$. T^f is the temperature of the sorptive fluid. The quantity $\mathbf{\Phi} = \mathbf{\Phi}(\mathbf{x})$ is the potential energy per unit mass at location ($\mathbf{x}$) of the surface forces of the sorbent atoms. As this quantity normally is an unknown function of space coordinates ($\mathbf{x}$), so are the local thermodynamic quantities (p, ρ, T, μ). For given model function of the surface

*) This definition may be extended to include also electric loads, electromagnetic work, radiation energy and information [1.64, 1.65].

of the sorbent, the surface potential $\Phi = \Phi(\mathbf{x})$, and the sorptive fluid equations of state $\rho = \rho(p, T)$, $\mu = \mu(p, T)$, Eqs. (1.11-1.13) allow one on principle to calculate $p = p(\mathbf{x})$, $\rho = \rho(\mathbf{x})$ at the specific free enthalpy $\mathbf{g} = \mu(\mathbf{x})$ at any point $(\mathbf{x})$ within the sorbate phase. Hence we can assign a (total or absolute) mass to the sorbate as

$$m^a = \int_{V^a = V(A)} \rho(\mathbf{x})\, dV \qquad 1.14$$

Here $V^a = V(A)$ is the volume filled by the set (A) of all molecules of the adsorbate, i. e. all molecules of the sorptive fluid influenced by or exerted to the surface forces of the sorbent material.*) Assuming the sorptive fluid far away from the sorbent material to approach an equilibrium state of constant density

$$\rho^f = \lim_{x \to \infty} \rho(\mathbf{x}), \qquad 1.15$$

we also can define the so-called Gibbs excess mass as

$$m^a_{GE} = \int_{V^a = V(A)} (\rho(\mathbf{x}) - \rho^f)\, dV \qquad 1.16$$

*) This is basically an energetic criterion deciding whether a molecule from the fluid is adsorbed on the surface of a solid material or not. It can be specified as follows: Let the energy (E) of the molecule be

$$E = E_{kin} + E_p^f + E_p^s.$$

Here E_{kin} is the kinetic energy of the translational and possibly internal degrees of freedom of the molecule. Let also (E_p^f) and (E_p^s) be the potential energies of interaction of the molecule with other molecules from the fluid phase (f) and the surface atoms or molecules of the solid (s). If

a) $|E_p^s| >> E_{kin}$ and

b) $|E_p^s| \geq |E_p^f|$,

we consider the molecule to be adsorbed even if it is not in direct contact with the surface of the solid. Otherwise the molecule is considered to be not adsorbed. Consequently, molecules from the fluid phase may change their status of being adsorbed or desorbed between two molecular collisions and the "boundary" between adsorbed and not adsorbed molecules from the fluid phase naturally will become fuzzy. In practice this boundary has to be defined indirectly by an experimental procedure allowing one to actually measure the absolute or total mass adsorbed. An example for such a procedure will be given in Section 5.2.

This is the difference between (m^a) and the mass of the sorptive fluid that would be included in the volume V^a if there were no (attractive) surface forces of the sorbent material, i. e. we have in view of Eq. (1.14), [1.1-1.3]

$$m^a_{GE} = m^a - \rho^f V^a \tag{1.17}$$

These concepts can be easily elucidated by a simple example sketched in Figure 1.20 referring to a dense, square sorbent material of mass (m^s) and volume (V^s) with only one of its plane surfaces interacting with the sorptive gas of total mass (m^*) and volume (V^s). The local density $\rho = \rho(x)$ of the adsorbed molecules becomes a function of only one coordinate (x), directed perpendicular to the plane surface of the sorbent. As the surface forces of the sorbent normally are attractive, the sorbate density $\rho(x)$ will be larger than the density of the sorptive gas far away from the sorbent, i. e. we have

$$\rho(x) > \rho^f.$$

A qualitative curve of $\rho(x)$ is sketched below the sorption system in Figure 1.20.

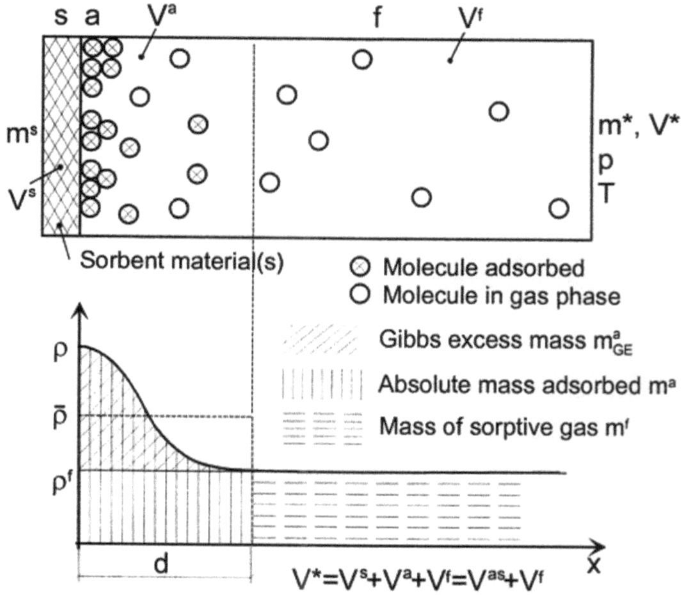

Figure 1.20. Plane surface sorption system (PSSS) in a box of total volume (V^*) including a certain mass of sorptive gas (m^*) part of it being adsorbed on the surface of the sorbent, the sorbate having the absolute mass (m^a) (layer model) and the Gibbs surface excess mass (m^a_{GE}).

If one is interested only in the Gibbs surface excess mass defined by Eq. (1.16), one has to assume the sorptive gas phase to extend to the surface of the sorbent material, i. e. to have a mass of

$$m^a_{GE} = \rho^f (V^* - V^s) \qquad 1.18$$

Then from the mass balance of the sorptive gas, i. e. equalization of gas mass prior to and after adsorption equilibrium has been attained,

$$m^* = m^a_{GE} + m^f_{GE} \qquad 1.19$$

it follows

$$m^a_{GE} = m^* - m^f_{GE} \qquad 1.20$$

Let us now introduce a reference gas density (ρ^{f*}), namely the density the sorption gas would have if no sorbent surface forces were acting on its molecules, i. e.

$$m^* = \rho^{*f} (V^* - V^s) \qquad 1.21$$

or

$$\rho^{f*} = \frac{m^*}{V^* - V^s} \qquad 1.22$$

Then we get for (m^a_{GE}) in view of (1.18), (1.20), (1.21)

$$m^a_{GE} = (\rho^{f*} - \rho^f)(V^* - V^s) \qquad 1.23$$

This formula also holds for more complex geometries of the sorbent material than that simple one sketched in Figure 1.20, cp. Figure 1.21 showing the structure of zeolite molecular sieve UOP/Linde MS 13X. Formula (1.23) allows one to calculate the Gibbs surface excess (m^a_{GE}) from measured data (ρ^f, V^*) if the volume (V^s) of the sorbent material (which is impenetrable for the gas molecules) is known. And here is where often difficulties start, as (V^s) normally is unknown, at least for most technical sorbents which are a mixture of the true sorbing material and some cement or glue and possibly some other components providing mechanical stability to sorbent pellets or grains. Hence one has to use some approximation for (V^s), the most common of which will discussed in brief in the next Section.

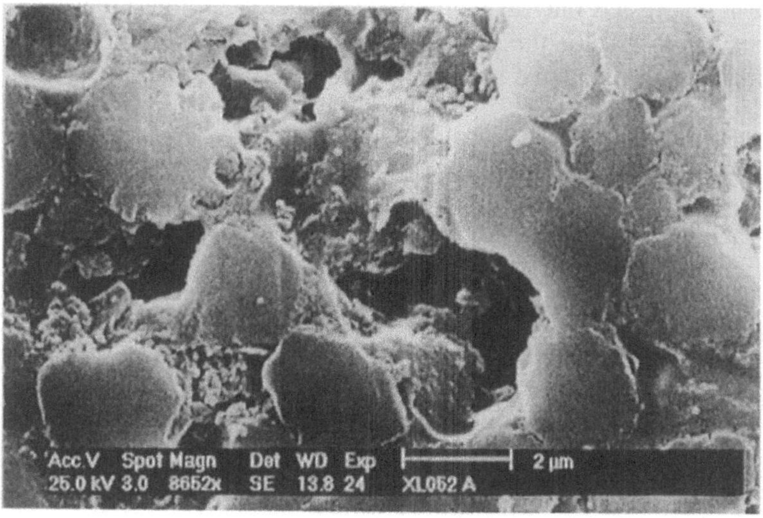

Figure 1.21. Surface structure of zeolite molecular sieve UOP/Linde MS13X. Macropores and capillaries supporting transport of sorptive gas molecules to and from meso- and micropores clearly can be recognized.

5.1 Models for the Void Volume (V^s) of a Sorbent Material and the Volume (V^a) of a Sorbate

In this Section we are going to present four different models for the volume (V^s) of a porous solid impenetrable to the molecules of a sorptive gas and also for the volume (V^a) of the absolute or total mass (m^a) of a sorbate phase [1.50].

The approximation of the void volume (V^s) of a sorbent material most often used in practice today is the so-called helium volume (V^s_{He}), i. e. the proposition

$$\text{P1:} \qquad V^s = V^s_{He} \qquad 1.24$$

V^s_{He} can be calculated from the mass balance (1.19) of helium sorption experiments, assuming helium not to be adsorbed, i. e. $m^a_{GEHe} = 0$. Then the mass balance Eq. (1.19) reads

$$m^*_{He} = m^f_{GEHe} = \rho^f_{He}(V^* - V^s_{He}) \qquad 1.25$$

leading to

$$V_{He}^{s} = V^{*} - \frac{m_{He}^{*}}{\rho_{He}^{f}} \qquad 1.26$$

Hence we have from Eqs. (1.23, 1.24, 1.26)

$$m_{GE}^{a} = (\rho^{f*} - \rho^{f})\frac{m_{He}^{*}}{\rho_{He}^{f}} \qquad 1.27$$

This relation allows one to calculate the Gibbs surface excess (m_{GE}^{a}) from mass and density measurements of the sorptive gas (m^*, ρ^{f*}, ρ^{f}) and accompanying helium measurements leading to calibration parameters ($m_{He}^{*}, \rho_{He}^{f}$). Details of the experimental procedure by volumetric / manometric measurements and several examples will be given in Chap. 2.

From a thermodynamic point of view (W. Schottky, M. Planck and others [1.63-1.65]) all the mass of gas included in the sorbent surface-near-layer of thickness (d) is adsorbed on the surface (Adsorption layer model [1.3], Fig. 1.20). Hence the volume (V^f) of the sorptive gas phase is the volume of the Gibbs excess sorptive gas phase (V_{GE}^{f}) reduced by the volume of the adsorbed phase (V^a), i. e. we have

$$V^{f} = V_{GE}^{f} - V^{a} \qquad 1.28$$

As the thickness (d) of the adsorbed layer is ill-defined we again have to use an approximation for (V^a) which may be related to the total mass adsorbed by

P2: $$V^{a} = \frac{m^{a}}{\rho_{0}^{L}} \qquad 1.29$$

Here (ρ_0^L) is an approximation for the average density ($\overline{\rho}$) of the sorptive layer, cp. Fig. 1.20, a quantity which at least for technical porous sorbents with a complicated, fractal-like surface structure cannot be measured up to date, cp. Figure 1.21, [1.3]. Actually, there are two possibilities so far to get physically reasonable numerical values for (ρ_0^L):

a) One can choose (ρ_0^L) as either the bulk density of the sorptive fluid in the liquid boiling state at ambient pressure, or as the density in the liquid triple state [1.66].

b) One may consider (ρ_0^L) as an optimization parameter which is to be determined by a least square optimization procedure of measured adsorption data. To realize this idea one has to introduce an analytic function for the masses adsorbed (m^a), i. e. a so-called adsorption isotherm, cp. Chap. 7,

$$m^a = m^a(p, T, m^s, A_1 ... A_M) \tag{1.30}$$

Here $A_1 ... A_M$ indicate certain parameters whose numerical values also have to be determined by the optimization procedure.

In view of the fact that the most often used measurement methods for gas adsorption equilibria, namely the volumetric/manometric method, Chap. 2 and the gravimetric method, Chap. 3, only allow one to measure instead of (m^a) or (m^a_{GE}) the so-called reduced mass (Ω), cp. Eq. (1.5),

$$\Omega = m^a - \rho^f V^{as} = m^a_{GE} - \rho^f V^s \tag{1.31}$$

with

$$V^{as} = V^s + V^a \tag{1.32}$$

indicating the total volume of the sorbent / sorbate system, cp. Figure 1.20, the optimization function is, in view of (1.24), (1.29)

$$\sum_{i=1}^{N} \left[\Omega_i - \left(1 - \frac{\rho_i^f}{\rho_0^L} \right) m^a(p_i, T, m^s, A_1 ... A_M) + \rho_i^f V_{He}^s \right] \rightarrow \text{Min} \tag{1.33}$$

Here $(\Omega_i, p_i, T, \rho_i^f = \rho^f(p_i, T), m^s)$ are measured data and $(A_1 ... A_M, \rho_0^L)$ are optimization parameters. Obviously, such a procedure only will lead to meaningful results if there are enough experimental data available, i. e. if $N > (M + 1)^2$. Also, the resulting numerical value for (ρ_0^L) will depend on the analytic form of the adsorption isotherm (1.30) actually chosen, cp. Chap. 7.

Once the density (ρ_0^L) has been determined, the absolute mass adsorbed (m^a) can be calculated from Eqs. (1.24), (1.29), (1.31) as

$$m^a = \frac{\Omega + \rho^f V_{He}^s}{1 - (\rho^f / \rho_0^L)} \tag{1.34}$$

Likewise we have from Eqs. (1.31) combined with (1.27)

$$m^a = \frac{m^a_{GE}}{1-(\rho^f/\rho^L_0)} = \frac{\rho^{f*}-\rho^f}{1-(\rho^f/\rho^L_0)} \bullet \frac{m^*_{He}}{\rho^f_{He}} \qquad 1.35$$

From equations (1.34), (1.35) the total mass adsorbed (m^a) can be calculated from density measurements and calculations of the sorptive gas (ρ^{f*}, ρ^f) and given data of the corresponding helium experiments (m^*_{He}, ρ^f_{He}) and after choosing or calculating the reference density (ρ^L_0) of the sorbate phase.

It should be noted that according to Eq. (1.30) at low gas densities, $\rho^f << \rho^L_0$ or, equivalently low gas pressures, $p < p_s$ ($T < T_c$), the absolute mass adsorbed (m^a) and the Gibbs surface excess (m^a_{GE}) are nearly equal

$$\rho^f << \rho^L_0,\ p \rightarrow 0:\ m^a \cong m^a_{GE} \qquad 1.36$$

However, for high gas densities ($\rho^f \rightarrow \rho^L_0$) or, equivalently, high gas pressures both quantities are becoming different. Consequences of this are demonstrated in Figure 1.22 below showing (m^a_{GE}, m^a)-data of nitrogen adsorbed on activated carbon Norit R1 Extra [1.50].

Both concepts of masses of an adsorbate discussed so far – the Gibbs surface excess (m^a_{GE}) based on proposition (P1) and calculated by Eq. (1.27) and the absolute mass adsorbed (m^a) based on proposition (P2) and calculated by either Eq. (1.34) or (1.35) – do have their physical limitations. Hence it is desirable to mention other possibilities to define and to measure masses of adsorbates in gas adsorption systems.

A fairly simple possibility to do this is to realize that contrary to a PSSS, Fig. 1.20, the volume (V^s) of an irregular porous sorbent being impenetrable for molecules of a sorptive gas, generally will depend on the size of the gas molecules, cp. Figs. 1.1, 1.21. This is well known in the mathematical theory of fractal surfaces [1.67] and thermodynamic phases of fractal dimension [1.68]. Also the range of surface forces $V^a = V(A)$, cp. (1.14), of a sorbent material is limited in space*) but also depend on the sorptive gas molecules. Hence one may consider the volume of the joint sorbent/sorbate phase (V^{as}) as

*) The boundaries of the region (A) of surface forces of the sorbent can be defined as surface on which the potential energy of the surface forces is equal to the (average) energy of the thermal fluctuations of a sorptive gas molecule. Naturally these boundaries will depend on the type of sorptive gas considered and – predominantly – on its temperature.

an unknown system parameter which is to be determined by a least square optimization procedure of measured adsorption data; i. e. we propose

$$\text{P3:} \qquad V^{as} \equiv V^s + V^a = \text{const} \qquad 1.37$$

To realize this idea, one again has to introduce an analytic adsorption isotherm, Eq. (1.30). Restricting to volumetric/manometric and gravimetric measurements, i. e. observing that only reduced masses (Ω), (1-31) can be measured by these methods, the respective optimization function is

$$\sum_{i=1}^{N}\left[\Omega_i - m^a(p_i, T, m^s, A_1 ... A_M) + \rho^f(p_i, T)V^{as}\right] \rightarrow \text{Min} \qquad 1.38$$

Here (Ω_i, p_i, T_1, m^s) are measured data and ($A_1 \ldots A_M$, $V^{as} = \text{const}$) are optimization parameters. The number (N) of experimental data sets available should exceed the square of the number of parameters to be determined, i. e. $N > (M + 1)^2$. Once the volume ($V^{as} = \text{const}$) has been determined from (1.38), the masses m^a and m^a_{GE} can be calculated from Eq. (1.31) as

$$m^a = \Omega + \rho^f V^{as} \qquad 1.39$$

$$m^a_{GE} = \Omega + \rho^f V^s_{He} \qquad 1.40$$

It also should be noted that within procedure (P3), (1.37) the density of the sorbate phase

$$\bar{\rho} = \frac{m^a}{V^a} \qquad 1.41$$

can be calculated from (1.32) with $V^s = V^s_{He}$ and (1.39) as

$$\bar{\rho} = \frac{\Omega + \rho^f V^{as}}{V^{as} - V^s_{He}} \qquad 1.42$$

This quantity can provide information on phase changes within the sorbate layer, for example the transition from a lattice gas type structure, cp. Fig. 1.2, to a patch-liquid, or a dense mono-layer structure [1.3].

Another proposition for the volume of the combined sorbent/sorbate phases (V^{as}) is, to consider explicitly the mass dependent volume of the sorbate $V^a = m^a/\rho_0^L$ and to determine the (sorptive gas dependent) void volume of the sorbent (V^s) by an optimization procedure:

P4:*) $$V^{as} = V^s + \frac{m^a}{\rho_0^L} \qquad 1.43$$

Here ρ_0^L should be chosen as bulk liquid density in a reference state (triple state or liquid boiling state at ambient pressure) representing an approximate value of the average sorbate density $(\bar{\rho})$. The optimization function for this model then reads in view of Eqs. (1.31), (1.43):

$$\sum_i^N \left[\Omega_i - m^a(p_i, T, m^s, A_1 ... A_M)\left(1 - \frac{\rho^f(p_i, T)}{\rho_0^L}\right) + \rho^f(p_i, T)V^s \right] \to \text{Min} \qquad 1.44$$

Here again (Ω_i, p_i, T, m^s) are experimental data and ($A_1 \ldots A_M$, $V^s = \text{const}$) have to be determined from the minimization procedure.

To elucidate consequences of all four different model assumptions ($P_1 \ldots P_4$) for the volume of a porous sorbent/sorbate system, adsorption data (Ω) of nitrogen N_2 (5.0) on activated carbon (AC) Norit R1 Extra which have been measured gravimetrically at 298 K, and the adsorbed masses calculated from these data are shown in Figure 1.22, [1.50]. The lowest curve presents the Ω-data. Note that these decrease at high gas densities as then seemingly the increase in buoyancy of the sorbent/sorbate sample is larger than the uptake of nitrogen gas, cp. Eq. (1.31). The other curves present data as follows:

1) These are Gibbs excess masses (m_{GE}^a) calculated by using proposition (P1), i. e.. using the helium volume approximation for (V^s), (1.24) and formula (1.27).

It should be noted that the numerical values of these data (■) decrease at high gas densities, i. e. for $\rho_{N2}^f > 120 \text{kg}/\text{m}^3$ we have

*) The model (1.43) easily can be generalized to multicomponent adsorption systems by

$$V^{as} = V^s + \sum_i \frac{m_i^a}{\rho_{i0}^L} \qquad 1.45$$

with m_i^a, i = 1, 2, 3,... being the mass adsorbed of component i and ρ_{i0}^L being the density of pure component (i) in a reference liquid state [1.69].

$$\left(\frac{\partial m^a_{GE}}{\partial \rho^f}\right)_T < 0 \qquad 1.46$$

This is in contradiction with the requirement for thermodynamic stability

$$\left(\frac{\partial m^a_{GE}}{\partial \rho^f}\right)_T > 0 \qquad 1.47$$

which itself is a consequence of the Second Law of thermodynamics [1.64, 1.68]. Hence this model proposition (P1) should not be used for high pressure gas adsorption data.

The solid line next to data (1) is the best fit we could get for correlating the data to a somewhat generalized Langmuir adsorption isotherm (AI) in the ($m^a_{GE}, \rho = \rho(p,T)$) –space,

$$m^a_{GE} = m^a_{\infty GE} \frac{(b\rho)^\alpha}{1+(b\rho)^\alpha} \qquad 1.48$$

with asymptotic mass ($\rho \to \infty : m^a_{GE} \to m_{\infty GE}$) and the exponent ($\alpha$) being related to energetic heterogeneity of the sorbent surface, cp. Chap. 7, Sect. 2.1. Obviously the correlation is poor and unacceptable. For data dispersions cp. Table 1.7.

2) Data present absolute masses adsorbed (m^a) calculated from Eq. (1.34), i. e. assuming $V^s = V^s_{He}$ and $V^a = m^a / \rho^L_0$ with $\rho^L_0 = 0.600$ g/cm^3 having been determined via the optimization procedure (1.33) using the Langmuirian adsorption isotherm (AI)

$$m^a = m^a_\infty \frac{(b\rho)^\alpha}{1+(b\rho)^\alpha}, \qquad 1.49$$

parameters having a similar meaning as in (1.48). Though m^a-data reasonably could be correlated by the adsorption isotherm (1.49), it should be noted that again for gas densities $\rho^f_{N2} > 130$ kg/m^3 we have

$$\left(\frac{\partial m^a}{\partial \rho^f}\right)_T < 0,$$

indicating thermodynamic instability which actually could not be observed in any of the several experiments performed with this system. Hence also proposition (P2) should not be used at high sorptive gas pressures.

3) These data present absolute amounts of adsorbed masses calculated from Ω-data by using proposition (P3), i. e. optimization procedure (1.38) with the Langmuirian isotherm (1.49). The data fit is reasonably well, cp. Table 1.7. Also data increase monotonously with increasing gas density, i. e. the stability condition (1.47)

$$\left(\frac{\partial m^a}{\partial \rho^f}\right)_T > 0,$$

holds.

4) These data also present absolute masses of N_2 adsorbed on the AC Norit R1 which have been calculated by proposition (P4) (1.43), i. e. taking the (changing) volume of the adsorbate phase $(V^a = m^a / \rho_0^L)$ into account. The volume (V^s) of the AC impenetrable to the N_2 molecules has been calculated by the optimization procedure (1.44) using the adsorption isotherm (1.49). Data can easily be fitted by the Langmuir isotherm (1.49). Also the thermodynamic stability condition (1.47) holds.

As one can recognize from the curves (1-4), all model assumption lead to nearly the same results for low gas densities ($\rho_{N2}^f < 20$ kg/m^3). However, big differences occur for high gas pressures ($p \leq 14$ MPa). Also, thermodynamics inconsistencies occur for high gas pressures if the helium approximation for the void volume ($V^s = V_{He}^s$) is used as in propositions (P1,P2). No such inconsistencies could be observed for propositions (P3, P4) using a value of (V^s) fitted to Eq. (1.49) and theexperimental Ω-data. This can be more clearly seen from the respective data of all the fits included in Table 1.7.

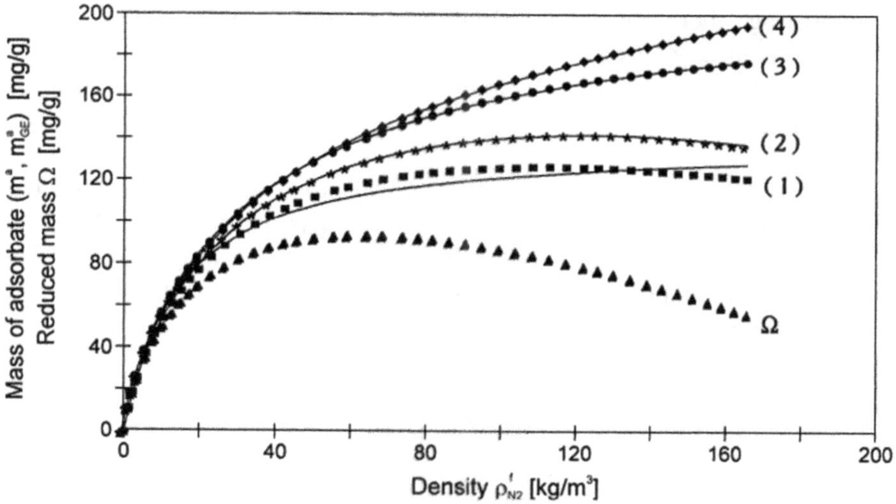

Figure 1.22. Adsorption equilibria data of nitrogen N_2 (5.0) on activated carbon (AC) Norit R1 Extra at T = 298 K for pressures up to 14 MPa equivalent to gas densities up to 170 kg/m³ [1.50].

$\Omega = m^a - \rho^f V^{as}$: Reduced masses measured gravimetrically;

(1) Gibbs surface excess masses (m^a_{GE})
$V^s = V^s_{He}$, $V^a = 0$;

(2) Absolute masses adsorbed (m^a)
$V^s = V^s_{He}$, $V^a = m^a/\rho^L_0$, $\rho^L_0 = 0.600$ g / cm³, optimization parameter (1.33), AI (1.49).

(3) Absolute masses adsorbed (m^a)
V^{as} = const.. .optimization parameter (1.38),
Langmuir adsorption isotherm (1.49).

(4) Absolute masses adsorbed (m^a)
V^s = const.. .optimization parameter (1.44), $V^a = m^a/\rho^L_0$,
$\rho^L_0 = \rho_s$ (77 K) = 0.803 g/cm³, Langmuir adsorption isotherm (1.49),

The parameters of all the data fitting procedures(1-4), i. e. numerical values for V^s, V^a, ρ^L_0, $m^a_{GE\infty}$, m^a_∞, b, α and also dispersions (σ_Ω) of Ω-data fitted to the AIs (1.48), (1.49) respectively are given in Table 1.7. As one can see, method (1) referring to Gibbs surface excess masses (m^a_{GE}) does not lead to physically acceptable results in the pressure range considered ($0 < p < 14$ MPa). The quality of correlation of procedures (2-4) is nearly equal, with all AI-parameters having nearly the same numerical values. From a physical point of view procedure (4) seems to be most satisfying though we cannot see any real arguments against procedures (2, 3).

Table 1.7. Adsorption equilibria of nitrogen N_2 (5.0) on activated carbon (AC) Norit R1 Extra at T = 298 K. Parameters of data correlation for different models of $V^{as} = V^s + V^a$ using a generalized Langmuir adsorption isotherm Eqs. (1.41, 1.42).

No.	Model		Optimization Parameters				Dispersion
	$V^s=$	$V^a=$		m_∞ g/g	b m^3/kg	α	σ_Ω %
1	$V^s_{He} = 0.3835 \frac{cm^3}{g}$	0	—	0.136	0.0692	1.016	4.92
2	$V^s_{He} = 0.3835 \frac{cm^3}{g}$	$\frac{m^a}{\rho^L_0}$	$\rho^L_0 =$ 0.600g/m³	0.244	0.0293	0.860	0.66
3	V^s=const	V^a=const	$V^{as}=$ 0.728 cm³/g	0.260	0.0286	0.863	0.68
4	V^s=const	m^a/ρ^L_0 $\rho^L_0 = 0.80 \frac{g}{cm^3}$	$V^s=$ 0.474 cm³/g	0.248	0.0292	0.861	0.66
	$V^{as} = V^s + V^a$		$\sigma^2_\Omega = \frac{1}{N}\sum_{i=1}^{N}\left(\frac{\Omega_{exp} - \Omega_{cal}}{\Omega_{exp}}\right)^2_i$				

Summarizing the analytic statements and the experimental results presented in Figure 1.22 above, it seems to be safe to state that

a) at low sorptive gas pressures $((p/p_s(T)) < 0.3)$ the helium volume approximation of the void volume of a porous sorbent $(V^s \approx V^s_{He})$ impenetrable to the gas molecules, can be used to calculate both the Gibbs surface excess (m^a_{GE}) and the total or absolute amount of gas adsorbed (m^a) by Eqs. (1.27) and (1.34), (1.35) respectively;

b) at higher pressures $((p/p_s(T)) > 0.3)$ the helium volume approximation may lead to thermodynamically inconsistent results again for both the Gibbs surface excess and the absolute mass adsorbed. Hence it is recommended to replace it by another (approximate) value for (V^s) which may be determined by a Gauss-minimization procedure fitting measured data of the reduced mass (Ω) to an appropriately chosen adsorption isotherm, cp. (1.43), (1.44);

c) for comparison of experimental data with either analytic model adsorption isotherms and/or numerical results of molecular simulation procedures, always the absolute mass adsorbed (m^a) should be used and not the Gibbs surface excess (m^a_{GE}). The later is only part of the total mass adsorbed, cp. (1.17), and may become zero or even assume negative values for sorbent materials with very narrow pores or inkbottle-like pores [1.70].

d) In view of the ambiguity of curves (3), (4) in Figure 1.22 it is still desirable to have an experimental method at hand allowing one to measure the absolute amount of a fluid adsorbed on the surface of a porous solid without introducing any hypotheses on the void volume of the sorbent (V^s) or the volume of the sorbate phase (V^a). Indeed such a method exists [1.54]. Its thermodynamic principle will be outlined below. However, the method is very laborious from an experimental point of view. Hence only few data exist today [1.54]. Also no commercial instruments allowing automated and reliable measurements are available – up to now.

5.2 Outline of Calorimetric-Dielectric Measurements of Absolute Masses of Adsorbates

Absolute masses of adsorbates defined by Eq. (1.14) in principle can experimentally be determined by combined dielectric and calorimetric measurements. In this section we only will present the basic thermodynamic idea of this method and give an example. Details can be found in the literature [1.54].

If an electric field is applied to a sorbent material exerted to a sorptive gas, the dielectric polarizability of the combined sorbate/sorbent system can be measured, cp. Chap. 6 and literature cited herein. Combining these measurements with dielectric measurements of the empty sorbent material in vacuum, the dielectric polarizability (Ω^a_{DE}) of the sorbate phase itself can be calculated. This is a thermodynamic quantity of state, i. e. it can be represented by its dielectric equation of state (DEOS) as function of the gas pressure (p), temperature (T), mass of sorbent (m^s) and mass of sorbate (m^a):

$$\Omega^a_{DE} = \Omega^a_{DE}(p, T, m^s, m^a) \qquad 1.50$$

As (Ω^a_{DE}) is an extensive quantity it always can be written as

$$\Omega^a_{DE} = \alpha^a\left(p, T, \frac{m^a}{m^s}\right) m^a \qquad 1.51$$

Here (α^a) is the specific polarizability of the sorbate within the sorbent/sorbate system considered.

Similarly, the change in enthalpy (ΔH) of sorbate phase upon desorption

$$\Delta H = H^f - H^a > 0 \tag{1.52}$$

can be measured in a calorimeter, cp. [1.3] and Chap. 2, Sect. 5. It also is a quantity of state, i. e. can be represented as

$$\Delta H = \Delta H\ (p, T, m^s, m^a) \tag{1.53}$$

In view of the extensivity property of this quantity we also have

$$\Delta H = \Delta h \left(p, T, \frac{m^a}{m^s} \right) m^a \tag{1.54}$$

with Δh being the specific enthalpy of desorption of the adsorbate.

From the specific polarizability of the sorbate

$$\alpha^a = \alpha^a \left(p, T, \frac{m^a}{m^s} \right) \tag{1.55}$$

a representation of the pressure (p) as a function of (α^a, T, m^a/m^s) can be derived, i. e. we have

$$p = p \left(\alpha^a, T, \frac{m^a}{m^s} \right) \tag{1.56}$$

Inserting this function in (1.54) we get

$$\Delta H = \Delta h \left(p \left(\alpha^a, T, \frac{m^a}{m^s} \right), T, \frac{m^a}{m^s} \right) m^a \tag{1.57}$$

which in view of (1.51), the relation $\alpha^a = \Omega^a_{DE} / m^a$ can be written as

$$\frac{\Delta H}{m^a} = \Delta h\left(\frac{\Omega^a_{DE}}{m^a}, T, \frac{m^a}{m^s}\right) \tag{1.58}$$

This is an algebraic equation for the unknown mass (m^a) of the sorbate phase, as ($\Delta H, \Omega^a_{DE}, T, m^s$) are measured and hence well known quantities. Hence we get by algebraic inversion of (1.58)

$$m^a = m^a(\Delta H, \Omega^a_{DE}, T, m^s) \tag{1.59}$$

from which numerical values of (m^a) can be calculated once ($\Delta H, \Omega^a_{DE}$) have been measured. Of course, as the explicit structure of the EOSs (1.50), (1.53) often is not known, it is necessary to use series expansions, for example virial expansions in pressure (p) and mass ratio (m^a/m^s). To determine respective coefficients quite a number of ($\Delta H, \Omega^a_{DE}$)-data will be needed. Hence measurements will be quite laborious and take time. A few combined dielectric-calorimetric measurements of the adsorption of carbon dioxide CO_2 (4.5) on Wessalite (DAY) at T = 298 K have been performed during 2001-2003 at IFT, U Siegen. A schematics of the experimental installation is shown in Figure 1.23. It consists of a gas storage vessel combined with an adsorption chamber provided with thermopiles and including a capacitor which holds the sorbent material. Gas expansion experiments allow to determine the Gibbs surface excess in the helium volume approximation. Combined dielectric-calorimetric measurements allow one to determine the absolute mass adsorbed. Results of 5 measurements are shown in Figure 1.24.

The absolute masses are always somewhat larger than the Gibbs excess masses as it should be according to the underlying physical concepts. However, we should emphasize that the results presented only are preliminary, as the data base is fairly small and much more data would be needed to show the absolute – and the Gibbs excess-adsorption isotherm – over a considerable range of pressure in order to finally answer the question in what range of pressure the later is still a good approximation of the former. Experiments in a so-called sensor gas calorimeter (SGC) are presently being performed at IFT and should be published in due time, cp. also Chap. 2, Sect. 5.

Differences in the volumetrically and gravimetrically measured Gibbs excess masses depicted in Fig. 1.24 are partly due to different activation procedures of the zeolite samples used in the experiments. This of course should be avoided, but as a matter of fact happened in our experiments and therefore is brought to reader's attention.

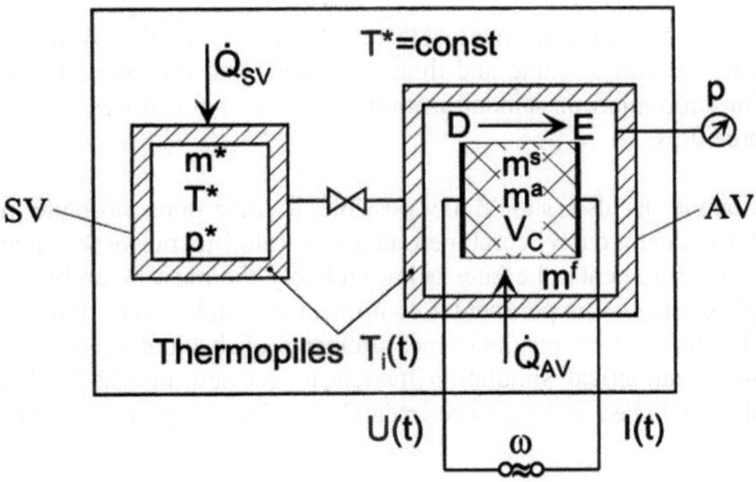

Figure 1.23. Experimental setup for calorimetric-dielectric measurements of pure gas-adsorption-equilibria (impedance calorimeter). © IFT U Siegen, 2001, German Patent DE 10019122 A 1.

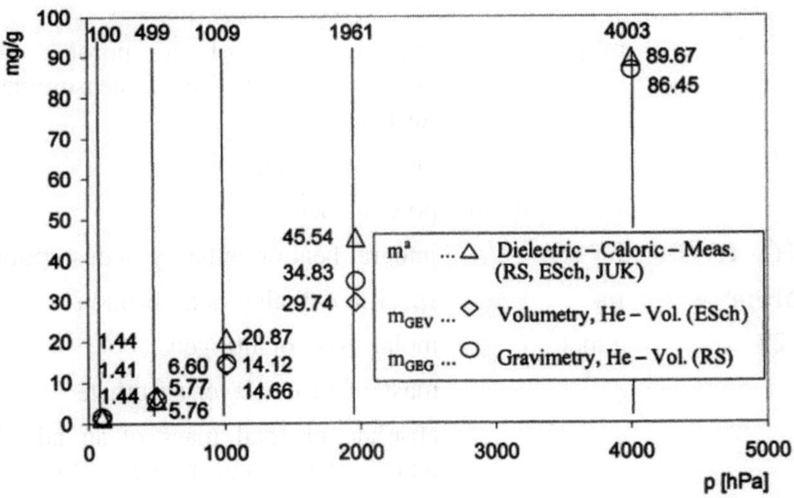

Figure 1.24. Absolute masses (m^a) measured by combined dielectric-calorimetric experiments and Gibbs surface excess masses (m^a_{GE}) measured manometrically and also gravimetrically of carbon dioxide CO_2 (4.5) adsorbed on Wessalite (Degussa DAY) at T= 298 K, [1.54].

In conclusion we want to emphasize that in principle it is possible to measure the absolute amount of a sorptive gas adsorbed on porous solids without using the so-called helium volume approximation by combined dielectric and calorimetric measurements. However, as already mentioned, experiments are cumbersome and time consuming and probably only will be needed for calibration or standardization purposes or for processes operating at high pressures.

In principle it also should be possible to determine absolute masses adsorbed by analogously combined magneto-calorimetric or by magneto-dielectric measurements, the later being probably the most favorable in view of the highly sensitive experimental equipment available today. It also would be desirable to compare results of measurements of absolute masses adsorbed with those of numerical simulation models to get new insight for both the analytical model used and the experimental technique proposed here [1.54].

6. LIST OF SYMBOLS

A list of most symbols (SI-units) and abbreviations used in this Chapter is given.

A_{BET}	m^2/g	surface area of a porous solid determined by BET-measurements, Sect. 4.3
De	1	Deborah number
$d = 2r$	$1 nm = 10^{-9}$ m	pore diameter
$\Delta H = H^f - H^a > 0$	J	integral heat or enthalpy of desorption
$\Delta h = \Delta H/m^a$	J/g	specific enthalpy of desorption
$M_{N2} = 28$	g/mol	molar mass of nitrogen
m_l	g	mass of a monolayer adsorbate
m^a	g	absolute or total mass of an adsorbed phase or adsorbate, cp. Fig. 1.20
m_{He}	g	total mass of helium gas in an adsorption system
$m^f = m^f_{GE} - \rho^f V^a$	g	mass of sorptive gas phase which is outside of the adsorbed phase
$m^a_{GE} = m^a - \rho^f V^a$	g	Gibbs surface excess mass of a sorbate phase, cp. Fig. 1.20

m^f_{GE}	g	mass of Gibbsian sorptive gas phase extending up to the surface of a sorbent material, cp. Fig. 1 .20
m^s	g	mass of sorbent material
PSSS		plane surface sorption system, Fig. 1 .20
p	Pa	pressure
p_0	Pa	saturation pressure of a sorptive gas at temperature (T_0)
r	$1\ nm = 10^{-9} m$	pore radius
$R_{He} = 4.003$	g/mol	molar mass of helium
T	K, °C	temperature
T_c	K, °C	critical temperature of a sorptive gas
t_c	s	cycle period of an industrial adsorption process
V^*	cm^3	total volume of an adsorption chamber including a sorbent material and a sorptive gas, Fig. 1 .20
V^a	cm^3	volume of a sorbate phase as limited by the atoms of a sorbent material and by the molecules of a sorptive gas, cp. Fig. 1.20
$V^{as} = V^s + V^a$	$cm^3 = 10^{-6}\ m^3$	volume of a sorbate/sorbent system as seen by the molecules of a sorptive gas
V^f	cm^3	volume of sorptive gas phase outside of the adsorbed phase
V^f_{GE}	cm^3	volume of Gibbsian sorptive gas phase extending up to the surface of a sorbent material, cp. Fig. 1 .20
V^s	cm^3	void volume of a porous sorbent material impenetrable to molecules of a sorptive gas
V^s_{He}	m^3	volume of a sorbent material seen by helium gas molecules (helium volume of (m^s))
x_α, $\alpha = 1, 2, 3$	m	space coordinates

α^a	Asm^2/Vg	specific dielectric polarizability of the adsorbed phase
$\Phi = \Phi(\mathbf{x})$	J/g	potential energy of surface forces of a sorbent material per unit mass of adsorbate
μ	J/g	chemical potential of a sorbate phase
$\rho = \rho(\mathbf{x})$	g/cm^3	local density of a sorbate phase at position ($\mathbf{x} = (x_1, x_2, x_3)$)
$\bar{\rho}$	g/cm^3	average or equivalent mass density of an inhomogeneous sorbate phase
ρ^f	g/cm^3	mass density of a sorptive gas
$\rho_{He} = \rho_{He}(p, T)$	kg/m^3	mass density of helium gas
ρ_0^L	g/cm^3	density of sorptive fluid in a reference liquid state
σ	Pa m	surface tension of a liquid
$\sigma_{N2} = 0.162$	$(nm)^2 = 10^{-18}\ m^2$	cross section area of a N_2-molecule
$\Omega \equiv m^a - \rho^f V^{as}$	kg	reduced mass of an adsorbate
Ω^a_{DE}	Asm/(V/m)	macroscopic dielectric polarizability of the adsorbate phase

REFERENCES

[1.1] **Gregg S. J., Sing K. S. W.**
Adsorption, Surface Area and Porosity, Academic Press, London etc., 1982.

[1.2] **Ruthven D. M.**
Principles of Adsorption and Adsorption Processes,
J. Wiley & Sons, New York etc., 1984.

[1.3] **Rouquerol F., Rouquerol J., Sing K.S.W.**
Adsorption by Powders and Porous Solids,
Academic Press, San Diego, USA, 1999.

[1.4] **Do D. D.**
Adsorption Analysis: Equilibria and Kinetics,
Imperial College Press, London, 1998.

[1.5] **Dabrowski A.**
Adsorption – from theory to practice,
Adv. in Colloid and Interface Sci., 93 (2002), 135-224.

[1.6] **Rudzinski W., Panczyk T.**
Phenomenological Kinetics of Real Gas-Adsorption-Systems: Isothermal Adsorption,
J. Non.-Equilib. Thermodyn., 27 (2002), p. 149-204.

[1.7] **Kärger J., Ruthven D. M.**
Diffusion in Zeolites and Other Microporous Solids,
J. Wiley & Sons, New York etc., 1992.

[1.8] **Chen N. Y., Degnan T. F., Smith C. M.**
Molecular transport and Reaction in Zeolites,
VCH-Wiley, New York etc., 1994.

[1.9] **Staudt R., Rave H., Keller J. U.**
Impedance Spectroscopic Measurements of Pure Gas Adsorption Equilibria on Zeolites,
Adsorption, 5 (1999), p. 159-167.

[1.10] **Staudt R., Editor**
Adsorption by Porous Solids,
Festschrift in Honor of Prof. Dr. Jürgen U. Keller,
Fortschritt-Berichte VDI, Nr. 555, VDI-Verlag, Düsseldorf, 1998.

[1.11] **Steele W.**
The Interaction of Gases with Solid Surfaces,
Pergamon, New York, 1974.

[1.12] **Adamson A. W., Cast A. P.**
Physical Chemistry of Surfaces,
J. Wiley & Sons, 6th Ed., New York, 1997.

[1.13] **Beenacker J. J. M. et al.**
Chem. Physics Letter 232 (1995), 379.

[1.14] **Yang R. T.**
Gas Separation by Adsorption Processes,
Imperial College Press, London, 1997.

[1.15] **Kast W.**
Adsorption aus der Gasphase, Ingenieurwissenschaftliche Grundlagen und technische Verfahren, Verlag Chemie, Wertheim, Germany, 1988.

[1.16] **Bathen D., Breitbach M.**
Adsorptionstechnik, VDI-Buch, Springer, Berlin, New York etc., 2001.

[1.17] **Kaneko K., Setoyama N., Suzuki T., Kuwabara H.**
Ultramicroporosimetry of Porous Solids by He Adsorption,
Fundamentals of Adsorption (FoA4), Proceedings of FoA4 Conf., Kyoto 1992,
p. 315-322, M. Suzuki, Ed., Kodansha, Tokio, 1993.

[1.18] **Sing K.S.W. et al.**
Reporting Physisorption Data for Gas/Solid Systems with Special Reference to the Determination of Surface Area and Porosity, IUPAC Recommendations 1984, Pure & Appl. Chem., 57 (1985), 603-619.

[1.19] **Sing K.S.W. et al. (Eds.)**
IUPAC Recommendations 1994 for Reporting Physisorption Data,
Pure & Appl. Chem., 66 (1994), 1739.

[1.20] **Setoyama N., Kaneko K.**
Density of He adsorbed in Micropores at 4.2 K,
Adsorption, 1 (1995), p. 165-173.

[1.21] **Kaneko K., Murata K., Myiawaki J.**
Fundamental problems in high pressure adsorption of supercritical gases in microporous systems, Fundamentals of adsorption 7, p. 664-671.

[1.22] **Toennies P., Hollricher K.**
Das kalte Phänomen,
Bild der Wissenschaften, 1998 (7), p. 105-106.

[1.23] **Robens E.**
Some Remarks on the Interface Ice/Water
Proceedings VIII Ukrainian-Polish Symposium: Theoretical and Experimental Studies of Interfacial Phenomena and their Technological Applications, September 2004, Odessa, Ukraine, B. Kats et al., Eds., in preparation (2004).

[1.24] **Kestin J.**
A Course in Thermodynamics, Vols. I, II,
Blaisdell Publ. Comp., London, 1968.

[1.25] **Kaneko K., Kanoh H., Hanzawa Y.**
Fundamentals of Adsorption 7, Proceedings of FoA 7 Conference, Nagasaki, May 2001, IK International, Chiba, Japan 2002.

[1.26] **Schüth F., Sing K. S. W., Weitkamp J.**
Handbook of Porous Solids, Vols 1-5, p. 3141,
Wiley-VCH, Weinheim etc., 2002, ISBN 3-527-30246-8.

[1.27] **Yang R. T.**
Adsorbents, Fundamentals and Applications,
Wiley-Interscience, Hoboken, New Jersey, 2003.

[1.28] **Kienle H. von, Bäder E.**
Aktivkohle und ihre industrielle Anwendung
F. Enke Verlag, Stuttgart, 1980.

[1.29] **Rodriguez-Reinoso F., Me Enaney B., Rouquerol J., Unger K.**
Characterization of Porous Solids VI, Proceedings of the 6th Int. Symposium on the Characterization of Porous Solids, (COPS-VI), Alicante, May 2002, Studies in Surface Science and Catalysis, Vol. 144 Elsevier, Amsterdam etc., 2002.

[1.30] **Talu O., Ed.**
Proceedings of the 8th International on Fundamentals of Adsorption, FoA 8, May 2004, Sedona, USA, Adsorption, Kluwer, Boston etc., in preparation, 2004.

[1.31] **Suzuki M.**
Adsorption Engineering, Elsevier, Amsterdam etc., 1990.

[1.32] **Fleming H. L.,**
Adsorption on aluminas – current applications.
Article in: A. Dabrowski (Ed.), Adsorption and its Application in Industry and Environmental Protection, Vol. 1: Applications in Industry, Studies in Surface Sciences and Catalysis, Vol. 120 A, p. 561-585, Elsevier, Amsterdam, 1999.

[1.33] **Schepers B.**
Aluminiumoxid,
Artikel in Ullmanns Encyklodädie der technischen Chemie,
4. Auflage, Bd. 7, p. 293, Verlag Chemie, Weinheim, 1970.

[1.34] **Ray, M. S.**
Adsorptive and Membrane-type Separation:

A Bibliographical Update (1998),
Adsorption Science & Technology, 17 (1999), 205.

A Bibliographical Update (1999),
Adsorption Science & Technology, 18 (2000), 439.

A Bibliographical Update (2000),
Adsorption Science & Technology, 19 (2001),821.

[1.35] **Ullmann's**
Encyclopaedia of Industrial Chemistry, Carbon, 5. Activated Carbon,
Wiley-VCH, Weinsheim, 6th Edition, 2001.

[1.36] **He R., Zimmermann W, Keller J. U., Roehl-Kuhn B., Jakob J., Heil V., Kümmel R.**
Characterization of Impregnated Activated Carbon SC-44,
Poster 320 C presented at AIChE Annual Meeting 2003, San Francisco, Nov. 2003.

He R., Keller J. U.
Imprägnierung mikroporöser Substanzen aus der überkritischen Phase von CO_2, Teilprojekt 2 „Characterisierung von Aktivkohlen“ (46 Z/2)
Report AIF-46 Z (2003), TZB-Hannover, Hannover 2003.

[1.37] **Robens E., Wenzig J.**
Water storing and releasing properties of peat,
Int. Peat Journal, 6 (1996), 88-100.

[1.38] **Grumpelt H.**
C. Deilmann AG, Bad Bentheim, BRD,
Peat, Artikel in Ullmanns Encyclopädie der Technischen Chemie,
6. Auflage, Wiley – VHC, Weinheim, 2001.

[1.39] **Sircar S.**
Drying Processes
Article in [1.26], Vol. 4, p. 2533-67.

[1.40] **Baerlocher Ch. et al.**
Atlas of Zeolite Framework Types,
5th Ed. rev., Elsevier, N. Y., 2001,
www.iza-online.org/
(Homepage of the International Zeolite Society),
cp. „Databases / Zeolite Structures / Atlas of..."

[1.41] **Tomás Alonso F., Orenes Fernandez I., Palacios Latasa J. M.**
Preparation and characterization of porous sorbents for hot gas desulfurization, article in [1.29], p. 663-670.

[1.42] **Dabrowski A., Robens E., Klobes P., Meyer K., Podkościelny P.**
Standardization of Methods for Characterizing the Surface Geometry of Solids, Particle & Particle Systems Characterization, 20 (2003) 5, p. 311-322.

[1.43] **Rigby S. P.**
New methodology in mercury porosimetry, article in [1.29], p. 185-192.

[1.44] **International Standardisation Organization, Geneva (ISO)**
ISO 15901-1 (1998), Pore size distribution and porosity of solid materials – Evaluation by mercury porosimetry, Geneva, in preparation since 1998.

[1.45] **Sengers J. V., Kayser R. F., Peters C. J., White Jr. H. F., Eds.**
Equations of State for Fluids and Fluid Mixtures, Part I
IUPAC Series in Experimental Thermodynamics, Vol. V,
Elsevier, New York etc., 2000.

[1.46] **Staudt R., Bohn St., Dreisbach F., Keller J. U.**
Gravimetric and Volumetric Measurements of Helium Adsorption Equilibria on Different Porous Solids,
Proceedings COPS IV Conference, Bath, UK, Sept. 1996, p. 261-266, Mc Ennany, et al., Eds., The Royal Society of Chemistry, Special Publ. No. 213, London, 1997.

[1.47] **Robens E., Keller J. U., Massen C. H., Staudt R.**
Sources of Error in Sorption and Density Measurements,
J. of Thermal Analysis and Calorimetry, 55 (1999), 383-387.

[1.48] **Bohn St.**
Untersuchung der Adsorption von Helium an Aktivkohle Norit Rl Extra und dem Molekularsieb 5 A.
Diploma Thesis, Inst. Fluid- and Thermodynamics, University Siegen, Siegen, 1996.

[1.49] **Gumma S., Talu O.**
Gibbs Dividing Surface and Helium Adsorption,
Adsorption, 09 (2003), p. 17-28.

[1.50] **Staudt G., Sailer G., Tomalla M., Keller J. U.**
A note on Gravimetric Measurements of Gas-Adsorption Equilibria
Ber. Bunsenges. Phys. Chem. 97, (1993), p. 98-105.

[1.51] **Llewellyn P. L., Rouquerol F., Sing K. S. W.**
Critical appraisal of the use of nitrogen adsorption for the characterization of porous carbons, Studies in Surface Sciences and Catalysis, Vol. 128,
K. K. Unger et al., Eds., p 421-427, Elsevier Sciences B. V., Amsterdam etc., 2000.

[1.52] **Sircar S.**
Measurements of Gibbsian Surface Excess,
AIChE Journal, 47 (2001), p. 1169-1176.

[1.53] **Sircar S.**
Role of Helium Void Measurement in Estimation of Gibbsian Surface Excess,
Fundamentals of Adsorption (FoA7), p. 656-663, K. Kaneko et al. Eds.,
IK International, Chiba, Japan, 2002.

[1.54] **Keller J. U., Zimmermann W., Schein E.**
Determination of Absolute Gas Adsorption Isotherms by Combined Calorimetric and Dielectric Measurements, Adsorption, 9 (2003) p. 177-188.

[1.55] **International Standardisation Organization, Geneva (ISO)**
ISO 15901-2 (2004), Pore size distribution and porosity of solid materials by mercury porosimetry and gas adsorption – Part 2: Analysis of meso-pores and macro-pores by gas adsorption, Geneva, in preparation (2004).

[1.56] **International Standardisation Organization, Geneva (ISO)**
ISO 15901-3 (2004), Pore size distribution and porosity of solid materials by mercury porosimetry and gas adsorption – Part 3: Analysis of micro-pores by gas adsorption, Geneva, in preparation (2004).

[1.57] **Brunauer S., Emmet P. H., Teller E.**
J. American Chemical Society, 60 (1938), 309.

[1.58] **Sweatman M. B., Quirke N.**
Characterization of porous material by Gas Adsorption: Comparison of Nitrogen at 77 K and Carbon Dioxide at 298 K for Activated Carbon,
Langmuir, 17, 5011-5020 (2001).

[1.59] **Dubinin M. M.**
Physical Adsorption of Gases and Vapors in Micropores, Carbon, Vol. 21 (1983).

[1.60] **Seelbach M.**
Experimentelle und analytische Untersuchungen von Adsorptionsprozessen technischer Gase an mikroporösen Sorbentien, PH- D Thesis, Institute of Fluid- and Thermodynamics, University of Siegen, Siegen, Germany, 2003.

[1.61] **Rouquerol J., Llewellyn P., Navarrete R., Rouquerol F., Denoyel R.**
Assessing Microporosity by immersion microcalorimetry into liquid nitrogen or liquid argon, article in [1.29], p. 171-176.

[1.62] **Everett D. H., Ottewill R. H., Eds.**
Surface Area Determination,
Butterworths, London, 1970.

[1.63] **Schottky W., Ulrich H., Wagner C.**
Thermodynamik (in German),
Reprint of the 1929 Edition, Springer, Berlin (West), 1973.

[1.64] **Keller J. U.**
Thermodynamik der irreversiblen Prozesse mit Aufgaben, Rechenweg und Lösungen, Teil I: Thermodynamik und Grundbegriffe, Kap. A 14,
W. de Gruyter, Berlin – New York, 1977.

[1.65] **Sandler S. I.**
Chemical and Engineering Thermodynamics,
John Wiley & Sons, New York etc., 3rd Ed., 1999.

[1.66] **Reid R. C., Prausnitz J. M., Sherwood T. K.**
The Properties of Gases and Liquids,
3rd Ed., Mc Graw Hill, New York etc., 1977.

[1.67] **Mandelbrot B.**
The Fractal Geometry of Nature,
Freeman, New York, 1983.

[1.68] **Keller J. U.**
Equations of State of Adsorbates with Fractal Dimension,
Physica A 166(1990), p. 180-192.

[1.69] **Dreisbach F.**
Untersuchung von Adsorptionsgleichgewichten methanhaltiger Gasgemische an Aktivkohle als Grundlage zur Auslegung technischer Adsorptionsanlagen PHD-Thesis, IFT University of Siegen, Siegen, 1998, cp. also Fortschritt-Berichte VDI, Reihe 3, Verfahrenstechnik, No. 547, VDI-Verlag, Düsseldorf, 1998.

[1.70] **Murata K., Migawocki J., Li D., Kaneko K.**
Fundamental Problems in High Pressure Adsorption of Supercritical Gases in Microporous Systems,
article in [1.25], p. 664-671.

Chapter 2

VOLUMETRY / MANOMETRY

Abstract The physical principles and experimental techniques of pure gas- and multi-component gas adsorption measurements by the volumetric (or manometric) method are outlined. Examples are given. Thermovolumetric and combined volumetric–calorimetric measurements are presented. Pros and cons of the method are discussed. References. List of Symbols.

1. INTRODUCTION

Volumetry or, preferably "Manometry", cp. Sect. 2.1, is the oldest method to investigate sorption of gases in solids. Early experiments had already been performed by C. W. Scheele (1777), Chappuis (1881), W. Ostwald (1905) and J. Langmuir (1912), [1.1]. Some of these experiments were designed to determine the volume of geometrically complicated objects like cauliflower or sea-urchins. A special instrument for this purpose was designed by R. W. Pohl about 1940 and was called a "volumimeter" [2.1]. It served as a prototype of many of today's volumetric instruments.

The physics of volumetric gas adsorption experiments is simple: a given amount of sorptive gas is expanded into a vessel which includes a sorbent sample and which initially has been evacuated. Upon expansion the sorptive gas is partly adsorbed on the (external and internal) surface of the sorbent material, partly remaining as gas phase around the sorbent. By a mass balance, the amount of gas being adsorbed can be calculated if the void volume of the sorbent, i. e. the volume which can not be penetrated by the sorptive gas molecules is known – at least approximately.

The adsorbed mass per unit mass of sorbent is a characteristic quantity for a porous solid. Assuming a characteristic area per single molecule adsorbed on the surface of a solid – for N_2 at 77 K this area is 0.162 nm^2, for Kr at 87 K it is 0.152 nm^2 [2.2, p. 173] – the surface area of the solid can be calculated and from this the porous solid can be "characterized". Actually, many different types of instruments, normally operating around or below ambient pressure, have been designed for this purpose including static, gas flow and differential gas flow instruments. An overview is given in [2.2]; a list of commercial suppliers which in no way is meant to be exclusive or complete is given in Table 2.1 below, cp. also [2.3, 2.4, 3.7].

Table 2.1 Manufacturer of volumetric and carrier gas sorption measuring instruments and of gas pycnometers.

Manufacturer	A	P	C	D
Beckman Coulter UK Ltd., Oakley Court, Kingsmead Business Park, London Rd. High Wycombe Bucks. HP11 1JU, U.K. (including Omicron), Tel.: +44-149444-1181, Fax: -7558 beckmancoulter_uk@beckman.com, www.beckmancoulter.com	•	•	•	•
Thermo Finnigan, Thermoquest Italia, Strada Rivoltana, s.n., I-20090 Rodano, Milano, Italy, Tel.: +39-02950592-66, Fax: -56, woodwardc@finnigan.co.uk, www.thermo.com, www.porotec.de, info@porotec.de	•	•	•	•
Jouan Robotics - Gira, Rue des Bruyères, Z.I. Berlanne, F – 64160 Morlaas, France, Tel: +33-559-308-38-3, Fax: -4, g.chancel@gira-france.com, www.gira.fr	•	•	•	
Micromeritics, 1 Micromeritics Drive, Norcoss, GA 30093-1877, USA, Tel.: +1-770-662-36-60, Fax: -96 international@micromeritics.com, www.micromertics.com	•	•	•	•
Porous Materials, Inc., Cornell University Research and Technology Park, 83 Brown Road, Bldg. 4, Ithaca, NY 14850, USA, Tel.: +1-607-257-554, Fax: -5639, www.pmiapp.com, info@pmiapp.com	•	•	•	•
Quantachrome Corp., 1900 Corporate Drive, Boynton Beach, Florida 33426 USA, Tel.: +1-561-731-4999 Fax: -732-9888,www.quantachrome.com, webmaster@quantachrome.com	•	•	•	•
Zeton Altamira, 149 Delta drive, Suite 200 Pittsburgh, PA 15238,USA Tel: +1-412-9636-385, Fax: -485, www.zetonaltamira.com, altamira@compuserve.com	•			

A = specific surface area, P = pore size distribution, C = chemisorption, D = density

As we here are mainly interested in adsorption measurement techniques for industrial purposes, i. e. at elevated pressures (and temperatures), we restrict this chapter to volumetric instruments which on principle can do this for pure sorptive gases (N = 1), Sect. 2. Thermovolumetric measurements, i. e. volumetric/manometric measurements at high temperatures (300 K - 700 K) are considered in Sect. 3. In Section 4 volumetric-chromatographic measurements for multi-component gases (N>1), are considered as mixture gas adsorption is becoming more and more important for a growing number of industrial gas separation processes. In Section 5 we discuss combined volumetric-calorimetric measurements performed in a gas sensor calorimeter (GSC). Finally pros and cons of volumetry/manometry will be discussed in Sect. 6, and a list of symbols, Sect. 7, and references will be given at the end of the chapter.

2. VOLUMETRIC MEASUREMENT OF PURE GAS ADSORPTION EQUILIBRIA (N = 1)

2.1 Experimental

An instrument for volumetric measurements of pure gas adsorption basically consists of a gas storage vessel (volume V_{SV}) and an adsorption chamber (V_{AC}) being connected by a tube bearing a valve. Both vessels should completely be placed within a thermostat (water, oil, air etc.) and provided with tubes for gas supply and evacuation as well as with thermometers and manometers to measure the temperature (T) and pressure (p) inside the vessels, cp. Figure 2.1.

For pressures above ambient all vessels and tubes should be manufactured of stainless steel, inside surfaces should be electropolished or, preferably gilded. For pressures below ambient, i. e. vacuum systems glass should be preferred. Sealing materials should be chosen according to the sorptive gases to be used and the ranges of temperature and pressure of operation. For corrosive gases and high pressures the use of metallic sealings (silver, steel) always is recommended. The adsorption chamber includes a sample of sorbent material of mass (m^s), which prior to measurement should have been "activated" at higher temperature and in vacuum for several hours to reduce the amount of pre-adsorbed molecules. Standard procedures for activation have been described in [2.2, 2.5 – 2.7].

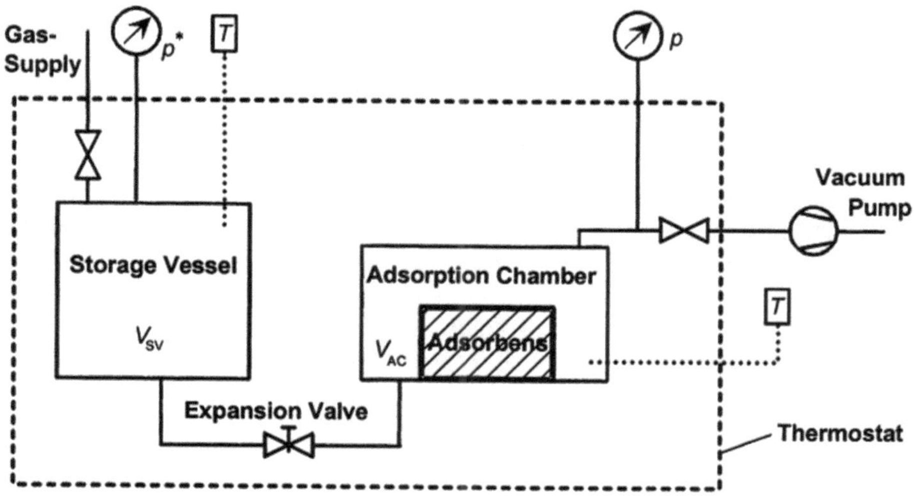

Figure 2.1. Experimental setup for (static) volumetric measurements of pure gas adsorption equilibria.

To measure adsorption a certain amount of gas of mass (m*) is prepared in the storage vessel and the adsorption chamber is evacuated. Upon opening the expansion valve, the gas expands to the adsorption chamber where it is partly adsorbed on the (external and internal) surface of the sorbent material. This process may last milliseconds, minutes, hours or even several days – as in case of helium on activated carbon (Norit R1) [2.8]. After thermodynamic equilibrium, i. e. constancy of pressure (p) and temperature (T) inside the vessels has been realized, these data can be taken as a basis to calculate the mass of the gas adsorbed on the sorbent (m^S). That is, volumetric adsorption experiments mainly result in pressure measurements. Hence the name "Manometry" for this method should be used [2.2].

2.2 Theory

To determine the mass (m^a) adsorbed in a volumetric experiment we consider the mass balance of the sorptive gas:

$$m^* = m^f + m^a \qquad 2.1$$

Here $m^* = m^*(p^*, T, V_{SV})$ is the mass of gas initially prepared in the storage vessel, m^f is the mass of the sorptive gas in the adsorption equilibrium state considered. It also can be calculated from its thermal equation of state (EOS)

$$pV^f = \frac{m^f}{M} RTZ(p,T) \tag{2.2}$$

In this equation M is the molar mass of the sorptive gas, R = 8.314 J/molK the universal gas constant and Z = Z (p, T) the compressibility factor or real gas factor of the gas which can be determined from p, V, T-measurements. Analytic expressions for Z are given for example in [2.9, 2.10], numerical values for a variety of technical gases over a considerable range of pressure and temperature can be found in [2.11 – 2.15]. For an ideal gas we have Z = 1 for any value of p, T. The volume (V^f) of the sorptive gas in Eq. (2.2) can be calculated from the relation

$$V^f = V_{SV} + V_{AC} - V^{as} \tag{2.3}$$

where V^{as} indicates the so-called void volume of the sorbent material and the sorbate phase which can not be penetrated by the sorptive's gas molecules. Equations (2.1-2.3) can be combined to present the result of volumetric measurements:

$$\Omega_V = m^a - \rho^f V^{as} \tag{2.4}$$

Here Ω_V has been introduced as an abbreviation being defined by

$$\Omega_V \equiv m^* - \rho^f (V_{SV} + V_{AC}) \tag{2.5}$$

with the sorptive gas density, cp. (2.2),

$$\rho^f = \frac{m^f}{V^f} = \frac{pM}{RTZ(p,T)} \tag{2.6}$$

It should be emphasized that the auxiliary quantity (Ω_V) only includes measurable quantities. Hence its numerical value is known from the gas expansion experiment. According to Eq. (2.4), Ω_V can be called "reduced mass" of the adsorbed phase as it is the difference between the mass of the adsorbed phase (m^a) and the mass of the sorptive gas that would be included in the void volume of both the sorbent and the sorbate phase (V^{as}) [2.3, 2.7]. As this quantity also is unknown, some model assumptions for it have to be introduced to calculate the adsorbed mass (m^a) from Eq. (2.4).

As outlined in Chapter 1, V^{as} is often approximated by the so-called helium volume,

$$V^{as} \cong V^{s}_{He} \tag{2.7}$$

that is the volume of the sorbent without that of the sorbate phase as "seen" by helium molecules which are assumed not to be adsorbed at all on the surface of the sorbent. Indeed under these conditions V^{s}_{He} can be determined from volumetric experiments performed with helium – preferably at room temperature or even at higher temperatures – leading via (2.1 - 2.3) with $m^{a}_{He} = 0$ to

$$V^{s}_{He} = V_{SV} + V_{AC} - V^{f}_{He}, \tag{2.3a}$$

with

$$V^{f}_{He} = \frac{m^{*}_{He}}{p_{He} M_{He}} RTZ_{He}(p_{He}, T) \tag{2.2a}$$

Here V^{f}_{He} is the volume of the helium gas of mass m^{*}_{He} at pressure p_{He}, temperature T and Z_{He} is the compressibility or real gas factor of helium [2.11, 2.12]. In view of the approximation (2.7), the sorbate mass m^{a} in Eq. (2.4) is the Gibbs surface excess mass (GSE) (m^{a}_{GE}) cp. Chap. 1. Inserting Eqs. (2.5 - 2.7) in (2.4) this mass can be represented as

$$m^{a}_{GE} = m^{*} - \frac{pM}{RTZ(p,T)}(V_{sv} + V_{AC} - V^{s}_{He}) \tag{2.8}$$

If the void volume (V^{as}) in (2.4) is approximated by

$$V^{as} = V^{s}_{He} + \frac{m^{a}}{\rho^{L}_{0}}, \tag{2.9}$$

cp. Chapter 1, m^{a} in Eq. (2.4) is the absolute or total mass adsorbed which equivalently to (2.8) via Eqs. (2.4 – 2.5, 2.8, 2.9) can be represented as

$$m^{a} = \frac{m^{a}_{GE}}{1 - (\rho^{f} / \rho^{L}_{0})} \tag{2.10}$$

with m^{a}_{GE} as given by (2.8) and ρ^{f} by (2.6).

It should be noted that for $\rho^f \rightarrow 0$, $m^a \cong m^a_{GE}$, and for $\rho^f \rightarrow \rho_0^L$, $m^a \rightarrow \infty$, which reflects the occurrence of a liquid phase outside the sorbent material, i. e. a vapor-liquid-state of the sorptive fluid.

Auxiliary remarks:

1. The quantity m^a in Eq. (2.1) on principle also includes the mass of gas adsorbed on the walls of the adsorption chamber and possibly desorbed from the walls of the storage vessel upon gas expansion (Δm^a_{wall}). These quantities should be determined by calibration experiments without using a sorbent sample, i. e. $m^s = 0$, cp. Chapter 4. If the surface of the walls of the adsorption chamber and the storage vessel is small compared to the BET surface of the sorbent sample, one may assume $(\Delta m^a_{wall}) << m^a$, i. e. neglect (Δm^a_{wall}) compared to (m^a).

2. The volumes V_{AC}, V_{SV} should be determined experimentally by using a sample, preferably a cylinder of calibrated volume (V_0) made of dense material (Ti, Au) and performing gas expansion experiments as described above. The amount of gas adsorbed on the surface of the sample normally can be neglected.**

3. Waiting times for thermodynamic equilibrium in the gas adsorption system are based on experimental experience which comes from gravimetric measurements, cp. Chap. 3. No general rules for these times are available today.

2.3 Uncertainties or Errors of Measurements

An important part of any kind of experimental work is to consider errors or uncertainties of measurements [2.16 – 2.18]. To keep these down, it is always recommended to repeat measurements of any experimental quantity at

** Suppose the sample (V_0) is placed in the adsorption chamber and a certain amount of nearly not adsorbing gas (He, Ar at $\simeq$ 300 K) has been prepared at pressure (p^*) in the storage vessel (V_{SV}). If after expansion to the adsorption chamber the gas is at pressure (p), the volume V_{AC} can be calculated via Eq. (2.2) by

$$V_{AC} = \left(\frac{p^* Z(T,p)}{p Z(T,p^*)} - 1\right) V_{SV} + V_0 \qquad (2.11)$$

least three times to guarantee reproducibility of data and reduce the uncertainty by a factor (1 / 3). To be more explicit let us consider briefly an arbitrary quantity (x) which has been measured $N \geq 3$ times, the result of the i-th measurement being

$$x_i \pm \Delta x_i \tag{2.12}$$

with the uncertainty or error $\Delta x_i > 0$. Then the statistical mean value of all measurements

$$\bar{x} = \frac{1}{N}\sum_{i=1}^{N} x_i \tag{2.13}$$

has, according to Gauss' law of propagation of error a dispersion (or variance or mean square deviation, MSD) of

$$\sigma^2(\bar{x}) = \sum_{i=1}^{N}\left(\frac{\partial \bar{x}}{\partial x_i}\right)^2 (\Delta x_i)^2 = \frac{1}{N}(\Delta x)^2 \tag{2.14}$$

We consider this quantity as uncertainty of the representive value $\bar{x}$ of the quantity x. In (2.14) it has been assumed that the uncertainties of all single measurements are equal ($\Delta x_i \equiv \Delta x$, $i = 1 \ldots N$), and that measurements themselves are statistically independent [2.16].

In view of space limitations we here consider only the uncertainty of the Gibbs surface excess mass adsorbed according to Eq. (2.8). The Gauss law of error propagation immediately leads to the expressions

$$\sigma^2_{mGE} = \sigma^2_{m*} + \sigma^2_{mf} \tag{2.15}$$

$$\begin{aligned} \sigma^2_{mf} = {} & \left(1 - \frac{p}{Z}\left(\frac{\partial Z}{\partial p}\right)_T\right)^2 \left(\frac{M}{RTZ}\right)^2 (V_{SV} + V_{AC} - V^s_{He})^2 \sigma^2_p \\ & + \left(1 + \frac{T}{Z}\left(\frac{\partial Z}{OT}\right)_p\right)^2 \left(\frac{pM}{T^2RZ}\right)^2 (V_{SV} + V_{AC} - V_{He})^2 \sigma^2_T \\ & + \left(\frac{pM}{RTZ}\right)^2 (\sigma^2_{VSV} + \sigma^2_{VAC} + \sigma^2_{sHe}), \end{aligned} \tag{2.16f}$$

$$\sigma_{m^*}^2 = (1 - \frac{p^*}{Z^*}(\frac{\partial Z^*}{\partial p^*})_T)^2 (\frac{V_{SV}M}{RTZ^*})^2 \sigma_{p^*}^2 + \\ + (1 + \frac{T}{Z^*}(\frac{\partial Z^*}{\partial T})_{p^*})^2 \cdot \frac{\sigma_T^2}{(T^2Z^*)^2} + (\frac{p^*M}{RTZ^*})^2 \sigma_{VSV}^2 \qquad 2.16^*$$

Analysis of numerical data shows that measurements of pressures (p, p*), i. e. values of $(\sigma_p^2, \sigma_{p^*}^2)$ are the most crucial for accurate measurements of m_{GE}^a followed of course by temperature measurements (σ_T^2) and those for basic data of the equipment $(\sigma_{VSV}^2, \sigma_{VAC}^2)$ and of the sorbing material (σ_{sHe}^2). To achieve useful results, i. e. $(\sigma_{mGE} / m_{GE}^a) < 5\%$, limiting values of variances should hold as follows [2.7]: $\sigma_p < 10^{-4} p, \sigma_T < 10^{-3} T$, $\sigma_{VSV} \simeq \sigma_{VAC} < 10^{-6}(V_{SV} + V_{AC})$.

2.4 Example

In Figure 2.2 equilibrium adsorption data for carbon dioxide (CO_2) on zeolite Na 13X (Linde, UOP) are presented for temperatures 298 K and 303 K. The mol numbers of the Gibbs surface excess amounts per unit mass of sorbent are depicted as function of the sorptive gas pressure and temperature. Relative uncertainties of measurements are about $(\sigma_{mGE} / m_{GE}^a) \leq 2\%$. The subcritical isotherms are in the range of pressure measured of Type I – IUPAC classification [2.20].

However, the adsorption isotherms exhibit Type II structure if the saturation pressure (p_0 (T)) is approached, cp. Chapter 3. Prior to adsorption, the zeolite has been "activated" at 623 K for 3 hours under vacuum conditions (p < 15 Pa). Some characteristic data of the zeolite also have been measured in our laboratory [2.19], namely the BET surface at 77 K being 577 m^2 / g (UOP-date: 572.5 m^2 / g), the total pore volume being 420 cm^3 / kg, the He-expansion density being 2.51 g / cm^3.

Equilibration times varied from 15 minutes at low pressures up to (nearly) 1 hour at the higher pressures considered, i. e. p $\simeq$ 0,03 p_0 (T). This information was gained from time dependent temperature signals produced by the sorbent sample and gained simultaneously with adsorption equilibria measurements [2.23], cp. Sect. 5.

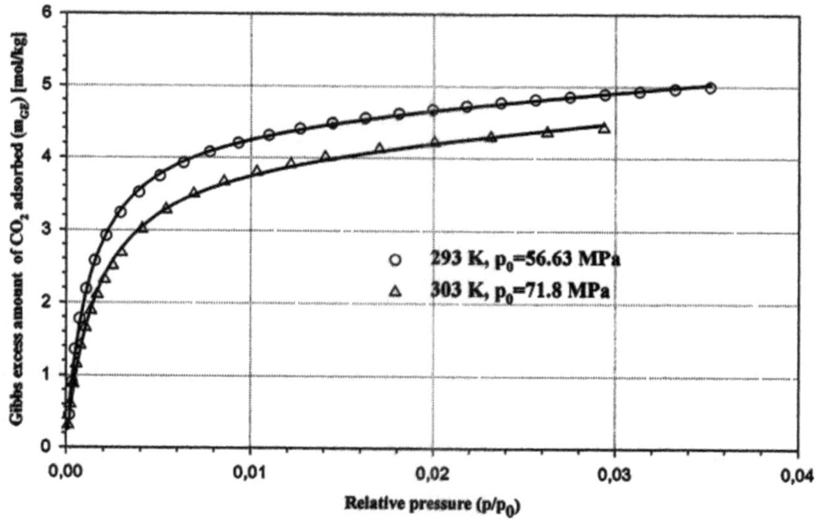

Figure 2.2. Adsorption equilibria of CO_2 on zeolite Na13X (Linde, UOP) at 293 K and 303 K for pressures up to 0.2 MPa. Lines indicate correlation of data by a generalized Langmuir type adsorption isotherm, cp. Chap. 7.

3. THERMOVOLUMETRY

Sorption of gases in highly porous and thermally stable materials like zeolites at high temperatures are becoming important in view of an increasing number of technical applications with considerable economic potential. Two examples of these are

i) Sorption enhanced chemical reactions [2.21]
ii) Volumetric receivers of solar radiation in solar-thermal energy converters [2.22].

In view of some inherent problems of traditional thermogravimetric measurements, cp. Chapter 3, it seems to be worthwhile to consider also volumetric / manometric measurements of gas adsorption equilibria at higher temperatures ($T > 400$ K). An instrument of this type has been designed and built in the labs of the authors, the scheme of it being presented in Figure 2.3. A photo showing mainly the high temperature thermostat and auxiliary equipment is given below.

Adsorption and desorption experiments of nitrogen (N_2 (5.0)) on zeolite KE-H 555 (UOP) have been performed in the temperature range

293 K < T < 673 K and for pressures up to 5.5 MPa. Data are given in Figure 2.5 that show clearly that nitrogen is adsorbed even at high temperatures.

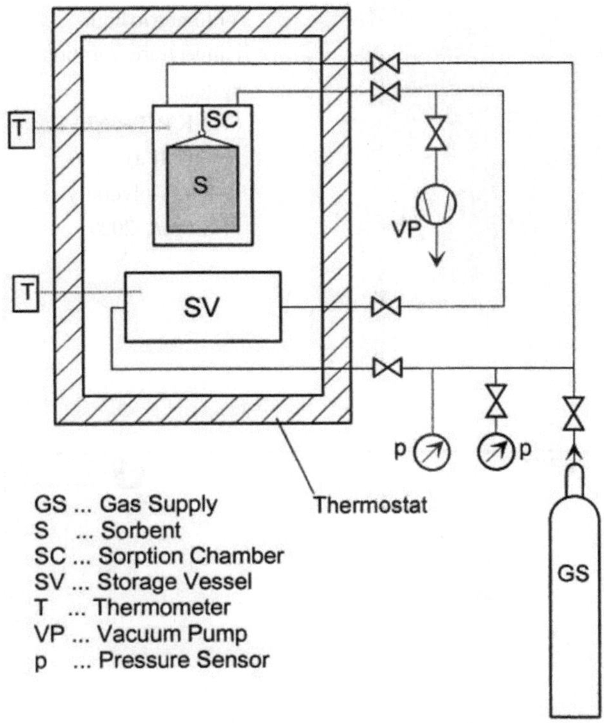

Figure 2.3.
Experimental set-up for volumetric high temperature adsorption / desorption experiments.

In calculating the adsorbed masses of nitrogen, the thermal expansion of the storage and of the adsorption vessel at higher temperatures must be taken into account. Also slight changes in the so-called helium-volume of the zeolite (V^s_{He}) depending also on temperature should be considered.

No coadsorption measurements at high temperatures using the volumetric / manometric method seem to have been performed yet. Some experiments with humid nitrogen, i. e. N_2 / H_2O mixtures are presently (2003) under way at the author's institute. Data will be reported in the homepage of IFT, cp. [4.15].

Figure 2.4. Volumetric high temperature sorption system ($293\ K < T < 673\ K$, $p < 10 MPa$) © IFT, University of Siegen, 2000.

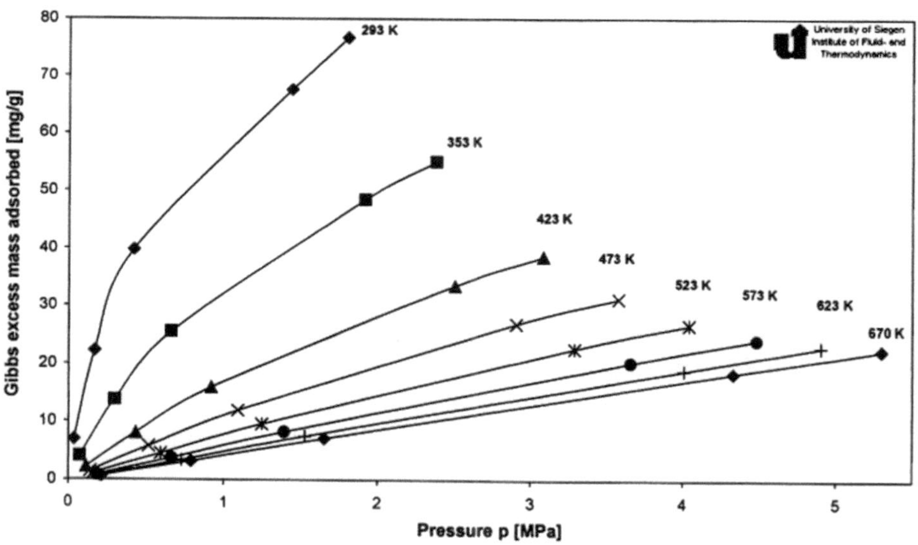

Figure 2.5. Adsorption isotherms of N_2 on zeolite KE-H 555 (UOP) at different temperatures (293 K – 673 K); helium volume $V^s_{He} = 0.45\ cm^3/g$.

4. VOLUMETRIC MEASUREMENT OF MULTICOMPONENT GAS ADSORPTION EQUILIBRIA (N > 1)

4.1 Experimental

Coadsorption equilibria of multi-component gases on porous materials today most often are measured by using volumetric / manometric instruments. The flow diagram of such a device is sketched in Fig. 2.6. It basically has the same structure as the instrument for pure gas adsorption measurements, Fig. 2.1. However, one now has to provide a gas circulation loop including a pump between the storage vessel and the adsorption chamber in order to maintain equal and constant sorptive gas concentrations everywhere in the system. Also a sample loop has to be added which may be outside the thermostat but in any case has to be connected to a gas chromatograph or mass spectrometer. No commercial suppliers of volumetric multi-component gas adsorption instruments are known to the authors, cp. [2.3, 2.4]. However, an instrument which easily can be used as such, after having been equipped with a standard gas chromatograph, is offered by BEL Japan, cp. Chapter 4. Materials of vessels and tubes should be chosen according to the recommendations given in Section 2.1. For the analysis of a sorptive gas sample in a gas chromatograph, helium at elevated temperatures (373 K – 473 K) should be used as a carrier gas. More information for design of a multicomponent volumetric gas adsorption instrument is given in [2.20].

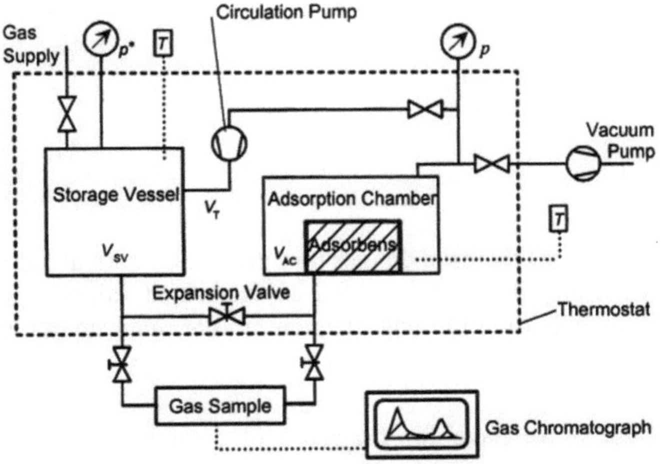

Figure 2.6. Experimental set-up for (static) volumetric / manometric measurements of multicomponent gas adsorption equilibria.

4.2 Theory

Let us assume a certain amount of gas mixture with known mass concentrations w_i^*, and molar concentrations y_i^*, i = 1...N is in the storage vessel (V_{SV}) of the instrument, Fig. 2.6. The adsorption chamber (V_{AC}) including the sorbent mass (m^s) has been evacuated. Upon opening the valve connecting both vessels, adsorption occurs. Recirculating the gas in the system will enhance equilibrium between the gas and the sorbate phase. If this is attained, a gas sample may be taken to determine its molar concentrations (y_i^f, i = 1...N). From this information the masses (m_i^a) of components i = 1...N of the adsorbate can be determined as follows: First we note the mass balances of all components

$$m_i^* = m_i^a + m_i^f, \qquad i = 1...N \tag{2.20}$$

Here m_i^* is the mass of component i of the sorptive gas filled to the storage vessel (V_{SV}). It can be calculated from the relation

$$m_i^* = \rho_i^* V_{SV} = M_i y_i^* n^*, \qquad i = 1...N \tag{2.21}$$

with the total number of mols (n^*) of the gas originally supplied being given by the thermal equation of state (EOS)

$$n^* = \sum_i^N \frac{m_i^*}{M_i} = \frac{p^* V_{SV}}{RTZ^*} \tag{2.22}$$

Here

$$Z^* = Z(p^*, T, y_1^* ... y_n^*) \tag{2.23}$$

is the compressibility or real gas factor of the gas mixture. Its numerical value can be calculated from an analytical EOS, [2.9-2.25] measured values from pressure (p^*) and Temperature (T) and known molar concentrations ($y_1^* ... y_N^*$) of the gas.

Similarly to (2.21-2.23) we have for the mass of component i in the sorptive gas

$$m_i^f = \rho_i^f V^f = M_i y_i^f n^f, \qquad i = 1...N \tag{2.24}$$

with the total mol number (n^f) again being obtained from the EOS

$$n^f = \sum_i^N \frac{m_i^f}{M^i} = \frac{pV^f}{RTZ^f} \tag{2.25}$$

including the compressibility factor

$$Z^f = Z(p, T, y_1^f ... y_N^f) \tag{2.26}$$

Inserting Eqs. (2.21-2.25) into the mass balance (2.20) and taking Eq. (2.3) into account we get the equations

$$\Omega_{Vi} = m_i^a - \rho_i^f V^{as}, \qquad i = 1...N \tag{2.27}$$

with

$$\Omega_{Vi} \equiv m_i^* - \rho_i^f (V_{SV} + V_{AC}), \qquad i = 1...N \tag{2.28}$$

being the "reduced mass" of component i in the sorbate phase, a quantity which according to its definition (2.28) can be calculated from measured experimental parameters of the adsorption system. In order to calculate (m_i^a) from (2.27) we must introduce an approximate value for (V^{as}). Choosing again the helium volume approximation (2.7), the adsorbed mass indeed is the Gibbs surface excess mass of component (i), cp. Chap. 1, for which we get from (2.27)

$$m_{iGE}^a = \Omega_{Vi} + \rho^f V_{He}^S \tag{2.29}$$

or in view of (2.28, 2.21, 2.24),

$$m_{iGE}^a = \frac{M_i}{RT}\left(\frac{y_i^* p^* V_{SV}}{Z^*} - \frac{y_i^f p}{Z}(V_{SV} + V_{AC} - V_{He}^s)\right), \qquad i = 1...N \tag{2.30}$$

Choosing for (V^{as}) in (2.27) instead of (2.7) the approximation (2.9) we get for the absolute adsorbed masses (m_i^a) of components i = 1...N in view of (2.21-2.25, 2.30), the relations

$$\sum_{k=1}^{N}(\delta_{ik} - y_i^f \frac{M_i p}{\rho_0^L RTZ}) m_k^a = m_{iGE}^a, \qquad i = 1...N \tag{2.31}$$

This is a system of N linear equations from which the absolute adsorbed masses (m_k^a) of all components k = 1...N can be calculated. The symbol δ_{ik} = 0 for $i \neq k$ and δ_{ii} = 1 for i = 1...N is the Kronecker Symbol. For the reference density of the adsorbed phase an approximate value

$$\frac{1}{\rho_0^L} = \sum_{i=1}^{N} \frac{w_i^a}{\rho_{0i}^L} \tag{2.32}$$

with – according to experience – roughly estimated values of the expected concentrations w_i^a is used. Of course results can be improved by iteration, i. e. using numerical solutions of (2.31) to get a new approximation for (2.32) etc.

From either the set (m_{iGE}^a), (2.30) or (m_k^a), (2.31) the concentrations of masses in the adsorbed phase $w_i^a = m_i^a / (\sum_k^N m_k^a)$, i = 1...N and of mole numbers $x_i^a = M_i m_i^a / (\sum_{k=1}^N M_k m_k^a)$, $i = 1...N$ can be calculated. Hence the *selectivities* of any two components i, k of the gas-adsorption-system

$$S_{ik} = \frac{x_i^a / y_i^f}{x_k^a / y_k^f}, \quad i,k = 1...N \tag{2.33}$$

can be determined. These are key quantities for adsorption separation processes.

4.3 Uncertainties or Errors of Measurements

We restrict this discussion to dispersions or mean square deviations (MSD) (σ_{miGE}^2) of the Gibbs excess masses of component i = 1...N of a multicomponent adsorbate as given by Eqs. (2.30). Applying the Gauss law of propagation of error or uncertainty we have in view of (2.28), (2,24), (2.27):

$$\sigma_{miGE}^2 = \sigma_{i*}^2 + \sigma_{if}^2, \tag{2.34}$$

$$\sigma_{i*}^2 = M_i^2[n^{*2}\sigma_{yi*}^2 + y_i^{*2}(\frac{\sigma_{m*}^2}{M^{*2}} + (\frac{m^*}{M^{*2}})^2 \sum_{i=1}^{N} M_i^2\sigma_{yi*}^2)] \qquad 2.34a$$

$$\sigma_{if}^2 = M_i^2[n^{f2}\sigma_{yif}^2 + y_i^{f2}(\frac{\sigma_{mf}^2}{M^{f2}} + (\frac{m^f}{M^{f2}})^2 \sum_{i=1}^{N} M_i^2\sigma_{yif}^2)] \qquad 2.34b$$

Here $n^* = m^* / M^*$ with $M^* = \sum_i^N y_i^* M_i$ is the mol number of the gas in the storage vessel prior to adsorption, likewise $n^f = m^f / M^f$ with $M^f = \sum_i^N y_i M_i$ is the mol number of the sorptive gas after equilibration. The dispersions σ_{yi*}, σ_{yif} refer to concentrations in the gas prior (y_i^*) and after adsorption (y_i^f). Approximate values of the dispersions σ_{yi*}, σ_{yif} referring to the total masses of the sorptive gas prior (m^*) and after adsorptionare (m^f) given by Eqs. (2.16*) and (2.16f) respectively. Formulae (2.33, 2.34 a, b) clearly show the profound influence of the dispersions σ_{yi*}, σ_{yif} of sorptive's gas molar concentrations y_i^*, y_i^f respectively. Measurements of these should be performed with meticulous care in order to get useful values of the masses (m_{iGE}^a) of the adsorbate components. In practice, concentration measurements (by GC or MS) should maintain uncertainties of $\sigma_{yi*} \approx \sigma_{yif} < 10^{-2}$. Provided uncertainties of all the other measured quantities are of the orders of magnitude given at the end of Sect. 2.3, one may expect relative uncertainties for binary (N = 2) mixture measurements of about $(\sigma_{iGE} / m_{iGE}^a) \leq 2\%$, and for ternary (N = 3) mixture measurements of $(\sigma_{iGE} / m_{iGE}^a) \leq 5\%$.

4.4 Example

In this section we present data for the coadsorption equilibria of CH_4 / N_2 gas mixtures on activated carbon (AC) Norit R 1 Extra taken at T = 298 K and total gas pressure p = 0.3 MPa. Measurements were performed using the volumetric – gas chromatographic method described above (N = 2). Standard activation procedures of the AC were applied, namely vacuum (10 Pa) at 400 K for 4 hours, then cooling down to ambient temperature under vacuum within 12 hours. First the system was investigated by M. Tomalla at the author's institution in 1993, [2.20]. Measurements were repeated using the same method by St. Dohrmann in the labs of O. Talu, Cleveland State University, Cleveland, USA in 1997. Results of both measurements are shown in Figures 2.7 and 2.8 below. Figure 2.7 shows the partial molar Gibbs surface excess amounts of CH_4 and N_2 respectively being adsorbed on the AC at constant temperature (298 K) and pressure (0,3 MPa) but at varying molar concentration of methane (y_{CH4}) in the gas phase. Data show that for an equal molar gas mixture ($y_{CH4} = y_{N2} = 0{,}5$) nearly three times as much methane than nitrogen is adsorbed ($S_{CH4,N2} \cong 3$). The dark full circles represent data taken at Cleveland. The open symbols indicate data taken 4 years earlier at Siegen. Differences between the data basically are due to the fact that AC is not a very stable material and the sample used may have changed somewhat over the period of 4 years. Also slightly different activation procedures may have been applied. In Figure 2.8 the corresponding Mc Cabe – Thiele diagram of the equilibria data is presented. Here the data seem to be nearly identical, thus proving experimental consistency.

Today there is a considerable amount of binary and also of some ternary coadsorption data of technical gas mixtures on a variety of sorbens materials available in the literature. The interested reader is referred to Journals like "Adsorption", "Adsorption, Science & Technology", "Langmuir" etc.

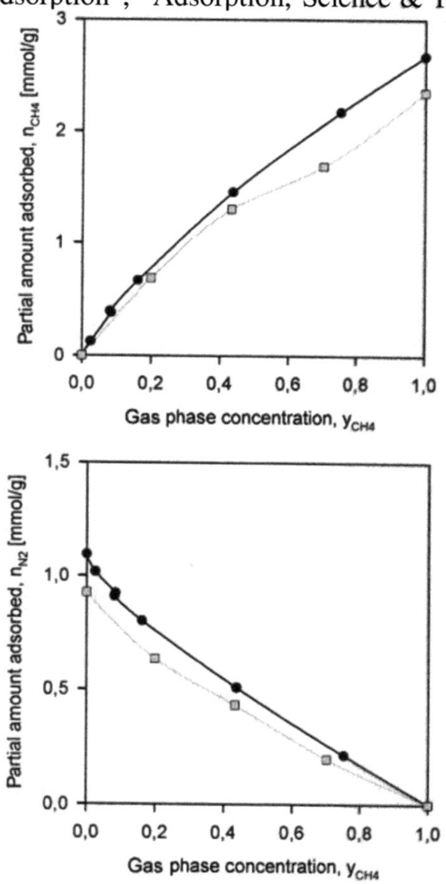

Figure 2.7.
Coadsorption equilibria of CH_4 / N_2 gas mixtures of AC Norit R1 Extra at 298 K, 0.3 MPa. The data indicate partial molar Gibbs excess amounts of CH_4 (upper figure) and N_2 (lower figure) and their dependence on the sorptive gas concentration (y_{CH4}).

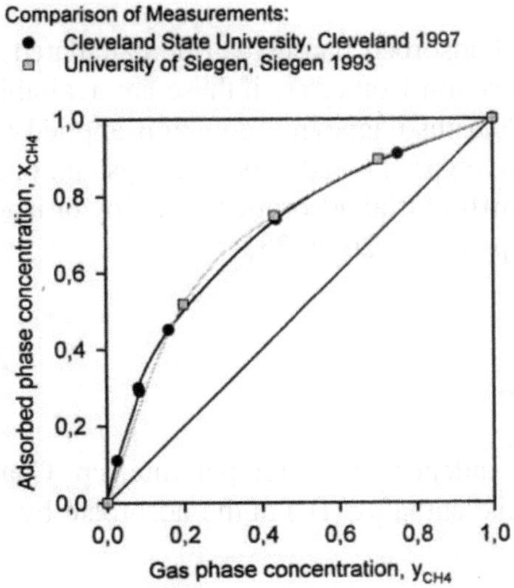

Figure 2.8.
Mc Cabe – Thiele diagram of coadsorption equilibria of CH_4 / N_2 gas mixtures on AC Noit R1 at 298 K, 0.3 MPa. Full circles (●) indicate data taken at Cleveland in 1997, shaded symbols (■) refer to measurements done at Siegen in 1993, [2.20].

5. VOLUMETRIC – CALORIMETRIC MEASUREMENTS THE SENSOR GAS CALORIMETER (SGC)

Assuming the adsorbed phase to be a thermodynamic system in the sense of W. Schottky [2.24], cp. Chap. 1, we can assign it an (integral) enthalpy (H^a) which can be considered as function of either the sorptive gas pressure (p) and temperature (T) or to depend on the mass of the adsorbate (m^a), its temperature (T) and the mass of the sorbent (m^s), i. e.

$$H^a = H^a(p, T, m^s) = H^a(m^a, T, m^s) \qquad 2.35$$

This is an important quantity likewise for single and for multi-component adsorbates which characterizes the energetic state of the adsorbed molecules. It normally is determined by measuring the difference (ΔH) between the enthalpies of the mass (m^a) in the gaseous state (H^f) and the adsorbed state (H^a), i.e. the so-called adsorption enthalpy

$$\Delta H = H^a(m^a, T, m^s) - H^f(p, T, m^a) < 0 \qquad 2.36$$

Today there are many experimental techniques and instruments at hand to perform measurements of this type both for characterization of the sorbent material, cp. for example [2.25], but also for industrial purposes [2.26]. A good overview of adsorption calorimetric measurement methods is given in [2.2].

On principle calorimetric data of adsorbed phases can be calculated from adsorption equilibria data, i. e. adsorption isotherms, if these are available for different temperatures. From the Clausius-Clapeyron equation applied to the phase equilibrium in the ideal sorptive gas (f) – adsorbate (a) system, one can derive an equation for the isosteric differential adsorption enthalpy or isosteric differential heat of adsorption [2.2, p. 43], [2.26, p. 38]

$$h^a(m^a, T, m^s) = R\left(\frac{\partial \ln(p/p_0)}{\partial(1/T)}\right)_{m^a} < 0 \tag{2.37}$$

which here has been assumed to be independent of temperature, cp. Chap. 7, Fig. 7.1. This quantity is related to the enthalpy (H^a) of the adsorbate by

$$h^a = \left(\frac{\partial H^a(m^a, T, m^s)}{\partial m^a}\right)_{T, m^s}, \tag{2.38}$$

i. e. we have

$$H^a = \int_0^{m^a} h^a(m^a, T, m^s)\, dm^a \tag{2.39}$$

In practical applications of Eq. (2.37) the r.h.s. differential quotient has to be approximated by a difference quotient, i. e.

$$\begin{aligned} \frac{\partial \ln(p/p_0)}{\partial(1/T)} &\cong \frac{\ln(p_2/p_0) - \ln(p_1/p_0)}{(1/T_2)(-(1/T_1)} \\ &= \frac{\ln(p_2/p_1)}{T_1 - T_2} T_1 T_2 \end{aligned} \tag{2.40}$$

This may cause considerable deviations of numerical values for (h^a) calculated in this way via (2.37) from measured data of (h^a) or (H^a), which may add up to a 100 % or even more [2.23]. Hence it always is recommended to measure integral adsorption enthalpies (H^a) or integral heats of adsorption (ΔH), (2.36) and to determine the differential heats of adsorption (h^a) by differentiating analytic expressions for $H^a = H^a(m^a, T, m^s)$ to the mass adsorbed, cp. (2.38).

In order to measure simultaneously the mass and the enthalpy of an adsorbed phase, the adsorption vessel in the volumetric instrument, Fig. 2.1, has to be replaced by a calorimeter vessel. Traditionally this vessel is provided with a large number of thermocouples allowing to measure changes in temperature due to transfer processes of the heat of adsorption flowing from the sorbent/sorbate system to the (isothermal) heat bath covering the sorption system. We here are not going to present this technique in detail as this has been done in literature several times, [2.2], [2.25], [2.27], [2.28] and the literature cited therein. Instead we would like to describe a new type of calorimeter avoiding thermocouples but using instead an inert gas as sensor medium. Hence, we will call it in what follows a sensor gas calorimeter (SGC). In Sect. 5.1 the basic experimental information is provided. The theory of the instrument is outlined in Sect. 5.2 followed by examples of calibration measurements. Examples of simultaneous mass and heat of adsorption measurements are presented in Sect. 5.3. These are compared to results of analytical calculations of the heat of adsorption via Eq. (2.37) and discussed to a certain extend.

5.1 Experimental

In 1994, W. Langer proposed to use a newly designed calorimeter for simultaneous measurements of heats and isotherms of gas adsorption or desorption processes [2.29]. This instrument may be called a sensor gas calorimeter (SGC), as the heat flowing during a gas adsorption process from the sorbent/sorbate system to the sorptive gas will then pass a gas jacket surrounding the adsorption vessel. This gas, normally He or N_2, acts as a sensor. The heat flux introduced changes its temperature and, as the gas volume is constant, also its pressure. Actually, the time-dependent pressure signal can be correlated with the heat flux passing the sensor gas and after integration over time gives a measure of the total heat of adsorption released from the adsorption vessel to the thermostat. A schematics of the instrument is given in Fig. 2.9, followed by a photo of the instrument as designed and built by W. Zimmermann in Lab PB-A 320, IFT, University of Siegen [2.23].

The sensor gas calorimeter (SGC) basically consists of a classical volumetric gas adsorption device complemented by two gas thermometers (cp. Fig. 2.9). The core of the instrument is an adsorption vessel which is placed within a second vessel, the sensor gas jacket. This jacket vessel is filled with gas at pressure (p_{SG}) acting as a sensor via a capillary (1) connecting the vessel with a difference manometer (P3). Additionally, a reference vessel also filled with the (same) sensor gas at pressure (p_{RG}) is placed in the thermostat and connected via capillary (2) to the manometer (P3). Upon opening the valves (V7, V7A) the pressures (p_{SG}, p_{RG}) of the sensor gases in the jacket vessel and the reference vessel are equalized, i. e. we have $p_{SG} = p_{RG}$. Thermal equilibrium at temperature (T^*) in the system provided, these pressures remain to be equal and constant even after closing valves (V7,V7A). Consequently, in thermal equilibrium there is a constant signal at the difference manometer (P3).

However, if for any reason the temperature of the sensor gas in the jacket vessel is changed, so will be its pressure (p_{SG}). Hence the gas pressures on both sides of the manometer (P3) will be different. This pressure difference signal ($\Delta p = p_{SG} - p_{RG}$) can be displayed and/or monitored by a supporting data acquisition system. As the sensor gas pressure ($p_{SG} = p_{RG} + \Delta p$) is directly related to the temperature (T_{SG}) of the sensor gas via its thermal equation of state it provides a direct indication of any change of this temperature. Such a change my be caused by heat flowing from (or to) the adsorption vessel to (or from) the thermostat fluid via the sensor gas.

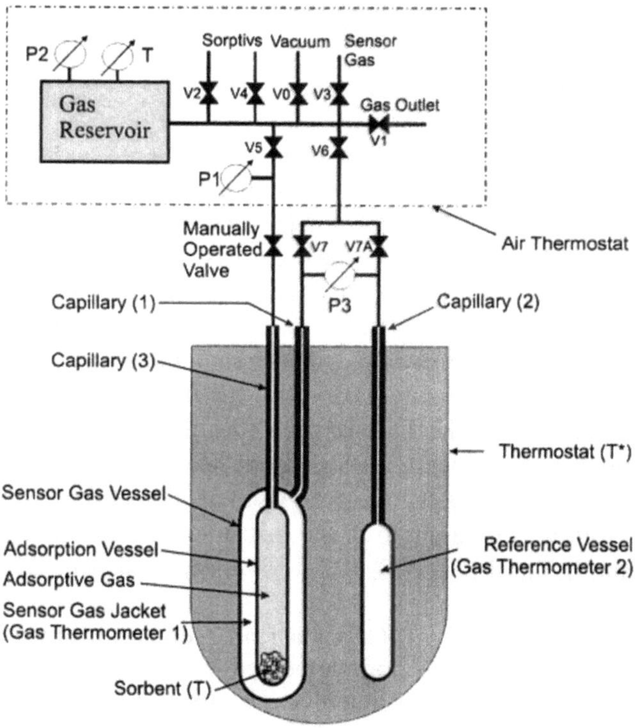

Figure 2.9. Schematic diagram of a sensor gas calorimeter (SGC), [2.23].

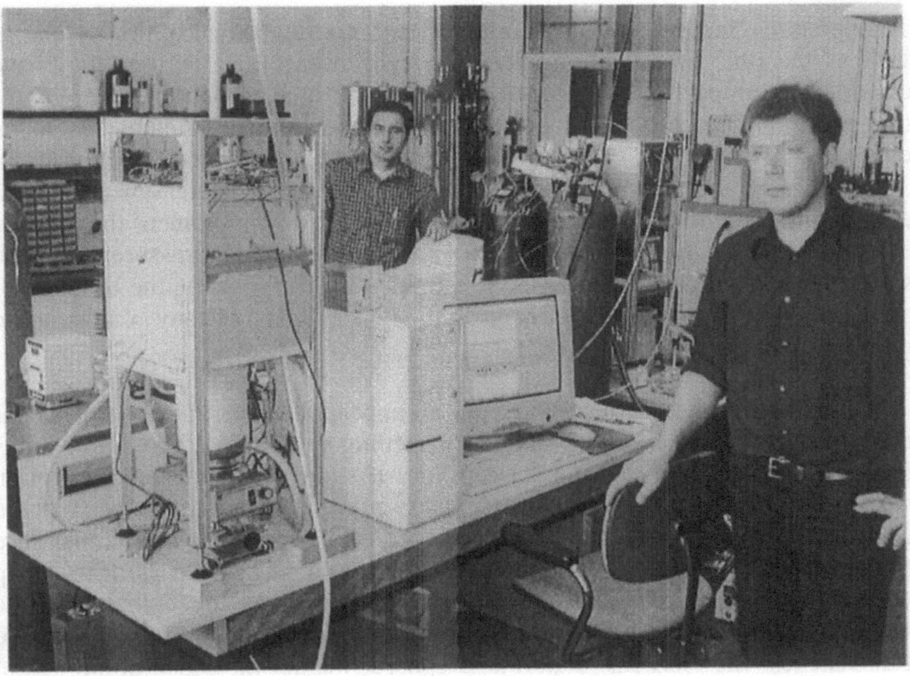

Figure 2.10. Sensor gas calorimeter (SGC) for simultaneous measurements of adsorption isotherms and enthalpies. © IFT, University of Siegen, 2003. E-mail:zimmermann@ift.maschinenbau.uni-siegen.de

The instrument is complemented by a gas reservoir and several solenoid valves for the dozing of the gases coming from the gas reservoirs in order to adjust the sorptive gas pressure (p) inside the adsorption vessel. The capillary (3) connecting the adsorption vessel with valves and the sorptive gas reservoir outside the thermostat has to be chosen carefully: diameter and length should be such that the sorptive gas coming from the reservoir assumes the temperature of the thermostat (T*), but also evacuation of the adsorption vessel, which for example is necessary for activation processes of the sorbent material, does not take too long or even becomes practically impossible.

Auxiliary equipment like vacuum system, thermostat, a data acquisition system including a PC and safety installations also have to be provided. Depending on the specific surface and the density of the material used for examination an amount between 0.5 g and 2g has to be filled into the adsorption vessel. At the beginning of an experiment the sample material is activated by simultaneous evacuation and heating up of the adsorption vessel outside the thermostat. For activated carbon sorbent materials temperatures about (100°C - 150°C) are recommended. For zeolite sorbent materials often activation temperatures about (400°C) and even higher may be needed.

Temperature gradients during heating up and cooling down of the sample as well as the time of activation should be chosen carefully as they may have considerable impact on the properties of the activated sorbent material. After this pre-treatment procedure is completed the whole instrument has to be evacuated. Then sensor gas should be provided to the gas thermometers (1,2). After this the instrument is placed within the thermostat. When thermal equilibrium is reached valves (7,7A) are closed. At this moment the reading of the difference pressure gauge corresponding to equal pressures on both sides of the instruments membrane is a stable signal providing the baseline of any pressure signal measurement and hence also of any calorimetric measurement.

After filling the gas reservoir with an appropriate amount of sorptive gas the adsorption process can be started by opening valve (V5). The sorptive gas is flowing from the reservoir via capillary (3) into the (evacuated) sorption vessel. During this process it is important to make sure that the velocity of the gas flow is small enough to allow thermal equilibration of the gas temperature to thermostat temperature (T*). Otherwise it would become difficult to set up a concise energy balance of the adsorption process as the enthalpy of the incoming sorptive gas would not be known exactly. As soon as the sorptive gas reaches the sorbent material prepared inside the adsorption vessel adsorption occurs and the heat of adsorption is released to the sorptive gas and then transferred via the wall of the adsorption vessel to the sensor gas inside the sensor gas vessel, cp. Fig. 1. As the temperature of the sensor gas is increased due to this heat transfer so is its pressure (p_{SG}) compared to the pressure in the reference vessel (p_{RG}). Hence there is a pressure difference $\Delta p = p_{SG} - p_{RG} \neq 0$ which is monitored and registered by the difference pressure manometer (P3). Additionally, pressures at manometers (P1) and (P2) also are monitored allowing to determine the amount of gas adsorbed i. e. to evaluate the adsorption isotherm.

A numerical example for a pressure difference signal recorded during a calibration process of an SGC is given in Fig. 2.11. It shows an increase in its first part reaching a maximum value after which it again approaches the baseline i. e. its initial value indicating thermal equilibrium in the overall system. After this, sorptive gas again can be added to the adsorption vessel increasing thus the inside pressure (p) and starting a new adsorption process which finally will lead to a new adsorption equilibrium at some higher pressure ($p' > p$).

5.2 Outline of Theory and Calibration

If heat is generated inside the adsorption vessel of the SGC, Fig. 2.9, it will be transferred via the sensor gas to the surrounding thermostat fluid. According to the Newton-Fourier Law of heat transfer we have for the total heat flow

$$\dot{Q} = K(T(t) - T^*) \quad 2.41$$

Here K is an instrument parameter to be determined by calibration experiments; T = T(t) indicates the time dependent average temperature of the sensor gas and T^* = const is the temperature of the thermostat fluid surrounding the sensor gas jacket.*) For an ideal sensor gas the temperature T easily can be related to its pressure ($p_{SG}(t)$) as

$$T = \frac{p_{SG} V_{SG}}{m_{SG} R / M_{SG}} \quad 2.42$$

In (2.42) V_{SG}, m_{SG} are the (constant) volume and mass of the sensor gas and M_{SG} its molar mass. Combining (2.41, 2.42) we get for the total heat released during a process

$$Q = \frac{K V_{SG}}{m_{SG} R / M_{SG}} \int_0^\infty (p_{SG}(t) - p^*) dt \quad 2.43$$

As $\Delta p(t) \equiv p(t) - p^*$ is the time dependent signal recorded by the pressure difference manometer (P3, Fig. 2.9), with $p^* = p_{RG}$ indicating the pressure in the reference gas thermometer, the heat (Q) can be calculated from this relation by simple integration. As an example pressure signals ($\Delta p(t)$) as responses to Ohm's heat inputs of (0.5, 1.0, 1.5...5.0) J inside the adsorption vessel are presented in Figure 2.11.

*) The Newton-Fourier Law (2.41) seems to be adequate for the heat transfer process from the sensor gas to the thermostat as long as there is no turbulent convection within the gas, i. e. for Grashof numbers $Gr < 10^6$, [2.30, 2.31]. For situations with $Gr > 10^6$ it has to be generalized taking aftereffects, i. e. the history of the sensor gas temperature $\{T(s), 0 \leq s \leq t\}$ into account. This can be done by using the theory of Linear Passive Systems (LPS), cp. Chap. 6 and the literature cited there. Details will be published in a forthcoming paper [2.32].

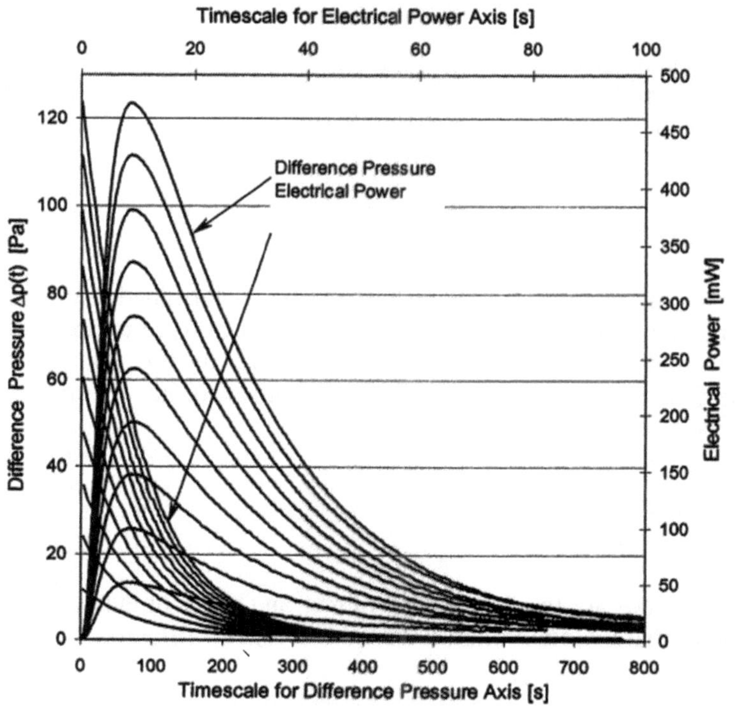

Figure 2.11. Calibration experiments of the sensor gas calorimeter (SGC). Ohm's heat of (0.5, 1.0, 1.5 ... 5.0) J is supplied by electric heating inside the adsorption vessel. The corresponding pressure signals increase in height and size with increasing heat input. The area below these curves is direct proportional to the total heat supplied to the SGC. Sensor gas used: N_2, 5.0, $p^* = 0.15$ MPa, $T^* = 298$ K.

The monotonously decreasing curves at the left side present the input of electric power in (mW), their time integral the total electric energy supplied to the system. The curves showing a maximum are the pressure signals corresponding to the respective electric energy / Ohm's heat signals. It should be observed that the time scales related to the electric power input (upper scale) and the pressure signal (lower scale) are not identical, but for sake of graphical presentation differ by a factor 8. As indicated by Eq. (2.43) the area

$$A \equiv \int_0^\infty \Delta p(t)dt \qquad 2.44$$

below the signal pressure curves ($\Delta p = \Delta p(t)$) and the total Ohm's heat supplied (Q) are linearly related, i. e. we have $Q \cong A$. The respective correlation of all curves depicted in Fig. 2.11 is shown in Fig. 2.12.

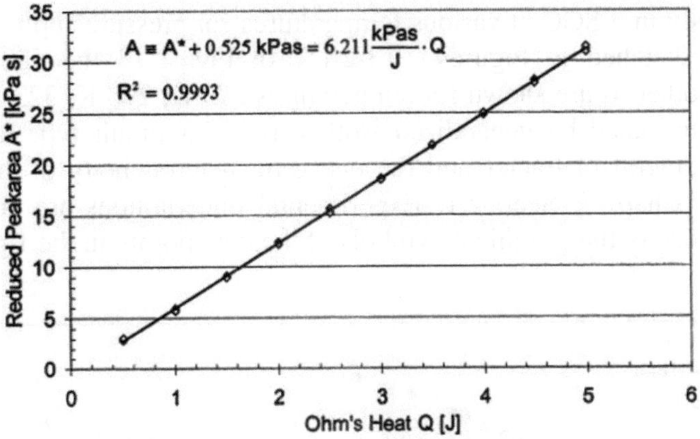

Figure 2.12. Linear correlation of the so-called peak area A, Eq. (2.44) and its actually measured reduced value $A^* = A - 0.525$ kPas, to the total Ohm's heat (Q) supplied during calibration experiments, [2.23]. The statistical, i. e. least square correlation function is indicated in the diagram together with the correlation coefficient (R). Sensor gas: N_2, 5.0, $p^* = 0.16$ MPa, $T^* = 298$ K, [2.23].

As data recording during all calibration experiments was always stopped after $t_E = 800$ s, only a certain portion (A^*) of the total peak area (A) could be observed, i. e. we have

$$A = A^* + \Delta A \qquad 2.45$$

At the time (t_E) all pressure signals Δp ($t \geq t_E$) were at the detection limit of the difference manometer. Hence one could expect ΔA to assume the same numerical value for all experiments performed. This holds true, indeed, and the least square minimization of all data taken led to $\Delta A = 0.525$ kPas for the set of experiments shown in Figure 2.11.

5.3 Example

Integral heats of adsorption and Gibbs excess masses of n-butane (C_4H_{10}) adsorbed on activated carbon (AC BAX 1100) have been measured simultaneously in a SGC at various temperatures for pressures up to 0.2 MPa. Results are sketched in Figures 2.13-2.15. In Fig. 2.13 the Gibbs excess adsorption isotherms are shown for temperatures 273 K, 298 K, 323 K, 343 K. Data were correlated by generalized isotherms of Langmuir type taking into account the fractal character and the energetic heterogeneity of the sorbent material, cp. Chap. 7, Sect. 2.1. Experimental uncertainties are about three times of the size of the graphical symbols of the data points in the figure.

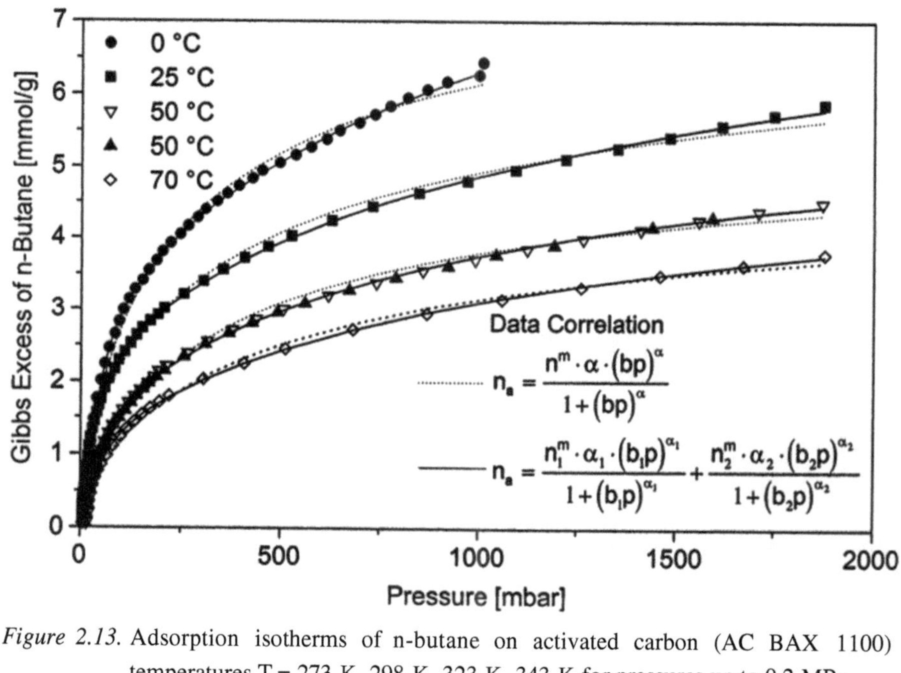

Figure 2.13. Adsorption isotherms of n-butane on activated carbon (AC BAX 1100) at temperatures T = 273 K, 298 K, 323 K, 343 K for pressures up to 0.2 MPa.

In Figure 2.14 the integral heats of adsorption (ΔH, cp. (2.36)) per unit mass of sorbent ($m^s = 1$ g) are shown as functions of the Gibbs excess mass of n-butane adsorbed. Data were taken for 4 temperatures. Measurements at T = 323 K were performed twice. Uncertainties of data are about three times the size of the graphical symbols presenting data points. The experiments performed with the GSC so far indicate that the sensitivity of measurements decreases with increasing temperature but increases with decreasing thermal conductivity of both the sorptive gas and the sensor gas used in the SGC.

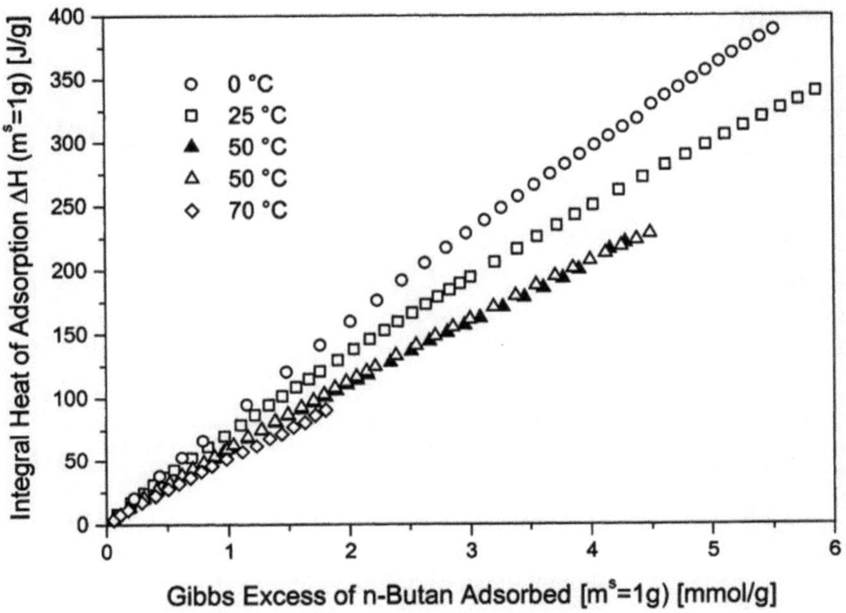

Figure 2.14. Integral heat of adsorption of n-butane (ΔH), per unit mass of sorbent (AC BAX 1100), (m^s = 1 g) as function of the Gibbs excess mass of n-butane adsorbed for temperatures T = 273 K, 298 K, 323 K, 343 K and corresponding pressures up to 0.2 MPa.

In Figure 2.15 integral and differential heats of adsorption (ΔH, (2.36), h^a, (2.38)) of n-butane on activated carbon (AC BAX 1100) are presented as function of the Gibbs excess amount of n-butane adsorbed per unit mass of sorbent. Measurements refer to T = 25 °C and p ≤ 0.2 MPa. Data of the integral heat of adsorption (Δ) were correlated by a polynom of order 5, leading by differentiation to the correlation line of the differential heats of adsorption (o).

Uncertainties of **ΔH-data** are about 3 times the size of the graphical symbols (Δ) representing the data, but of **h^a-data** about 6 times the size of the respective symbols (o).

The differential heats of adsorption (h^a) decrease with increasing amounts of n-butane (C_4H_{10}) adsorbed approaching values of about 40 kJ/mol near saturation pressure p_s (298 K) = 0.22 MPa which is about twice the value of the heat of condensation (r_{C4H10} = 20.9 kJ/mol). Approximate values of h^a

have been calculated via Eq. (2.37) from adsorption isotherm measurements at two different temperatures.

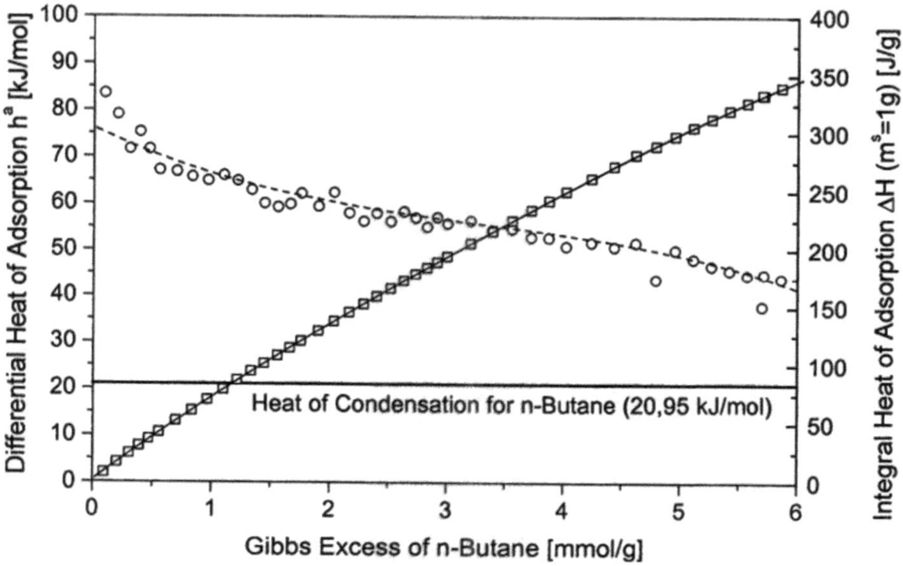

Figure 2.15. Differential and integral heats of adsorption (h^a, (2.38); ΔH, (2.36)) of n-butane on activated carbon (AC BAX 1100) at T = 298 K for pressures up to 0.2 MPa. The correlation line for **h^a-data (o)** is calculated by differentiation of the 5th order polynomial correlation function of the integral heats of adsorption (Δ).

Resulting data are deviating from direct measurement data by up to 30 % at low and 10 % - 20 % at higher adsorption loads. Hence we relinquish to present them here but recommend direct calorimetric measurements if accurate and reliable enthalpy data are needed [2.2, 2.28].

In conclusion it can be said, that the sensor gas calorimeter (SGC) is a very useful instrument for simultaneous measurements of adsorption isotherms and (integral and differential) heats of adsorption. Also hints on the kinetics of the gas adsorption process can be gained from the time dependence of the pressure signal curve, cp. Fig. 2.11. However, to achieve high sensitivity and accuracy of measurements, type and amount of the sensor gas have to be chosen very carefully. At low temperatures (77 K) helium is recommended at reference pressures of about (0.1 – 0,2) MPa. At higher temperatures (298 K) nitrogen should be preferred at the same pressures, [2.23, 2.29].

6. PROS AND CONS OF VOLUMETRY / MANOMETRY

In this section we will discuss some of the advantages and disadvantages of the volumetric or manometric method of measuring gas adsorption equilibria as they have appeared to us after more than 20 years of experimental work in this field.

6.1 Advantages

1. Simplicity
   Volumetric gas adsorption instruments are fairly simple and do not require sophisticated high tech equipment, cp. Figs. 2.1, 2.6. Also the experiment in itself is simple - opening of a valve between gas storage and adsorption chamber. Therefore, volumetric instruments easily can be automated and run without permanent supervision for hours and days [2.3, 2.4]. To achieve this it is recommended that every instrument is equipped with its own personal computer (PC) for data handling, storage and transmission. For screen based operation and control the program "Lab View" (Version 6.1) has proved to be very useful. Combined with the program "PC Anywhere" it could allow remote operation of the measuring instrument from nearly any place around the world.

2. Measuring techniques
   Volumetric measurements of gas adsorption equilibria reduce, if the mass of the sorbent sample used has been determined, to measurements of pressures and temperatures in gas phases. For this a variety of high precision measuring instruments operating in a fairly wide range and partly also in corrosive environment are available today. Of course these instruments prior to measurement have to be calibrated with meticulous care which may be laborious and even cumbersome. For pressure measuring devices, calibration with pressure maintaining valves of Desgranges & Huot has proved to be successful.

6.2 Disadvantages

1. Amount of sorbent material
   For volumetric measurements of technical gas-sorbent equilibria a certain amount of sorbent material is needed, usually several grams. This is caused by the fact that only then considerable changes in the gas pressure which are due to adsorption can be observed. If only tiny

amounts – say several milligrams – of the sorbent are available, gravimetric measurements are strongly recommended.

2. Approach to equilibrium
   Gas adsorption processes may last for seconds, hours or – sometimes – even days. Therefore one never can be sure whether mermodynamic equilibrium in a volumetric experiment has been realized. Hence the time which should elapse between opening the expansion valve and reading of instruments, especially thermometer and manometer has to be chosen according to experience or accompanying gravimetric measurements which – contrary to volumetry / manometry – also provide information on the kinetics or the sorption process, cp. Chap. 3.

3. Wall sorption
   Upon expansion from the storage vessel the sorptive gas may not only be adsorbed on the surface of the sorbent material but also on the walls of the adsorption vessel and the tube connecting both vessels. This may cause additional uncertainties in measurement. These often but not always can be reduced by performing complementary experiments with gas expansion to the *empty* adsorption chamber including no sorbent material at all. To reduce wall adsorption electropolishing of all inner surfaces is recommended. An experiment allowing to determine wall adsorption is described in Chapter 4, Sect. 3.6.

4. Uncertainties in step-up experiments
   In step-up pressure experiments, i. e. gas expansion and adsorption processes with remnant gas in the adsorption chamber, cp. Figs. 2.1, 2.6, the uncertainties of the adsorbed mass accumulate due to the algebraic structure of the sorptive gas mass balance equation

$$\Delta m^*_{n+1} + m^f_n + m^a_n = m^f_{n+1} + m^a_{n+1}, \quad n = 0,1,2... \qquad 2.35$$

   Here Δm^*_{n+1} indicates the mass of gas added to the storage vessel in step (n+1) at closed expansion valve. The quantities m^f_n, m^a_n are the masses of the gas in the sorptive phase and of the adsorbed phase at the end of the n-th experiment respectively. Note that for step-up adsorption experiments $\Delta m^*_{n+1} > 0$, whereas for step-down *desorption* experiments $\Delta m^*_{n+1} < 0$. In practice the total number of steps should not exceed 3-4 as then uncertainties of about 20 % and more typically occur.

5. In situ activation procedures
   Prior to adsorption experiments the sorbent sample should be activated by heating the adsorption chamber to (say) 420 K for activated carbons, 670 K for zeolites, degassing etc. During this procedure the mass of sorbent is changed by typically 1 % or more. This change normally can not be included in volumetric experiments but must be taken into account afterwards, at least approximately.

Volumetric / manometric adsorption experiments do not give information on the *kinetics* of the process; they are not useful at extreme low or high pressures as pressure measurements in these regions become difficult and accurate EOS of the sorptive gases not always are available. Also there may be problems with the thermostatization of the vessels as during expansion the gas (normally) is cooled down due to the Joule-Thomson-effect and it will take some time till heat transfer (and radiation) bring both sorbent and sorptive gas to the same temperature again. This is especially important for desorption experiments which will take even longer till thermal equilibrium is reached [2.7, 2.20].

7. LIST OF SYMBOLS

A collection of most of symbols used in this Chapter including their SI-units is given.

m^a_{GE}	kg	Gibbs excess mass of an adsorbed phase
m^*_i	kg	mass of component i = 1...N of a sorptive gas mixture supplied to the storage vessel prior to adsorption
m^a_i	kg	mass of component i = 1...N being adsorbed on a certain mass of sorbent (m^s)
m^f_i	kg	mass of component i = 1...N of a sorptive gas in an adsorption equilibrium state
m^a_{iGE}	kg	Gibbs excess mass of component i of an adsorbed phase

m^*	kg	mass of sorptive gas in storage vessel prior to adsorption
m^a	kg	mass of adsorbate on the surface of a certain mass (m^s) of sorbent
m^f	kg	mass of fluid, i. e. gaseous or liquid sorptive phase
$M_{He} = 4.0026$	g/mol	molar mass of helium
M_i	g/mol	molar mass of component i = 1...N of a sorptive gas mixture
n^*	mol	number of mols in a gas mixture filled in the storage vessel prior to adsorption
n^f	mol	number of mols of a sorptive gas mixture in a gas-adsorption equilibrium state
p	Pa	pressure of sorptive gas
R=8.314	J/Kmol	universal gas constant
S_{ik}	1	selectivity of component i relative to component k in a multi-component gas-adsorption equilibrium state
T	K	absolute temperature
V_{AC}	m^3	volume of sorption chamber
V^{as}	m^3	volume of the combined sorbate and sorbent system
V_{SV}	m^3	volume of storage vessel
V^s_{He}	m^3	volume of a (porous) sorbent measured by helium expansion experiments
$w_i^a = m_i^a / (\sum_k m_k^a)$	1	mass concentration of component i in an adsorbed phase

x	[x]	measurable physical quantity
$\bar{x}$	[x]	mean value of a measurable quantity x
x_i^a	1	molar concentration of component i in an adsorbed phase
Δx	[x]	experimental uncertainty of the numerical value $(x \pm \Delta x)$ of a measurement of the quantity (x)
y_i^f	1	molar concentration of component i = 1 ...N in a sorptive gas mixture
y_i^*	1	molar concentration of component i = 1 ...N in a sorptive gas mixture supplied to the storage vessel prior to adsorption
Z	1	compressibility or real gas factor
Z_{He}	1	compressibility or real gas factor of helium
ρ^f	kg/m^3	density of a sorptive gas
ρ_0^L	kg/m^3	density of a sorptive medium in a liquid reference state
ρ_{0i}^L	kg/m^3	density of pure component i of a sorptive in a reference liquid state
σ_x^2	$[x^2]$	dispersion, variance or mean statistical deviation (MSD) of a measurable quantity (x)
$\Omega_V = m^a - \rho^f V^{as}$	μg	reduced mass of adsorbed phase measured volumetrically, cp. eqs. (2.4, 2.5)

REFERENCES

[2.1] **Pohl R. W.**
Experimentalphysik, Vol. 1, Mechanik, Akustik, Wärmelehre
Springer, Berlin etc., 1954.

[2.2] **Rouquerol F., Rouquerol J., Sing K.S.W.**
Adsorption by Powders and Porous Solids, Academic Press, San Diego, USA, 1999.

[2.3] **Keller J. U., Robens E., von Hohenesche Cedric du Fresne**
Thermogravimetric and Sorption Measurement Techniques / Instruments,
Proceedings of 6th Int. Symposium on the Characterization of Porous Solids (COPS VI), Alicante, Spain, May 8-11, 2002, Studies in Surface Sciences and Catalysis, Vol. 144, F. Rodriguez-Reinoso et al., Eds., p. 387-394, Elsevier,
New York etc., 2003.

[2.4] **Keller J. U., Robens E.**
A Note on Sorption Measuring Instruments,
Proceedings of the 29th Int. Conf. on Vacuum Microbalance Technology, Lexica Ltd.,
J. of Thermal Analysis and Calorimetry, 71 (2003), 37-45.

[2.5] **Keller J. U., Dreisbach F., Rave H., Staudt R., Tomalla M.**
Measurement of Gas Mixture Adsorption Equilibria of Natural Gas Compounds on Microporous Sorbens,
Adsorption, 5 (1999), 199-214.

[2.6] **Dreisbach F., Staudt R., Keller J. U.**
High Pressure Adsorption Data of Methane, Nitrogen, Carbon Dioxide and their Binary and Ternary Mixtures on Activated Carbon
Adsorption, 5 (1999), 215-227.

[2.7] **Staudt R.**
Gasadsorption, Ingenieurwissenschaftliche Grundlagen, Messmethoden zur Bestimmung und Modelle zur Beschreibung von Adsorptionsgleichgewichten reiner Gase und Gasgemische an porösen Stoffen,
Habilitationsschrift, IFT, University of Siegen, Siegen, 2000.

[2.8] **Staudt R., Bohn S., Dreisbach F., Keller J. U.**
Gravimetric and Volumetric Measurements of Helium Adsorption Equilibria on Different Porous Solids, p. 261-266,
Proceedings of IV Conference on Porous Solids, COPS IV, Bath UK, 1996, B. Mc Enany et al., Eds., The Royal Society of Chemistry, Special Publ. No. 213, London, 1997.

[2.9] **Reid R.C., Prausnitz J. M., Poling B. E.**
The properties of Gases and Liquids
Mc Graw Hill, New York etc., 1983.

[2.10] **Sengers J. U., Kayser R. F., Peters C. J., White H. J., Eds.**
Equations of state for Fluids and Fluid Mixtures, 2 Vs.
IUPAC Physical Chemistry Division,
Commission on Thermodynamics, Elsevier, Amsterdam etc., 2000.

[2.11] **International Union of Pure and Applied Chemistry**
Physical Chemistry Division, Commission of Thermodynamics,
International Thermodynamic Tables of the Fluid State – Nos. 1, 2, 3..., Pergamon Press, Oxford, UK, 1970 - ...

[2.12] **DECHEMA, e. V.**
DETHERM, DECHEMA Chemistry Data Series
Database, Version 1.4 (December 2000)
Frankfurt Main, Germany, 2000, Cp. also: www.dechema.de

[2.13] **Wagner W., de Reuck M.**
International thermodynamic tables of the fluid state – 13, methane, Blackwell Science, Oxford, 1996.

[2.14] **Wagner W., Pruß A.**
The IAPWS formulation 1995 for the thermodynamic properties of ordinary water substance for general and scientific use. J. Phys. Chem. Ref. Data 31(2002), 387-535.

[2.15] **Span R., Wagner W.**
Equations of state for technical applications.
I. Simultaneously optimized functioned forms for nonpolar and polar fluids. Accepted for publication in Int. J. Thermodynamics 23 (2002).
II. Results for nonpopular fluids. NAL.
Accepted for publication in Int. J. Thermodynamics 23 (2002).

[2.16] **International Organization for Standardization, ISO**
Guide to the expression of uncertainty in measurement,
2nd Printing, 1995, CH – 1211 Geneva

[2.17] **American Society Mechanical Engineers, ASME**
Instruments and Apparatus, Pt. 2, "Measurement Uncertainty"
ANSI / ASME PTC 19.1, New York, 1985.

[2.18] **Gränicher W. H. H.**
Messung beendet - was nun?
Einführung und Nachschlagewerk für die Planung und Auswertung von Messungen,
Hochschulverlag AG an der ETH Zürich,
B. G. Teubner Stuttgart, 2. Aufl., 1996.

[2.19] **Schein E.**
Optimierung eines Wärmeleitungsgasdruckkalorimeters zur simultanen Messung der Gasadsorption und Wärmetönung für 298,15 K,
Diploma Thesis, Dept. Chemistry, University of Siegen, 2001,
pdf-file available from the author.

[2.20] **Tomalla M.**
Experimentelle Untersuchung der Koadsorptionsgleichgewichte von CH_4/N_2 und CO_2/CH_4-Gasgemischen an Aktivkohle bei T = 298 K im Bereich p = 0 - 12 MPa,
PHD-Thesis, Institute of Fluid- and Thermodynamics, University of Siegen, 1994.

[2.21] **Hufton J.R., Mayorga S., Sircar S.**
Sorption-Enhanced Reaction Process for Hydrogen Production,
AIChE Journal, 45 (1999), p. 248, also as paper 157 c at 1998 Annual AIChE Meeting, Miami Beach, Fl., USA.

[2.22] **Buch R. et al.**
Solar-Hybrid Gas Turbine-Based Power Tower System (REFOS),
Proceedings of Solar Forum 2001, Solar Energy: The Power to Choose, April 21st –25th, 2001, Washington D. C.

[2.23] **Zimmermann W., Keller J. U.**
A new calorimeter for simultaneous measurements of isotherms and heats of adsorption, Thermochimica Acta, 405 (2003), p. 31-41.

[2.24] **Schottky W., Ulrich H., Wagner C.**
Thermodynamik (in German),
Reprint of the 1929 Edition, Springer, Berlin (West), 1973.

[2.25] **Guillot A., Stoeckli F., Banguil Y.**
The Microporosity of Activated Carbon Fibre KF 1500 Assessed by Combined CO_2 Adsorption and Calorimetry Techniques and by Immersion Calorimetry,
Adsorption, Science & Technology, 18 (2000), p. 1-14.

[2.26] **Kast W.**
Adsorption aus der Gasphase, Ingenieurwissenschaftliche Grundlagen und technische Verfahren, Verlag Chemie, Weinheim, Germany, 1988.

[2.27] **Webb P. A., Orr C.**
Analytical Methods in Fine Particle Technology,
Micromeritics Inc., Norcross, GA, 1997.

[2.28] **Hemminger W., Höhne G.**
Grundlagen der Kalorimetrie,
Verlag Chemie, Weinheim, Germany, 1979.

[2.29] **Langer W.**
Ein Wärmeleitungs-Gasdruck-Kalorimeter und die simultane Messung von Isothermen und Wärmen der Adsorption von N_2 an SiO_2, Ph-D thesis, Dept. of Chemistry & Biology, University of Siegen, Siegen, 1994.

[2.30] **Schlichting H.**
Boundary Layer Theory, 7th Ed. Mc Graw Hill, New York etc., 1979.

[2.31] **Baehr D., Stephan K.**
Wärme- und Stoffübertragung (in German)
2nd Ed., Springer, Berlin etc., 1996.

[2.32] **Keller J. U., Zimmermann W.**
The Gas Sensor Calorimeter, Theoretical Analysis and Calibration Experiments,
Thermochimica Acta, in preparation, 2004.

Chapter 3

GRAVIMETRY

Abstract The physical principles and experimental techniques of pure gas- and multicomponent gas adsorption measurements by the gravimetric method are outlined. Two beam microbalances and the magnetic suspension balance are considered. Several examples are given and uncertainties of data are discussed. Pros and cons of the gravimetric method are outlined. Lists of symbols. References.

1. INTRODUCTION

Gravimetry is a fairly new method for investigation of sorption phenomena of gases in porous solids, although comparing masses by weighing them in the gravity field of earth is a very old technique going back to biblical times [3.1]. The main reason for the newness of this has been the lack of highly sensitive balances able to measure small relative changes in the weight of a sorbent sample of – for example **(10^{-6} g/g)** – or even less. Though at the end of the nineteenth century fairly accurate mechanical balances were available in many European countries [3.1], it was only with the design of the electro-magnetically compensated two beam microbalance by Th. Gast, manufactured commercially by Sartorius, Göttingen, Germany during 1965-1995, that gravimetric gas adsorption measurements became feasible for non-corrosive gases within a certain range of pressures (< 15 MPa) and temperatures (< 450 K). The next milestone was the development of the single beam magnetic suspension balance (MSB) initiated also by Th. Gast but really accomplished by Lösch, Kleinrahm, and Wagner in the 1980s and today commercially available from Rubotherm GmbH, Bochum, Germany [3.2-3.6]. In a MSB the gas and sorbent filled vessel is completely mechanically decoupled from the microbalance thus allowing also measurements with highly corrosive sorptive gases in a broad range of pressures (< 100 MPa) and temperatures (< 1500 K). Today there are many companies around the world offering single beam and two beam microbalances operating both under vacuum and also high pressure conditions. A list of suppliers of vacuum balances is given in Table 1 below [3.7-3.9]. Automated instruments including microbalances are available for various purposes among them water vapor adsorption/desorption measurements and thermogravimetry, cp. Tab. 2, [3.6, 3.10]. Technical progress in the design and application of microbalances is mirrored in a series of (small) tri-annual conferences on vacuum microbalance techniques (VMT) whose proceedings are a very useful source of technical information in this field [3.11]. The physics of gravimetric sorption measurements is not as simple as it may occur on the first glance. As the sorbent material always is surrounded by a sorptive gas (or liquid, cp. Sect. 4.4), the microbalance can only register the difference between the sample's weight and its buoyancy force which includes the volume of the porous sample. It is this quantity which normally is not known exactly and therefore has to be approximated, similar to what is necessary in the volumetric/manometric method, Chap. 2.

Table 3.1. Manufacturer of vacuum balances*)

manufacturer	V	T	S	M
Beckman Instruments, Fullerton, CA 82834, USA		•		
Cahn Instruments, 16207 South Carmenita Rd., Cerritos, CA 90701 USA	•	•		
CI Electronics Ltd. Brunel Rd. Churchfields Salisbury, Wilts. SP2 7PX, U.K.	•	•		
Linseis GmbH, Viellitzer Str. 43, 95100 Selb, Germany		•		
Mettler-Toledo AG, CH-8606 Greifensee, Switzerland		•		•
Perkin-Elmer, 761 Main Ave, Norwalk CT 06859-0012, USA	•	•		
Rheometric Scientific, Surrey Business Park, Weston rd. Kiln Lane, Epson, Surrey KT 17 1JF, U.K.		•		
Rubotherm, Universitätsstr. 142, D-44799 Bochum, Germany			•	
SETARAM, 7 rue de l'Oratoire, F-69300 Caluire, France	•	•		
Sartorius AG, D-37070 Göttingen, Germany				•
TA Instruments, 109 Lukens Drive, New Castle, DE 19720-2795, USA		•		

V = vacuum balance, T = thermobalance, S = suspension balance, M = microbalance

*) This table in no way claims to be exclusive or complete.

Table 3.2. Manufacturers of gravimetric gas sorption measuring instruments*)

Manufacturer	Balance	S	P	C	W	D
Cahn Instruments, 5225 Verona Road, Bldg. 1, Madison, WI 53711-4495, USA	Cahn	•	•	•	•	
CI Electronics Ltd. Brunel Rd. Churchfields, Salisbury, Wiltshire, SP2 7PX, U.K.	CI			•	•	
Hiden Analytical, 420 Europa Bd., Gemini Business Park, Warrington WA5 5UN, U.K.	Cahn	•	•	•	•	
Rubotherm, Universitätsstr. 142, 44799 Bochum, Germany	Rubotherm			•	•	•
Surface Measurement Systems, P.O. Box 1933, Marlow, Bucks. SL7 3TS, U.K.	Cahn	•	•	•	•	
VTI Corp. 2708 W. 84th Street Hialeah, FL 33016, USA	Cahn or CI	•	•	•	•	

S = specific surface area, P = pore size distribution, C = chemisorption, W = water vapour, D = density

Gravimetric adsorption measurements are used today to characterize porous media [3.10, 3.12, 3.19], to measure gas adsorption equilibria [3.13, 3.14, 3.18, 3.55], and to investigate adsorption kinetics [3.14-3.18], [4.21-4.23]. We here will mainly discuss methods to measure pure gas and mixture gas adsorption equilibria gravimetrically. Other aspects can only be taken into account by presenting few complementary examples.

This chapter is organized as follows: In Sect. 2 we consider pure gas adsorption measurements by both two beam and single beam balances. Section 3 is devoted to thermogravimetry. In Section 4 multicomponent gas adsorption equilibria are discussed. Finally in Sect. 5 pros and cons of gravimetry especially compared to volumetry/manometry are elucidated. A list of symbols and abbreviations used is given followed by references cited.

*) This table in no way claims to be exclusive or complete.

2. GRAVIMETRIC MEASUREMENTS OF PURE GAS ADSORPTION EQUILIBRIA (N = 1)

2.1 Two Beam Balances

2.1.1 Experimental

Two beam microbalances can be used to measure adsorption equilibria of pure non-corrosive gases like (He, Ar, Ne, N_2, CH_4, etc.) on porous solids. Also the uptake (or release) of moisture (H_2O) from (or to) humid air in hygroscopic substances can be determined fairly well (Hiden-Balance). A schematic diagram of an instrument designed for this purpose is given in Figure 3.1 followed by a photo taken in lab PB-A 318 of IFT, 2003. As clearly can be seen, the two vessels beneath the balance including a sorbent sample and a ballast or tare fixed to the balance respectively are placed in a thermostat. The vessels are also provided with tubes for gas supply and evacuation as well as with manometer(s) and thermometer(s) to measure the pressure (p) and temperature (T) inside. A gas circulator can be helpful in enhancing adsorption and reducing waiting times for adsorption equilibria but is not mandatory for pure gas experiments. On the other hand the circulator may cause problems for fine grained sorbent materials like powders or activated carbon fibers which may be removed or simply blown away from the sorbent vessel. By this the sorbent may be distributed all over the instrument and causes problems – especially in valves.

For design and building a gravimetric adsorption instrument the technical hints regarding choice of materials etc. given here should be observed, cp. also Chap. 4 (2.1), [3.20].

A gravimetric adsorption experiment normally is started by mounting a sample of the sorbent material in the microbalance. Then activation by evacuation, heating and flushing with helium follows. At the end the balance again has to be evacuated to measure the mass of the sorbent sample. After this the sorptive gas can be introduced leading to a change of the balance's recording due to

a) adsorption of gas on the sorbent's surface and
b) buoyancy effects of the sample in the surrounding gas.

As the sorption process of the gas in the sample sometimes may last hours and even days, one has to define "technical equilibrium" as a practical term to finish the measurement.

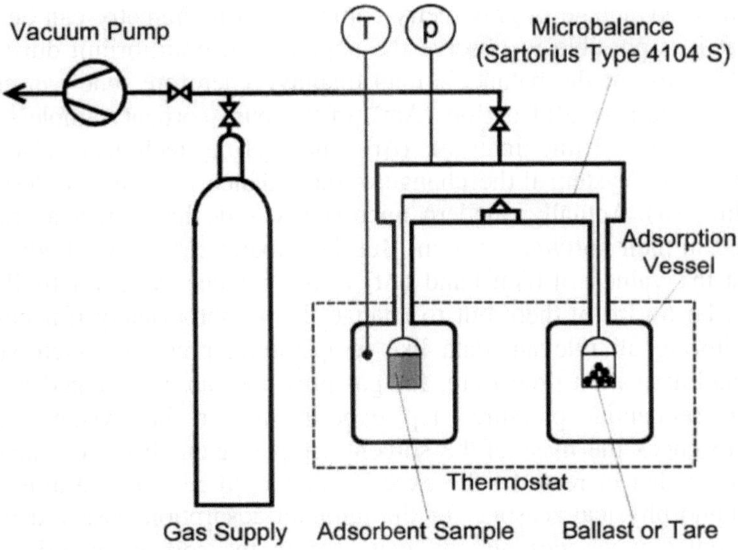

Figure 3.1. Experimental setup for gravimetric measurements of pure gas adsorption equilibria using a two beam balance. For commercial instruments cp. Tab. 3.2.

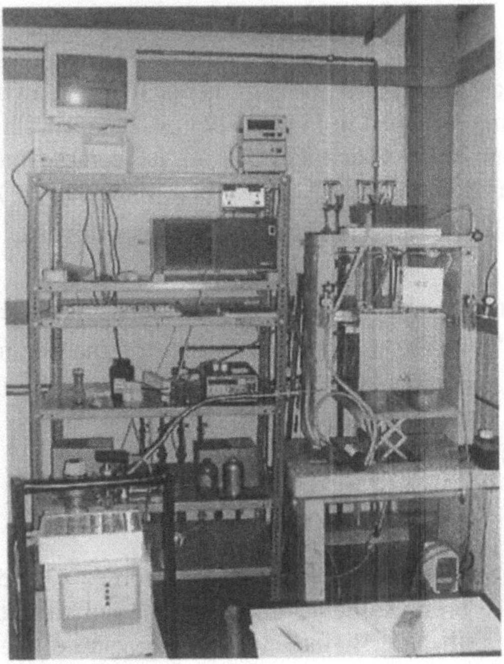

Figure 3.2.

Instrument for gravimetric measurements of pure gas adsorption equilibria using a two beam balance (Sartorius; Th. Gast, 4104 S).

Here the main advantage of gravimetry compared to manometry can be used, namely that it is possible to observe the approach to equilibrium during the adsorption process at the balance's data display. Therefore, one can choose arbitrarily a certain small fraction (Δm^s) of the pure sorbent sample's mass (m^s) and a certain time interval (Δt) and define technical adsorption equilibrium of the system if the change of the balance recording is less than (Δm^s) within (Δt). Actually, modern microbalances do have such a criterion incorporated in their software system. But the experimenter is well advised to check what the values of (Δm^s) and (Δt) in his balance system actually are, not only to be aware of them but to change them appropriately if necessary. After monitoring all relevant data like temperature, pressure, microbalance recording and activation procedure, the gas pressure can be changed in order to perform sequential pressure step experiments or the system can be evacuated to check the mass of the sorbent sample again. If it has changed – normally somewhat increased – the experiment should be repeated unless one can give sound physical reasons for the remnant adsorption observed or rely on experimental experience already gained with the sorbent material used. Indeed remnant or permanent adsorption occurs in nearly all types of sorbent materials if they are exposed to helium for the first time in order to determine their so-called helium-volume. Examples are given in Sect. 2.3, cp. [3.21].

2.1.2 Theory

To elucidate the physics of gas adsorption measurements using a two beam balance we consider the scheme sketched in Fig. 3.3 below. It shows the two beams of the balance with an electromagnetic coil in its center and equipped with two baskets containing a sample of the sorbent material (m^s) and a ballast or tare of approximately the same mass ($m^k \simeq m^s$) respectively.

The electric current in the electromagnetic coil is chosen such that the resulting mechanical moment (M_f^{sa}) of the magnetic forces keeps the beam exactly horizontal. Then the balance of moments is

$$\begin{aligned} -M_f^{sa} &= M_s + g(m^s + m^a)l_s - (B_s + g\rho^f V^{as} l_s) \\ &\quad -(M_k + g m^k l_k) + B_k + g\rho^f V^k l_k \end{aligned} \qquad 3.1$$

Here M_s and M_k are the mechanical moments of the gravity forces of the beam, wire, and basket of the balance on its sorbent sample site (s) and ballast or tare site (k) respectively. Likewise B_s and B_k indicate the moments of the buoyancy forces of beam, wire, and basket on the sorbent site (s) and ballast or tare site (k) respectively. The other quantities and parameters in Eq. (3.1) are: l_s, l_k ...length of the two beams of the balance, m^a ... mass of adsorbate; V^{as}, V^k ... volume of sorbent / sorbate system and ballast or tare, g ... gravity of earth, and ρ^f ... density of sorptive gas.

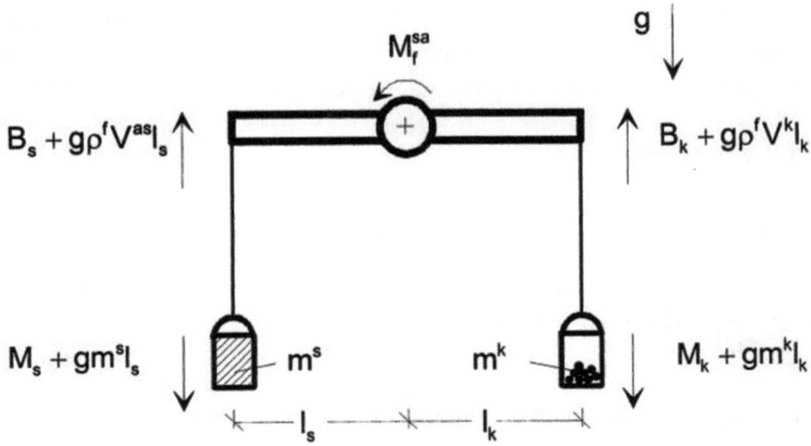

Figure 3.3. Schematics of a two beam balance with electromagnetic compensation (M_f^{sa}) (Th. Gast, Sartorius) loaded with a sorbent sample of mass (m^s) and a ballast or tare of mass (m^k).

In order to calculate the mass adsorbed (m^a) from the momentum balance (3.1) three additional measurements are necessary:

(1) Calibration measurement of the empty balance, i. e. the balance without a sorbent sample and a ballast or tare, in vacuum (M_0). In this case we have : $m^s = 0$, $m^k = 0$, $\rho^f = 0$, $B_s = B_k = 0$. Hence equation (3.1) reduces to

$$-M_0 = M_s - M_k \qquad 3.2$$

(2) Measurement of the loaded balance in vacuum (M_0^s). Since $\rho^f = 0$, eq. (3.1) reads

$$-M_0^s = M_s + \rho m^s l_s - (M_k + \rho m^k l_k) \qquad 3.3$$

(3) Measurement of the empty balance in the sorptive gas (M_f) to determine the moment of the buoyancy forces of the balance parts, i. e. beams, wires, baskets etc. From (3.1) we get for $m^s = m^k = 0$, $\rho^f \neq 0$:

$$-M_f = M_s - B_s - (M_k - B_k) \qquad 3.4$$

By combining the four measurements, i. e. eqs. (3.1 – 3.4) we can calculate the reduced mass of the adsorbate (Ω) defined as the difference between the mass adsorbed (m^a) and the buoyancy related to the volume of the sorbent / sorbate system (V^{as}) in the sorptive gas of density (ρ^f), i. e. the quantity

$$\Omega = m^a - \rho^f V^{as} \qquad 3.5$$

According to eqs. (3.1 – 3.4) Ω is given by

$$\Omega = \frac{1}{gl_s}\left[-M_f^{sa} + M_f - M_0 + M_0^s\right] - \rho^f V^k \frac{l_k}{l_s} \equiv \Omega_G \qquad 3.6$$

which is abbreviated by the symbol (Ω_G), the letter "G" indicating the gravimetric measurement method. This quantity can numerically be calculated from the results of the four measurements mentioned above, i. e. numerical values of the moments ($M_f^{sa}, M_f, M_0, M_0^s$) and some additional balance parameters (l_k, k_s, V^k, ρ^f, g). It is interesting to note the special form of eq. (3.6) in case there is no sorbent sample ($m^s = 0$) but only a ballast or tare at the balance ($m^k \neq 0$). Then according to (3.5), $\Omega = 0$ and (3.6) delivers the relation

$$\rho^f V^k \frac{l_k}{l_s} = \frac{1}{gl_s}\left[-M_f^0 + M_f - M_0 + M_0^0\right] \qquad 3.7$$

Here the new moments (M_f^0, M_0^0) refer to

a) measurement without a sorbent sample ($m^s = 0$) in sorptive gas (ρ^f), cp. eq.(3.1),

$$-M_f^0 = M_s - B_s - (M_k + gm^k l_k) + B_k + g\rho^f V^k l_k, \qquad 3.8$$

b) measurement without a sorbent sample ($m^s = 0$) in vacuum ($\rho^f = 0$), cp. (3.1) or (3.3),

$$M_0^0 = M_s - (M_k + \rho m^k l_k) \qquad 3.9$$

Combining eqs. (3.6), (3.7) we get for the reduced mass adsorbed no sorbent ($m^s = 0$)

$$\Omega_G = \frac{1}{g l_s}\left[-M_f^{sa} + M_f^0 + M_0^s - M_0^0\right] \tag{3.10}$$

vacuum ($\rho^f = 0$)

Also from eqs. (3.2), (3.3) we have the relation

$$-M_0^s + M_0 = \rho(m^s l_s - m^k l_k) \tag{3.11}$$

This can be used to calculate the exact values of balance's beams (l_s, l_k) if measurements are performed with well known calibration masses ($m^s = m_1^{cal}$, $m^k = m_2^{cal}$). It also can be used to determine the sorbent mass (m^s) if m^k already is known, cp. [3.1, 3.20, 3.22].

From eq. (3.5) the mass adsorbed (m^a) can be calculated if a model for the sorbent/sorbate volume (V^{as}) is introduced and all the necessary measurements have been performed, i. e. (Ω_G) is known according to eq. (3.10). As outlined in Chap. 1, V^{as} is often approximated by the so-called helium volume,

$$V^{as} \cong V_{He}^s \tag{3.12}$$

determined from either volumetric or gravimetric experiments performed with helium at room temperature. Indeed, assuming helium not to be adsorbed, i. e. $m_{He}^a = 0$, one has from (3.5),

$$V_{He}^s = -\Omega_{G,He} / \rho_{He}^f \tag{3.13}$$

with the reduced mass ($\Omega_{G,\,He}$) to be calculated from helium experiments via eq. (3.6) or (3.10) respectively. Inserting (3.12) in (3.5) we get the so-called Gibbs excess mass adsorbed as

$$m_{GE}^a = \Omega_G + \rho^f V_{He}^s \tag{3.14}$$

where (Ω_G) and (V_{He}^s) are calculated via eqs. (3.10), (3.13) respectively.

If the void volume (V^{as}) in (3.5) is approximated by

$$V^{as} = V^{s}_{He} + \frac{m^{a}}{\rho^{L}_{0}} \quad , \qquad 3.15$$

a model which takes the fictitious volume of the sorbate phase ($V^a = m^a / \rho_0^L$) into account, the mass (m^a) in (3.5) may be considered to be the total or absolute mass of the adsorbate. It is related to the Gibbs excess mass (3.14) by the relation

$$m^{a} = \frac{m^{a}_{GE}}{1-(\rho^{f}/\rho^{L}_{0})} \qquad 3.16$$

Here (ρ_0^L) is the (fictitious) density of the adsorbate. A physically reasonable assumption is to consider ρ_0^L to be either the density of the sorptive in its liquid triple state or in its liquid boiling state at system's temperature, if it exists. For low sorptive gas densities, i. e. $\rho^f << \rho_0^L$, $m^a \simeq m^a_{GE}$. But for high gas densities, i. e. $\rho^f \to \rho_0^L$, we have $m^a_{GE} << m^a \to \infty$, indicating the occurrence of a liquid phase outside the sorbent material according to vapor-liquid-equilibrium (VLE) of the sorptive fluid, cp. also Chaps. 1, 8.

Auxiliary remarks:

1. Calibration experiments with two nearly equal and well known calibration masses (m_1^{cal}, m_2^{cal}) prior and after gravimetric adsorption experiments are recommended. This can avoid systematic uncertainties in experimental data due to (slow) shifts of the zero-point-position of the balance.

2. Interchanges of the sorbent site and the ballast or tare site of the balance should be done from time to time. This can avoid unusual unsymmetrical effects in the balance as they may be caused by (unavoidable) shocks of the balance's base over the years.

3. If calibration masses of well known volume are available, the balance can be used to measure either the density (ρ^f) of the sorptive gas or, via an accurate equation of state, the gas pressure ($p = p(\rho, T)$) in the sorptive atmosphere. Use of two balances in parallel for combined sorption and pressure measurements is recommended in this case [3.1, 3.23-3.25].

2.1.3 Uncertainties or Errors of Measurements

As has been outlined in Chap. 2 uncertainties or errors of measured data constitute an important part of any kind of experimental work and hence always should be considered [2.18]. However, for sake of brevity we here provide the reader only with the formulae allowing one to calculate uncertainties or mean square deviations (MSD) (σ^2_{mGE}) of the Gibbs excess mass (m^a_{GE}) of an adsorbate which has been measured gravimetrically by using a two beam balance. This mass can be calculated from eq. (3.14) combined with eqs. (3.10) and (3.13). By using Gauss' law of propagation of uncertainties we have

$$\sigma^2_{mGE} = \sigma^2_{\Omega G} + V^{S2}_{He}\sigma^2_{\rho f} + \rho^{f2}\sigma^2_{VHe}, \tag{3.18}$$

with

$$\sigma^2_{\Omega G} = \left(\frac{1}{g^2}\sigma^2_g + \frac{1}{l^2_s}\sigma^2_{ls}\right)\Omega^2_G + \frac{4}{(gl_s)^2}\sigma^2_M. \tag{3.19}$$

As the thermal equation of state (EOS) of the sorptive gas always can be written as

$$pV^f = m^f RTZ(p,T) \tag{3.20}$$

with Z = Z(p, T) being the so-called real gas factor or compressibility which here is assumed to be a well known function of pressure (p) and temperature (T), we have for the MSD of the gas density $\rho^f = m^f/V^f$:

$$\sigma^2_{\rho f} = \left(\frac{1}{RTZ} - \frac{p}{RTZ^2}\left(\frac{\partial Z}{\partial p}\right)_T\right)^2 \sigma^2_p + \left(\frac{1}{T} + \frac{1}{Z}\left(\frac{\partial Z}{\partial T}\right)_p\right)^2 \left(\frac{p}{RTZ}\right)^2 \sigma^2_T \tag{3.21}$$

For the MSD of sorbent's helium volume $\left(V^s_{He}\right)$ we have according to (3.13)

$$\sigma^2_{VHe} = \frac{1}{\rho^f_{He}}\sigma^2_{\Omega GHe} + \left(\frac{V^s_{He}}{\rho^f_{He}}\right)^2 \sigma^2_{\rho f He} \tag{3.22}$$

with (cp. (3.19))

$$\sigma^2_{\Omega GHe} = \left(\frac{1}{g^2}\sigma_g^2 + \frac{1}{l_s^2}\sigma_{ls}^2\right)\Omega^2_{GHe} + \frac{4}{(gl_s)^2}\sigma_M^2 \cong \sigma^2_{\Omega G} \qquad 3.23$$

and (cp. (3.21))

$$\sigma^2_{\rho fHe} = \frac{1}{(RT)^2}\sigma^2_{pHe} + \frac{1}{T^2}\sigma_T^2. \qquad 3.24$$

Here we have assumed helium to be an ideal gas, i. e. $Z_{He} = 1$.

Quantities and parameters to be measured and their dispersions are listed in the table below:

g	l_s	$M^{\cdots}_{\cdots}$	p	T	p_{He}	ρ^f	
σ_g	σ_{ls}	σ_M	σ_p	σ_T	σ_{pHe}	$\sigma_{\rho f}$	3.25

Here the symbol $M^{\cdots}_{\cdots}$ indicates any of the electromagnetically measured mechanical moments occurring on the r. h. s. of eq. (3.10), the uncertainties of which (σ_M^2) having been assumed to be equal for all four types of measurements mentioned above, cp. Sect. 2.1.2.

From (3.18), (3.19) one can recognize that the MSD of the Gibbs excess mass (σ^2_{mGE}) will increase with increasing numerical values of the reduced mass of the adsorbate (3.5), (3.10) and – of course – with increasing MSD of the moments' measurements (σ_M^2). Hence it is very important to know system's parameters (g, l_s) fairly accurately, i. e. the relative uncertainties should be $(\sigma_g/g) < 10^{-6}$, $(\sigma_{ls}/l_s) < 10^{-4}$. Modern two beam microbalances, cp. Tab. 3.1, allow measurements with relative uncertainties of moments $\left|\sigma_M / M^{\cdot}_{\cdot}\right| < 10^{-6}...10^{-5}$. Combined with accurate measurements of sorptive's gas pressure $(\sigma_p / p_{max}) < 10^{-3}$, p_{max} indicating the maximum pressure measurable with the manometer considered, and accurate measurements of temperature $(\sigma_T / T) < 3 \cdot 10^{-2}$, they allow to determine Gibbs excess masses with relative uncertainties $\left(\sigma_{mGE} / m^a_{GE}\right) < 10^{-3}$, which normally is one order of magnitude better than results of volumetric / manometric measurements, cp. Chap. 2.

2.2 Single Beam Balances

2.2.1 Experimental

For measurement of adsorption equilibria of corrosive gases like (NH_3, SO_2, H_2S, NO_x etc.) on porous materials like catalysts etc. decoupling of the microbalance from the sorptive gas is mandatory. This can be achieved by using a magnetic suspension, i. e. a magnetic field acting on a permanent magnet which is coupled by a suspension rod to the sorbent load in order to keep it in balance against gravity, i. e. to make it float freely in the sorptive gas atmosphere. A schematics of such an instrument is given in Figure 3.4 followed by a photo taken in lab. PB-A0126/1 of IFT in 2002, Fig. 3.5. The sorbent sample is placed in a basket (stainless steel wire cloth) provided with a permanent magnet, coupled via an external magnetic field penetrating the adsorption vessel's walls to an electromagnet. The electric current in the electromagnet is always chosen such that the permanent magnet, basket, and sorbent sample (suspension) floats freely inside the adsorption vessel. In fact, in today's magnetic balances *) the suspension is oscillating at high frequency (ca. 10 kHz) with small amplitude around its mean position. Hence, in practice it can be considered to be at rest. As the current necessary for stable flotation of the suspension can be measured, its weight can be determined and, after appropriate calibration and vacuum measurements, the mass of sorptive gas adsorbed by the sorbent sample, cp. Sect. 2.2.2, [3.3, 3.4].

The procedure for pure gas adsorption measurement using the installation of Figs. 3.4, 3.5 is basically the same as with two beam balances which already has been described in Sect. 2.1.1. Nevertheless some additional remarks reflecting more than 10 years of practical experience with magnetic suspension balances (MSBs) seems to be appropriate:

1. MSBs offered today are fairly stable against small oscillations or motions of the basement, floor, or laboratory's table where they are mounted. The same is true as far as sneezing of the experimenter is concerned. However, one is well advised to avoid any unnecessary mechanical disturbance of a MSB installation as for example pneumatic (compressed-air) hammer activities in nearby laboratories etc.

*) Rubotherm GmbH, Bochum Germany. System's prize (2004) (subject to change): US $ 40,000.-

2. MSBs should be cleaned from time to time as small particles, for example activated carbon or zeolite powders may be transferred from the sorbent basket to the permanent magnet. Here they can change the magnetic field permanently and by this cause systematic uncertainties or errors of measurement. Once these are detected, they may be corrected by recalibrating the "zero load position" of the balance, cp. Fig. 4.12.

3. Electro smog as it may be caused by – undetected – wires and cables should be avoided. These electromagnetic fields, especially if enhanced by direct or nearby lightning, can seriously disturb measurements.

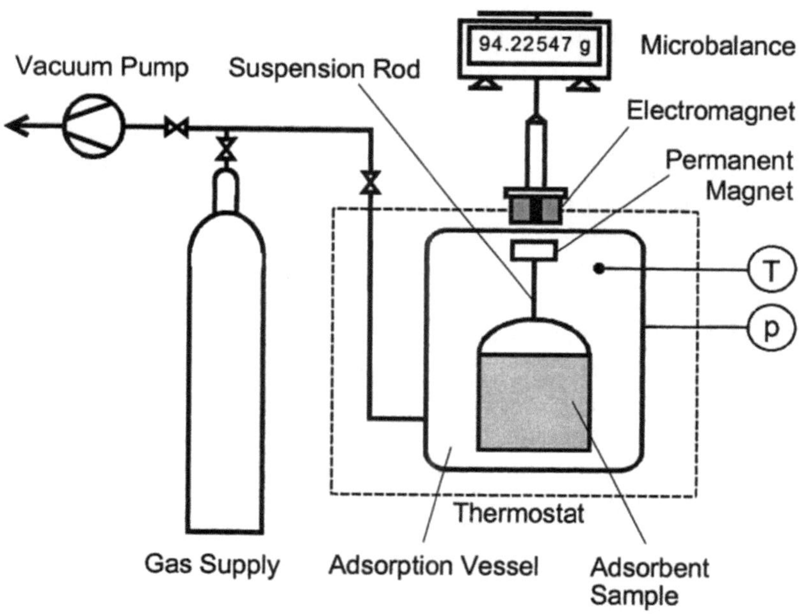

Figure 3.4. Schematic diagram of a gravimetric apparatus for gas adsorption measurements using a magnetic suspension balance (Rubotherm GmbH, Bochum, Germany).

4. It should be emphasized that MSBs today are used not only for gravimetric gas adsorption measurement but also for a still growing variety of other thermo-physical measurements of vertical forces at extreme pressures, temperatures, and chemical conditions, cp. Chap. 4 and [3.5, 3. 6, 3.26, 3.53, 3.54]. Besides, simple mechanical spring balances are still in use. They can give quick, but often only approximate data of adsorption equilibria, which however still may be useful for orientational purposes [3.1, 3.9].

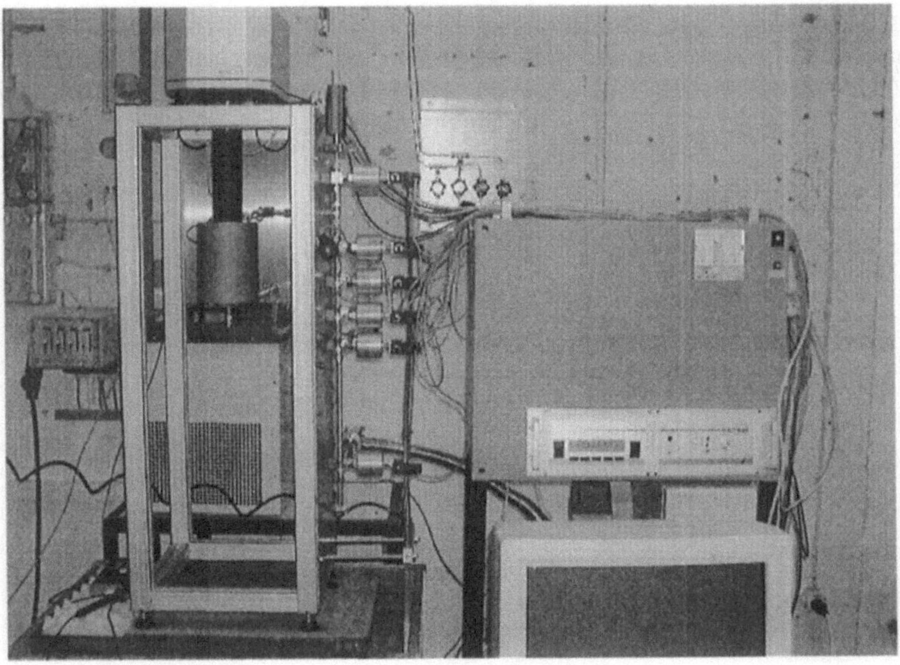

Figure 3.5. Magnetic suspension balance (Rubotherm GmbH) for automated gravimetric measurements of adsorption equilibria of pure corrosive gases. On the r. h. side of the balance 6 electro-pneumatic valves (Hoke, Frankfurt) for controlling the gas supply and evacuation of the installation clearly can be recognized.

2.2.2 Theory

To outline the physics of gas adsorption measurements using a magnetic suspension balance (MSB) let us consider the schematics, Fig. 3.6. It shows the adsorption chamber filled with sorptive gas at density (ρ^f), including the magnetic suspension consisting mainly of a permanent magnet of mass (m^{PM}), below of which a basket (mass: m^B) filled with sorbent material (mass: m^s) is fixed.

If the suspension is floating freely in the sorptive gas atmosphere, the balance of vertical forces is

$$G_f^{sa} = g(m^{PM} + m^R + m^B + m^s + m^a) - g\rho^f (V^{PM} + V^R + V^B + V^{sa}) \qquad 3.26$$

Here (G_f^{sa}) is the weight of the suspension floating freely in the sorptive gas, i. e. the vertical force measured by the microbalance, its zero point of calibration being assumed to refer to the "resting position" of the suspension at the bottom of the adsorption vessel. The masses and volumes of several parts of the suspension are indicated by m^{PM}, V^{PM}... permanent magnet, m^R, V^R... suspension rod, m^B, V^B... sorbent basket; m^s, m^a, V^{sa}... sorbent and sorbate. Also $g = 9.81\ m/s^2$ is the gravity of earth at laboratory's place and ρ^f is the density of the adsorptive gas.

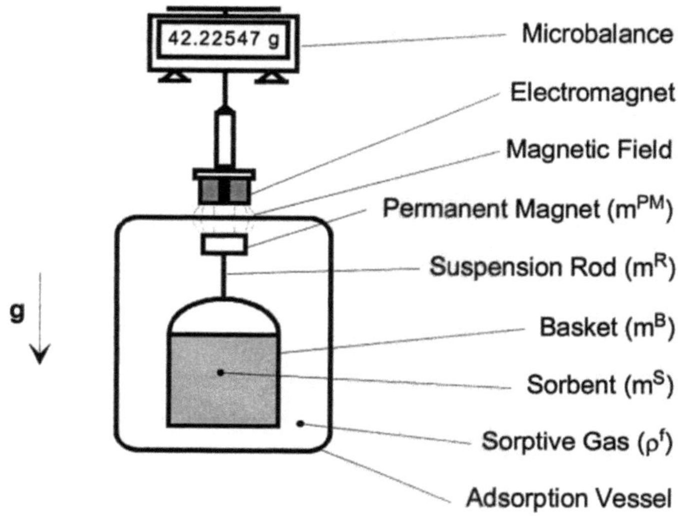

Figure 3.6. Schematics of a magnetic suspension balance (MBS) designed by Rubotherm GmbH, Bochum, Germany [3.3, 3.6].

In order to determine from eq. (3.26) the mass adsorbed (m^a), three additional measurements have to be performed:

(1) Calibration measurement (G_0) of the empty suspension ($m^s = 0$) in vacuum ($\rho^f = 0$). Then (3.26) reduces to the relation

$$G_0 = g(m^{PM} + m^R + m^B) \tag{3.27}$$

(2) Measurement (G_f) of the empty suspension ($m^s = 0$) in the sorptive gas atmosphere ($\rho^f \neq 0$):

$$G_f = g(m^{PM} + m^R + m^B) - g\rho^f(V^{PM} + V^R + V^B) \tag{3.28}$$

(3) Measurement (G_0^s) of the loaded suspension ($m^s \neq 0$) in vacuum. As $\rho^f = 0$, we have from (3.26)

$$G_0^s = g(m^{PM} + m^R + m^B + m^s) \tag{3.29}$$

From the measured quantities $(G_f^{sa}, G_f, G_0^s, G_0)$ we can calculate the reduced mass adsorbed, i. e. the quantity, cp. (3.5),

$$\Omega = m^a - \rho^f V^{sa}$$

via eqs. (3.26-3.29) as

$$\Omega = \frac{1}{g}(G_f^{sa} - G_0^s + G_0 - G_f) \equiv \Omega_G \tag{3.30}$$

The letter "G" in the abbreviation symbol "Ω_G" is indicating gravimetric measurements, contrary to volumetric/manometric measurements, cp. Chap. 2, eq. (2.5).

If the sorbent mass (m^s) in the basket is replaced by a dense ballast or tare or sinker of mass (m^K) and well known volume (V^K), weighing of the suspension in the sorptive gas delivers instead of (3.26) the result

$$\begin{aligned} G_f^K &= g(m^{PM} + m^R + m^B + m^K) - \\ &- g\rho^f(V^{PM} + V^R + V^B + V^K). \end{aligned} \tag{3.31}$$

Similarly, a vacuum measurement of this "sinker configuration" leads instead of (3.28) to

$$G_0^K = g(m^{PM} + m^R + m^B + m^K) \tag{3.32}$$

Combining eqs.(3.31), (3.32) and taking (3.27), (3.29) into account we can calculate the sorptive gas density by

$$\rho^f = \frac{1}{gV^K}\left(G_f^K - G_0^K + G_0 - G_f\right) \tag{3.33}$$

Introducing again the so-called helium hypothesis, i. e. approximating the volume (V^{sa}) in the reduced adsorbed mass (Ω), eq. (3.5) by the helium volume of the sorbent, i. e. $V^{as} \cong V_{He}^s$, cp. (3.13), we can calculate the Gibbs excess mass adsorbed (m_{GE}^a) from (3.30) and (3.33) as

$$m^a_{GE} = \Omega_G + \rho^f V^s_{He} \qquad 3.34$$

If the volume of the adsorbed phase (V^a) is taken into account, i. e. V^{sa} is approximated by eq. (3.15), the absolute mass adsorbed (m^a) again can be calculated from (m^a_{GE}) as given by (3.34) via eq. (3.16).

Hint: From the gas density ρ^f calculated by eq. (3.33) the sorptive gas pressure $p = p(\rho, T)$ inside the adsorption vessel can be calculated via an accurate thermal equation of state. That is, the MSB also can be used as a (fairly sensitive) manometer, especially at very high and very low pressures [3.6, 3.26].

2.2.3 Uncertainties or Errors of Measurements

The uncertainty of measurement represented by its dispersion or mean square deviation (MSD) of the Gibbs excess mass (σ^2_{mGE}) can be calculated from eq. (3.18) which results from eq. (3.14) or (3.34) by applying the Gauss law of propagation of uncertainties:

$$\sigma^2_{mGE} = \sigma^2_{\Omega G} + V^{s2}_{He}\,\sigma^2_{\rho f} + \rho^{f2}\sigma^2_{VHe}\,. \qquad 3.35$$

However, here the MSD $\sigma^2_{\Omega G}$ of (Ω_G) has to be calculated from eq. (3.30). Hence we have

$$\sigma^2_{\Omega GE} = \frac{\Omega^2_G}{g^2}\sigma^2_g + \frac{4}{g^2}\sigma^2_G\,. \qquad 3.36$$

The dispersions of the sorptive gas density ($\sigma^2_{\rho f}$) and of sorbent's helium volume (σ^2_{VHe}) occurring in eq. (3.35) are given by expressions (3.21) and (3.22), (3.23) respectively, cp. Sect. 2.1.3. For sake of clarity we list all parameters and quantities to be measured in a table:

g	$G^{\cdots}_{\cdots}$	p	T	p_{He}	ρ^f	V^s_{He}
σ_g	σ_G	σ_p	σ_T	σ_{pHe}	$\sigma_{\rho f}$	(σ_{VHe})

3.37

The symbol $G^{\cdots}_{\cdots}$ indicates any of the vertical forces defined by eqs. (3.26-3,29) and measured at the MSB. The uncertainties of measurement of all these forces (σ_G) have been assumed to be equal, cp. eq. (3.36). This assumption has to be checked at very high balance loads and / or high sorptive gas densities which may change the magnetic field providing coupling between the inner and the external part of the suspension balance.

What has already been said at the end of Sect. 2.1.3 is also valid here: To obtain accurate values of the Gibbs excess mass, it is essential to know system's parameters (g, V^s_{He}) and the sorptive gas density (ρ^f) fairly accurately which means inequalities $(\sigma_g / g) < 10^{-6}$, $(\sigma_{VHe} / V^s_{He}) < 10^{-4}$, and $(\sigma_{\rho f} / \rho^f) < 2 \cdot 10^{-4}$ to hold. Also relative uncertainties of forces $G^{\cdots}_{\cdots}$ should be less than $(\sigma_G / G^{\cdots}_{\cdots}) < 10^{-6}$. With these accuracies one normally can expect relative uncertainties of Gibbs excess masses $(\sigma_{mGE} / m^a_{GE}) < 10^{-3}$, cp. examples given in the next section.

2.3 Examples

Gravimetric measurements of pure gas adsorption equilibria on porous solids are normally performed for either of two purposes:

a) at low pressures (≤ 0.1 MPa) and temperatures to get information on the porous structure of the material, i. e. to "characterize" it, or
b) at higher pressures (≤ 2 MPa ... 10 MPa) and temperatures ($\simeq 300$ K) to determine the sorption capacity (and kinetic properties) of the material for industrial adsorption processes like air separation, natural gas purification etc.

In this section we will present several examples of gravimetrically measured gas adsorption data serving mainly purpose (b) without excluding (a):

1. Adsorption of CH_4, CO, CO_2, N_2 on activated carbon NORIT R1 EXTRA at 298 K for pressures ($p < 6$ MPa);

2. Adsorption of N_2, O_2 on zeolite Köstrolith SX6 at 293 K, 313 K, 333 K for pressures ($p < 0.15$ MPa);

3. Adsorption of N_2 and O_2 on zeolite Köstrolith SX6 with small amounts of water permanently presorbed at 313 K for pressures ($p < 0.12$ MPa); Adsorption of CO_2 on zeolite Na13X with presorbed water;

4. High pressure adsorption of N_2 and CO_2 on activated carbon NORIT Rl at 298 K-343 K for pressures ($p < 50$ MPa).
5. Adsorption of sub- and supercritical CO_2 on MS 5A.

Example 1
Adsorption of natural gas compounds on activated carbon

The sorbent material chosen is activated carbon NORIT R1 EXTRA made from peat by thermal and chemical activation. To determine its BET surface the adsorption isotherm (AI) of nitrogen (N_2) was taken at 77 K. It is presented in Figure 3.7A. It shows a very steep increase at low pressures indicating existence of micropores ($d < 2$ nm) and a slight hysteresis in the range ($0.4 < p / p_0 < 1$, $p_0 = 0.1$ MPa) reflecting mesoporous structures. The resulting BET surface area is 1430 m^2/g (ISO 9277: 1995). Other characteristic quantities are given below [3.27], [3.28]:

Elemental analysis given in (%atom):

C	O	N	Ca	S	Si	Others
93.7	4.8	0.4	0.2	0.2	0.7	0.3

Structure of C-based active sites given in (%):

$-CH_x$	-C-O-C-	-C=O	O=C-O	$-\underline{C}O_3$	-CX
70	13	6	5	4	2

The activated carbon was delivered by the manufacturer as cylindrical pellets of length 4 mm and diameter 1 mm having a bulk density of 0.435 kg/l.

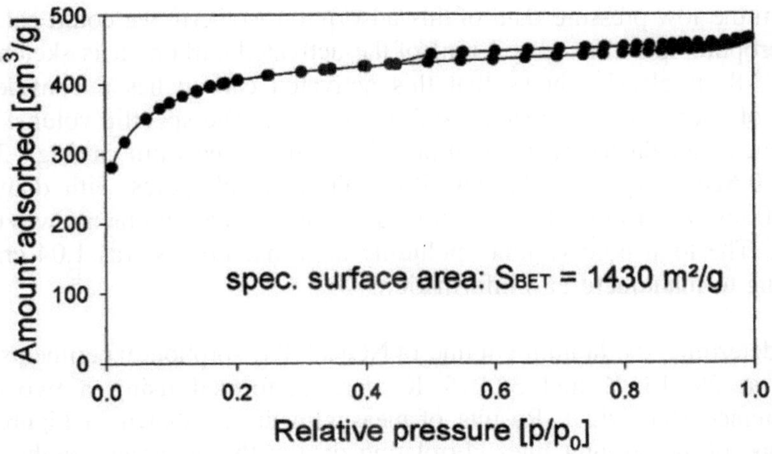

Figure 3.7A. Adsorption isotherm (AI) of pure nitrogen (N_2, 5.0) on activated carbon (AC) NORIT R1 EXTRA at 77 K, p_0 = 1 atm.

Due to the extreme steepness of the AI in Fig. 3.7A it was not possible to calculate the micropore spectrum from the data by the (standardized) Horwath-Kawazoe-Method, cp. [3.29].

In view of this we took similar measurements with Ar at 87 K. The resulting AI is sketched in Figure 3.7B.

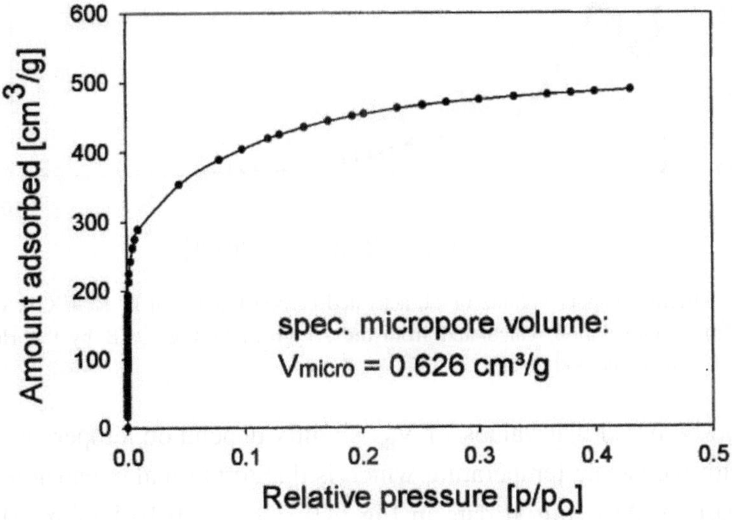

Figure 3.7B. Adsorption isotherm of pure argon (Ar, 5.0) on activated carbon (AC) NORIT R1 EXTRA at 77 K, p_0 = 1 atm.

From the low pressure data of this adsorption isotherm we could calculate the micropore spectrum (d < 2 nm) of the activated carbon. It is sketched in Figure 3.8. It clearly shows that this activated carbon has a considerable amount of micropores (diameter d < 2 nm = 20 A). The specific volume of all micropores calculated by integration of the pore spectrum in Fig. 3.8 is $V_{micro} = 0.626\ cm^3/g$ AC. The specific volume of all pores with diameters (d < 120 nm) determined from combined Ar- and N_2 measurements was 0.715 cm^3 / g. The total pore volume including also macropores was 1.04 cm^3 / g according to manufacturer's information.

To determine the helium volume of NORIT R1, sorption experiments with helium at 298.15 K and 323.15 K were performed using a two beam microbalance (Sartorius). Results of measurement are shown in Figure 3.9. The slope of the straight lines correlating data of the apparent weight of the sample as function of the density of the helium gas (ρ_{He}), is the negative value of sorbent's helium volume (V^s_{He}), cp. eq. (3.13).

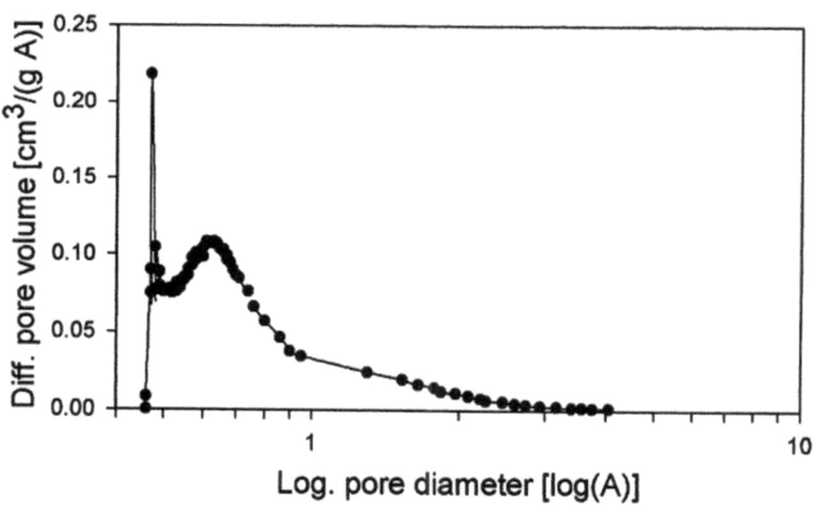

Figure 3.8. Differential pore volume of the activated carbon (AC) NORIT Rl EXTRA in the micropore region, calculated from the AI given in Fig. 3.7B by the Horvath-Kawazoe-method [3.29].

It should be noted that values of V^s_{He} slightly depend on temperature, i. e. increase with increasing temperature which is due to thermal expansion of the activated carbon. Also initial data in Fig. 3.9 at $\rho_{He} \cong 0.10\ kg / m^3$ show a jump or "primary adsorption" of helium of about $\Delta m = 0.15\ mg / g$ if the sorbent was cooled from 323 K to 298 K. This effect was irreversible, i. e. this amount of helium (Δm) was sorbed permanently in the sorbent and for other experiments could be considered as (constant) "presorbed gas" [3.30]. This effect of a spontaneous and irreversible uptake of helium at ambient or near ambient temperatures was observed in all kinds of sorbent materials, i. e. activated carbons, zeolites, metal foams and others, the specific amount of helium adsorbed depending strongly on the pretreatment of the sorbent material, i. e. its activation procedure etc., cp. Chap. 1, Sect. 4.2.

In Figure 3.10 equilibrium adsorption data of pure gases (CO_2, CH_4, CO, N_2) on activated carbon NORIT Rl EXTRA at pressures ($p \leq 6$ MPa) and temperatures 298.15 K are shown [3.27], [3.28]. The mol numbers of Gibbs excess amounts adsorbed per unit mass of sorbent are depicted as function of sorptive's gas pressure (and temperature). Relative uncertainties of data are about $(\sigma_{mGE} / m^{a}_{GE}) < 1$ %. Measurements were performed with a two beam balance (Sartorius 4104S), cp. Fig. 3.2. Prior to all measurements the sorbent was activated by the "standard procedure", i. e. by heating it up to 423 K for (4-5) h in vacuum ($p \leq 0.1$ Pa).

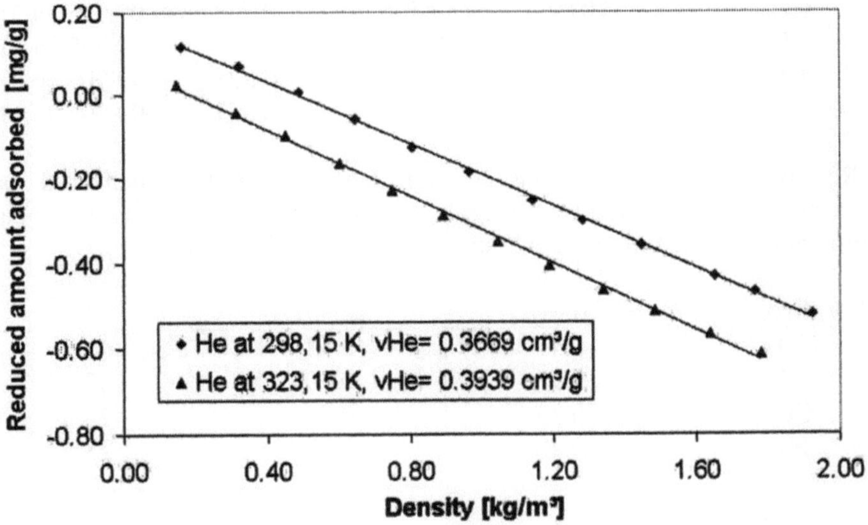

Figure 3.9. Adsorption equilibria of helium (5.0) on activated carbon (AC) NORIT Rl EXTRA at T = 298.15 K and T = 323.15 K. Data of the apparent weight of the sorbent sample are sketched as function of the density of the helium gas (ρ_{He}) within the region ($0 < \rho_{He} < 2$ kg / m^3) corresponding to the pressure range ($0 < p < 2.5$ MPa). Data were always taken 15 minutes after increase of helium gas pressure at a relative mass change rate of ($\Delta m / m^s \Delta t = 5.10^{-7}$ g / (gh)).

All the isotherms are in the range of pressure measured of Type I (IUPAC), [3.31-3.34]. Data have been correlated using an adsorption isotherm (AI) of the generalized Langmuir type, cp. Chap. 7:

$$n(p,T,m^s) = \alpha n_\infty(T,m^s)\frac{[b(T)p]^\alpha}{1+[b(T)p]^\alpha} \qquad 3.38$$

The quantity (αn_∞) is the (hypothetical) limiting amount adsorbed for $p \rightarrow \infty$. The parameter b(T) is the reciprocal of the "half-load-pressure", i. e. the pressure needed at given temperature (T) to load the sorbent with 50 % of its limiting load, i. e. $n(p = b^{-1}, T, m^s) = \alpha n_\infty (T, m^s) / 2$. The exponent $\alpha > 0$ can be related to the fractal dimension of the sorbent material and the diameter of the molecule adsorbed, cp. Chap. 7. Numerical values of parameters (n_∞, b, α) at T = 298.15 K for the sorptive gases investigated are:

Sorptive gas	n_∞ mmol/g	b $(Pa)^{-1}$	α (1)	Purity of gas (%)
CO_2	18.89	1.165	0.784	99.995 (4.5)
CH_4	13.38	0.695	0.725	99.995 (4.5)
CO	8.67	0.457	0.828	99.90 (3.0)
N_2	7.68	0.402	0.881	99.996 (4.6)

Relative absolute deviations (f) and dispersions (σ) of measured data (n_{exp}), and data calculated by fitting (n_{exp}) to (3.38) (n_{cor}) using a least square minimization procedure are given in the table below in percent (%):

Sorptive gas	f(%)	σ (%)	M
CO_2	0.47	0.54	17
CH_4	0.35	0.39	17
CO	0.33	0.41	25
N_2	0.40	0.55	23

$$f \equiv \frac{1}{M}\sum_{i=1}^{M}\frac{1}{n_{exp}}\left|n_{cor} - n_{exp}\right| \qquad 3.39$$

$$\sigma^2 \equiv \frac{1}{M}\sum_{i=1}^{M}\frac{1}{n_{exp}^2}(n_{cor} - n_{exp})^2 \qquad 3.40$$

The fit of experimental data (n_{exp}) shown in Fig. 3.10 can be improved, i.e. values of statistical deviations (f) and dispersions (σ) can be reduced if instead of the phenomenological isotherm eq. (3.38) a so-called two site Langmuir isotherm, cp. Chap. 7 is chosen. This isotherm consists of two terms analogously formed as the expression on the r. h. s. of eq. (3.38) having however different numerical values of the respective parameters (n_∞, α, b). These correspond to two energetically different types of adsorption places which may reflect two different types of pores, namely micro- and mesopores within the activated carbon. In view of space limitations we can not give more details here but refer to the literature [3.35].

To check quality and consistency of measured data it is recommended to represent them in a ln(p/n), n-diagram, especially those taken at low pressures. This diagram is very sensitive to statistical or systematic deviations of single data points or subsets of data relative to the other data. Examples for such deviation can be found in [3.36]. Figure 3.11 shows this diagram for data of Figure 3.10. As can be seen, all data measured fit fairly well to the correlation curves (3.38), as expected according to values of statistical deviations (f) and dispersions (σ) given above.

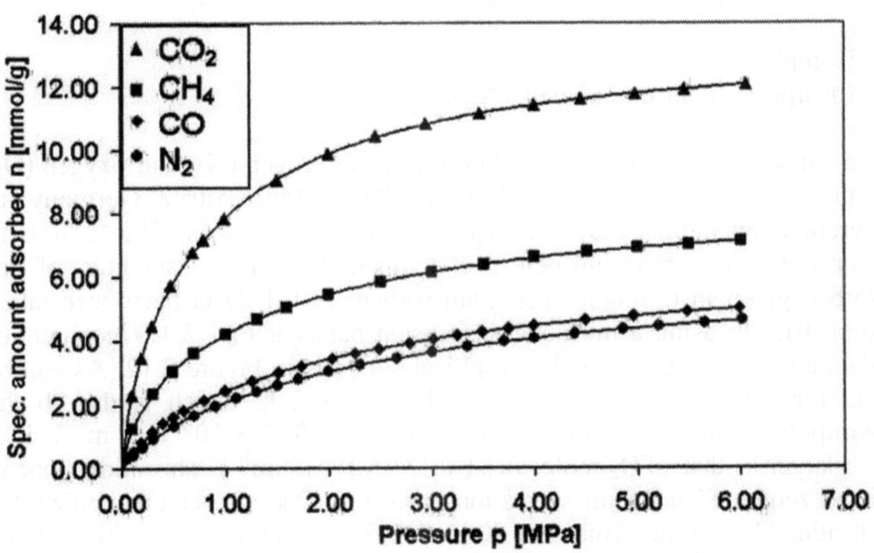

Figure 3.10. Gibbs excess adsorption isotherms of (CO_2, CH_4, CO, N_2) on activated carbon NORIT R1 EXTRA at T = 298.15 K, [3n.27]. Data have been correlated using an isotherm of the generalized Langmuir type (3.38).

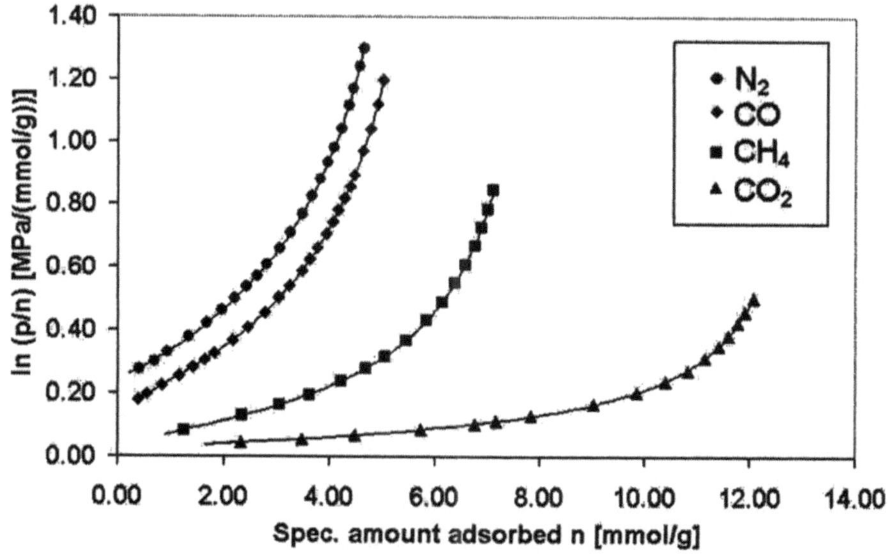

Figure 3.11. Logarithmic diagram (In (p/n), n) of adsorption equilibria data of (CO_2, CH_4, CO, N_2) on AC NORIT R1 EXTRA at T = 298.15 K, [3.27]. Data have been correlated by an AI of generalized Langmuir type , eq. (3.38).

Example 2
Adsorption of air compounds (N_2, O_2) on zeolite.

In Figure 3.12 adsorption equilibria data of nitrogen (N_2) and oxygen (O_2) on microporous zeolite Köstrolith SX6 (CWK, Bad Köstritz, Germany) at different temperatures (293 K, 313 K, 333 K) are shown. This is also to demonstrate the strong influence of temperature on the amounts of gas adsorbed given in milimoles per gram sorbent [3.16]. Data have been taken gravimetrically using a magnetic suspension balance, Fig. 3.4. Uncertainties of data are about the size of the graphical symbols in Figure 3.12. As can be seen, nitrogen is much more adsorbed than oxygen, which is due to the quadrupole moment of N_2-molecules which is ($-0.47 \cdot 10^{-34}$ $J^{1/2}$ $m^{5/2}$), i. e. much larger as that of O_2-molecules ($-0.13 \cdot 10^{-34}$ $J^{1/2}$ $m^{5/2}$). This allows one to use the zeolite SX6 for air separation processes. One also can recognize that by heating the zeolite from 293 K to 333 K, amounts of N_2 adsorbed are reduced to 40 % of their original values Hence one may argue that in technical adsorption processes with this zeolite thermal regeneration will be much more economical than pressure swing adsorption [3.37].

From the pure gas adsorption data presented in Fig. 3.12 the selectivities (S_{N2O2}) of (N_2, O_2)-mixtures can be determined – at least in the Ideal Gas

Adsorption Theory (IAST) approximation [3.38]. In this theory the selectivity of two gases simply can be represented as ratio of the Henry-constants (H_{N2}, H_{O2}), i. e. the steepnesses of the two pure gas adsorption isotherms at $p \rightarrow 0$. In our case we have

$$S_{N2O2} \equiv \frac{x_{N2}y_{O2}}{y_{N2}x_{O2}} = \frac{H_{N2}}{H_{O2}}, \tag{3.39}$$

i. e. for T = 293 K, $S_{N2O2} \cong 10$. Köstrolith SX6 is a zeolite molecular sieve of LiNaX-type with a (low) atomic ratio of Si/Al $\cong$ 1.01. Its BET surface measured with N_2 (5.0) at 77 K is 660 m^2 / g. The specific volume of macro- and (partly) mesopores determined by Hg-porosimetry is 0.280 m^3/g. The specific volume of the micropores determined by adsorption of CO_2 at 273 K and calculated by the Dubinin-Astarkhov-method [3.52] is 0.283 cm^3/g. The so-called helium volume of dry zeolite SX6 has been determined gravimetrically after activating the zeolite at 700 K for 6 hours as $V^s_{He}(293)K = 0.430\,cm^3/g$, $V^s_{He}(313)K = 0.432\,cm^3/g$, [3.16].

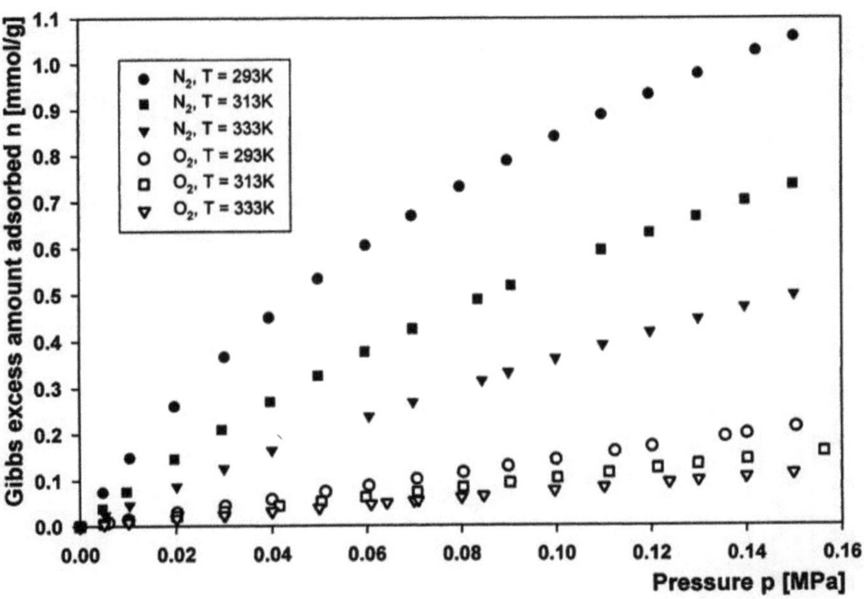

Figure 3.12. Gibbs excess adsorption amounts of N_2 and O_2 adsorbed per unit mass of dry zeolite Köstrolith SX6 at Temperatures 293 K, 313 K, 333 K, [3.16].

Example 3
Adsorption of air compounds (N_2, O_2) and carbon dioxide (CO_2) on zeolite with presorbed water.

Most zeolites are hydrophilic, i. e. attract water which usually is strongly adsorbed, i. e. only can be desorbed by heating the zeolite in vacuum or helium atmosphere up to 700 K. Hence one may conjecture that the amount of gas adsorbed on a zeolite at ambient temperature will depend on the amount of water which is permanently presorbed on the zeolite. This is indeed the case. To demonstrate, in Figures 3.13, 3.14 adsorption equilibria data of nitrogen (N_2) and oxygen (O_2) on three different samples of zeolite Köstrolith SX6, cp. example 2, including three different amounts of water are presented. The "wet zeolites" have been prepared by exposing the pure, i. e. freshly activated material to low pressure water vapor and weighing it again in vacuum. From the results of these measurements it could be concluded that the water at temperatures near ambient – and even up to 450 K – is permanently adsorbed on the zeolite. This was also the case after performing experiments with nitrogen and oxygen as the weight of the sorbent samples after desorption of (N_2) and (O_2) in vacuum turned out to be the same as prior to the (N_2, O_2)-experiments. Details are given in [3.16].

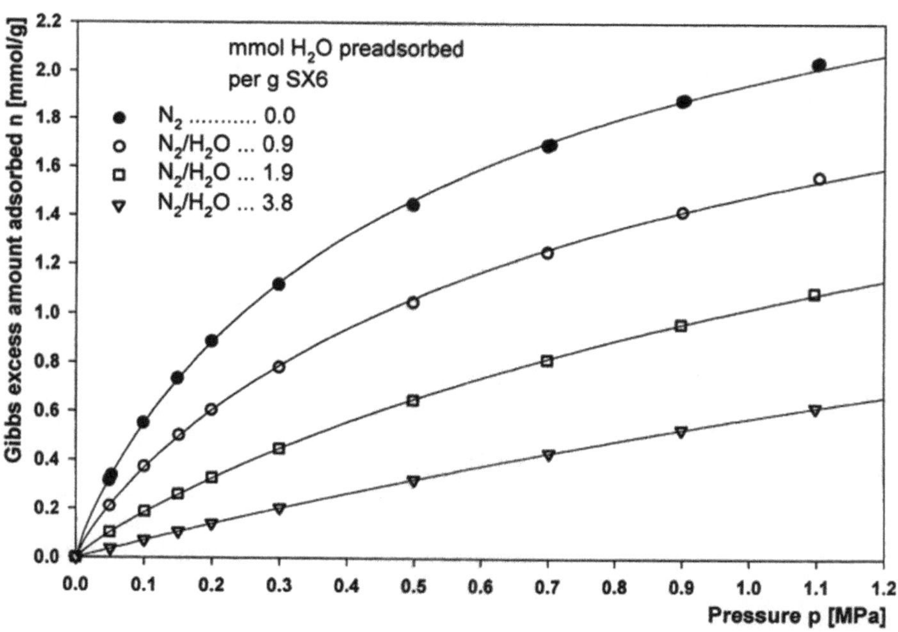

Figure 3.13. Adsorption isotherms of nitrogen (N_2) on zeolite Köstrolith SX6 without and with presorbed water of (0.91, 1.88, 3.81) mmol H_2O / g SX6 at 313.35 K.

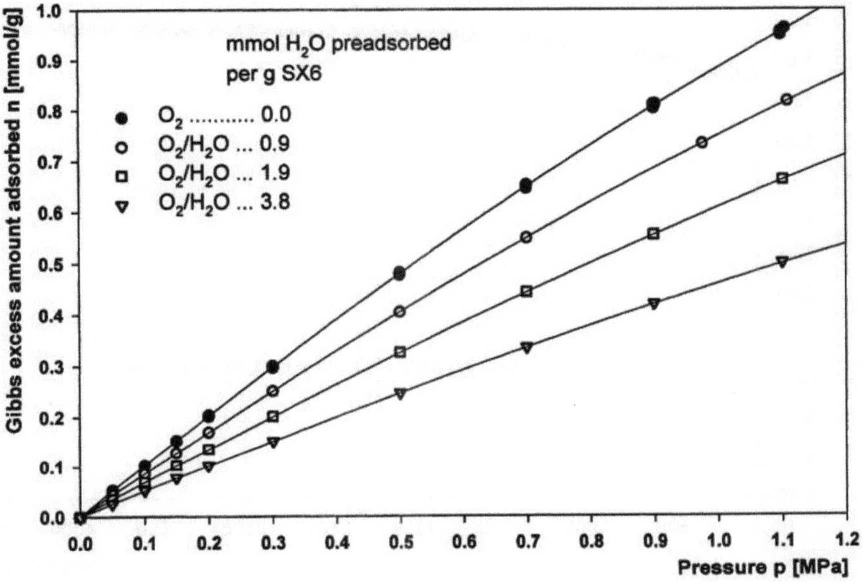

Figure 3.14. Adsorption isotherms of oxygen (O_2) on zeolite Köstrolith SX6 without and with pre-sorbed water of (0.89, 1.86, 3.79) mmol H_2O / g **SX6** at 313.35K.

The Gibbs excess amounts of N_2 and O_2 adsorbed on the zeolite SX6 have been calculated from the microbalance recordings following eq. (3.34). For data referring to the dry zeolite, the helium volume $V^s = V^s_{He} = 0.432\,cm^3/g$ measured in the dry zeolite has been used. For data referring to the "wet zeolite" including presorbed water, the helium volume corrected by the bulk volume of the adsorbed water, i. e. $V^s = V^s_{He} + (m_{H2O}/\rho_{H2O})$ has been used, (ρ_{H2O}) being the density of liquid water at T = 313.15K and p = 1.013 bar. As can be seen from the figures, the adsorption capacity of the zeolite due to the presorbed water is reduced much more for nitrogen than for oxygen. Hence it can be argued that the selectivity of the zeolite for air separation processes based on quadrupole induced adsorption of N_2 will be considerably reduced due to the presence of water. Indeed this has been experienced for various types of zeolites [3.37]. Actually, not only equilibria of (N_2, O_2) on SX6 are changed by presorbed water but also the kinetics of the adsorption process. This can be seen from Figure 3.15 showing uptake curves of nitrogen on dry and wet SX6 taken at 313.6 K. The adsorption process was initiated by increasing the gas pressure of N_2 (5.0) from 0.5 MPa to 0.7 MPa. Corresponding desorption curves were identical within experimental uncertainties, i. e. no hysteresis was found for this system and experiment.

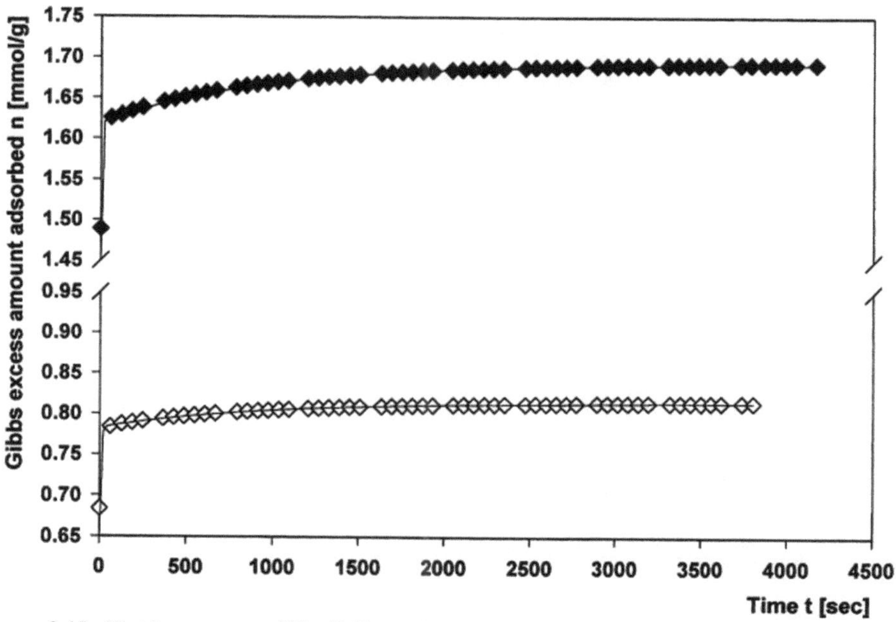

Figure 3.15. Uptake curves of N_2 (5.0) on dry zeolite Köstrolith SX6 (upper curve) (♦)and zeolite SX6 pre-loaded with 1.88 mmol H_2O/g SX6 (lower curve) (◊) at 313.6 K after increase of gas pressure from 0.5 MPa to 0.7 MPa.

Data have been taken using a magnetic suspension balance, Fig. 3.4 for about 1 hour. Data can be correlated using a superposition of two simple Langmuirian relaxation models [3.15, 3.16], i. e. representing the amount of (N_2) adsorbed by

$$n\,(p, t) = n_p(p, t) + n_s\,(p, t), \qquad 3.40$$

$$n_\beta\,(p, t) = n_{equ\beta}(p)\,(1\text{-}\exp\,((t\text{-}t_0)/\tau_\beta)) + n_{\beta 0}(p)\,\exp\,((t\text{-}t_0)/\,\tau_\beta), \qquad \beta = p, s \qquad 3.41$$

Here ($n_{\beta 0}$, $n_{equ\beta}$) are the initial ($t = t_0$) and the asymptotic ($t \rightarrow \infty$) value of the partial loads $n_\beta = n_\beta(t)$ adsorbed, the adsorption process being assumed to start at $t = t_0$.The index $\beta = p$, s indicates "primary" and "secondary" adsorption, cp. Chap. 7. Relaxation times τ_p, τ_s turned out to be nearly independent of the initial and the final gas pressure during the adsorption experiment, but were considerably different for dry and wet zeolite including presorbed water (H_2O/SX6 in mmol/g). Numerical values of τ_p, τ_s referring to experiments Fig. 3.15 are given below:

H_2O/SX6 (mmol/g)	τ_p/seconds	τ_s/seconds
0	2	950
1.88	8	850

As it was not possible to correlate experimental data of Fig. 3.15 using a single relaxation time model we argue that the uptake of nitrogen in the zeolite is a two step process. In the first step the N_2-molecules are quickly entering the macropores and also some mesopores (τ_p). In a second step the molecules slowly diffuse from the macropores to the micropores (τ_s). Presorption of water changes the first mechanism, i. e. uptake of N_2 in the wide pores is somewhat hindered by the water molecules. Contrary to this the secondary adsorption mechanism, i. e. transfer of N_2-molecules to the micropores does not seem to be much influenced by the presence of the permanently presorbed small amounts of water [3.16].*) To give another example of this influence of water on the sorption capacity of hydrophilic sorbents like (most of) zeolites, we present adsorption equilibria data of carbon dioxide (CO_2) on zeolite Na13X (Linde, UOP) taken at 313 K for pressures up to 1 MPa, Fig. 3.16 [3.16]. Measurements have been performed gravimetrically using a magnetic suspension balance, Fig. 3.4. Uncertainties of data are of the size of the graphical symbols in Fig. 3.16.

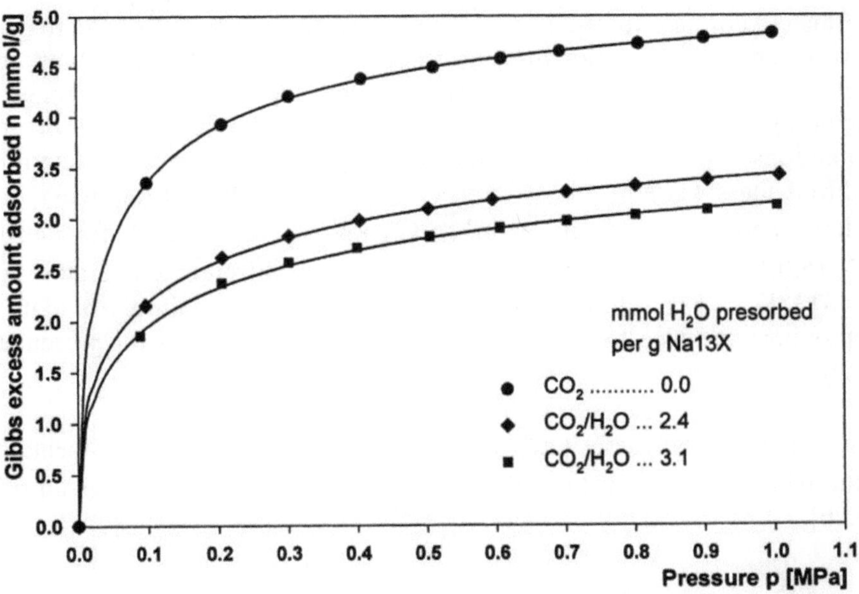

Figure 3.16. Adsorption equilibria of pure carbon dioxide (CO_2) on zeolite Na13X (Linde, UOP) without and with presorbed water of (2.4, 3.1) mmol H_2O/g Na13X at 313 K.

*) This is only true for temperatures (313.6 K) far below the desorption temperature of water (ca. 800 K).

Data have been correlated by using a generalized AI of Langmuir type, eq. (3.38). As can be seen from the figure, due to presorption of water of about 50 % of the limiting molar amount of CO_2, the capacity for CO_2-adsorption is reduced by ca. 40 %. We expect that part of the CO_2 adsorbed on zeolite with presorbed water will be dissolved in the "surface-water". In view of experimental difficulties we have not been able to do truly binary coadsorption measurements for CO_2/H_2O-mixtures at near ambient temperatures. Consequently we do not know the composition of the adsorbed phase for sure but leave this question open to the interested experimenter.

Example 4
High pressure adsorption of N_2 and CO_2 on activated carbon NORIT R1 EXTRA in the temperature range 298 K – 343 K for pressures up to 50 MPa.

Adsorption equilibria of the above kind have been measured at the Institute of Non-Classical Chemistry, University of Leipzig, Leipzig, Germany during 1999-2003. A high pressure version of a magnetic suspension balance allowing measurements up to 50 MPa was used. Experimental details of the instrument and the measurement procedure including data correlation are given in [3.26, 3.39]. In view of space limitations we only present graphically the gravimetrically determined reduced mass $\Omega_G = m^a - \rho^f V^{sa}$, cp. (3.5) and the Gibbs excess mass in the helium approximation, i. e. $m^a_{GE} = \Omega_G + \rho^f V^s_{He}$, cp. (3.14) for nitrogen (N_2, (5.0)) and carbon dioxide (CO_2, (4.5)) at four different temperatures (298 K, 313 K, 328 K, 343 K), Figs. 3.17 – 3.20. Uncertainties of data are approximately twice the size of the graphical symbols used for presentation [3.26].

The Gibbs excess masses (m^a_{GE}) have been calculated from experimental Ω_G-data using numerical values of the sorbent's helium volume as given below [3.26]:

T/K	298	313	328	343
V^s_{He} (cm^3/g)	0.462	0.469	0.473	0.478

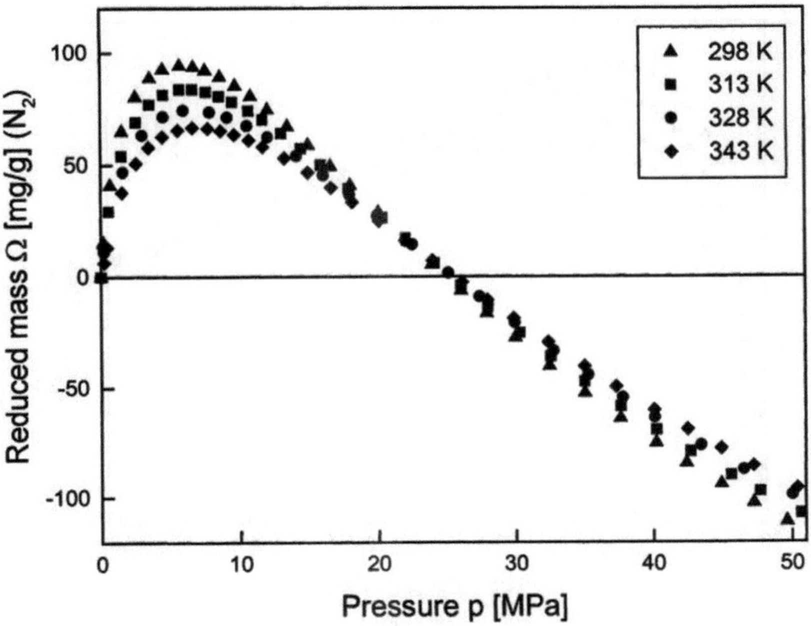

Figure 3.17. Reduced mass ($\Omega_G = m^a - \rho^f V^{sa}$) of nitrogen adsorbed on activated carbon NORIT R1 EXTRA at (298 K – 343 K) for pressures up to 50 MPa, [3.26].

Deviations of these data from the numerical values of (V^s_{He}) already given in Example 1 of this Section are due to different activation procedures of the NORIT R1 carbon applied in the labs at Siegen and Leipzig. As can be seen from Figs. 3.17, 3.19 the reduced mass adsorbed, a truly experimental quantity for both gases N_2 and CO_2, increases at low pressures with increasing pressure to reach a maximum after which it goes down to even negative values. This effect indicates that at high gas pressures, the density of the sorptive gas becomes so high that buoyancy effects surmount adsorption effects.

Indeed the nearly linear decrease of the Ω_G-signals for N_2 at pressures p > 20 MPa in Fig. 3.17 indicate that the amount of N_2 adsorbed tends to be constant, whereas the buoyancy based reduction of the sample weight increases nearly linearly with pressure. However, in case of CO2-adsorption, Fig. 3.19 this simple interpretation does not seem to be possible, or if so, only for pressures p > 40 MPa. Still for high pressures and supercritical temperatures (313 K, 328 K, 343 K) the buoyancy based reduction of the sorbent sample's weight again surmounts by far a possible increase in weight due to pressure induced adsorption.

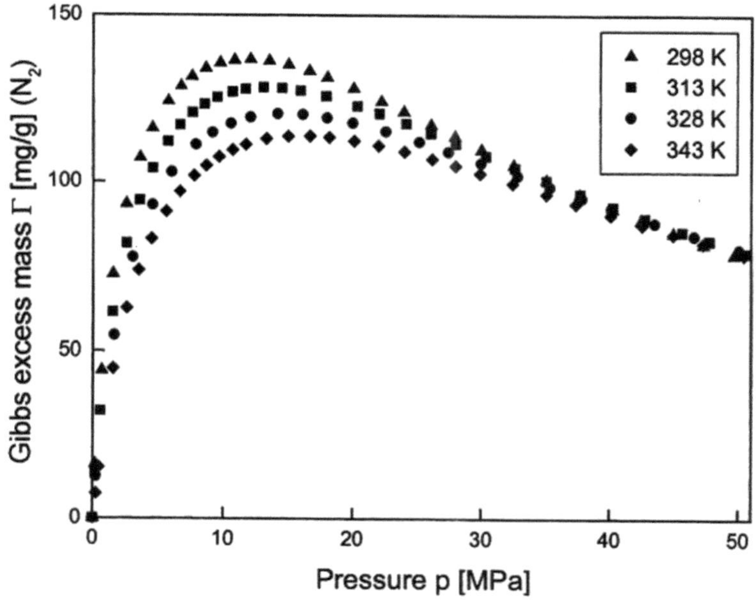

Figure 3.18. Gibbs excess mass ($m^a_{GE} = \Omega_G + \rho^f V^s_{He}$) of nitrogen adsorbed on activated carbon NORIT R1 EXTRA at (298 K – 343 K) for pressures up to 50 MPa, [3.26].

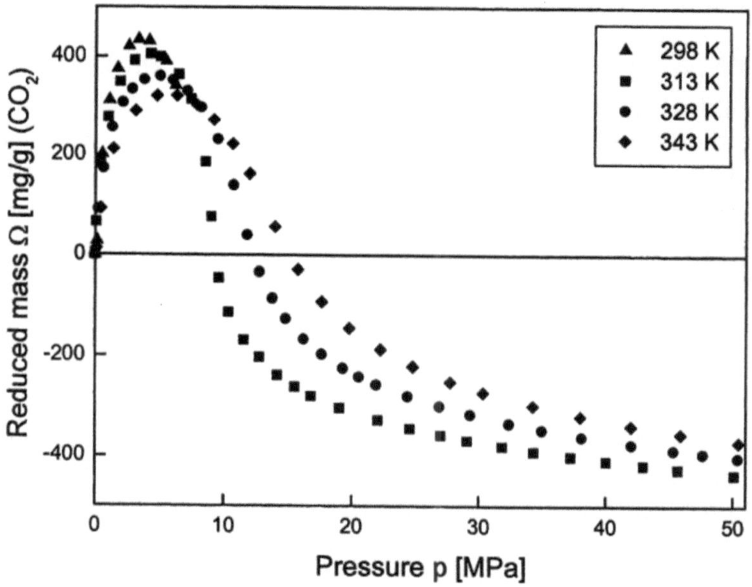

Figure 3.19. Reduced mass ($\Omega_G = m^a - \rho^f V^{sa}$) of carbon dioxide adsorbed on activated carbon NORIT R1 EXTRA at (298 K – 343 K) for pressures up to 50 MPa, [3.26].

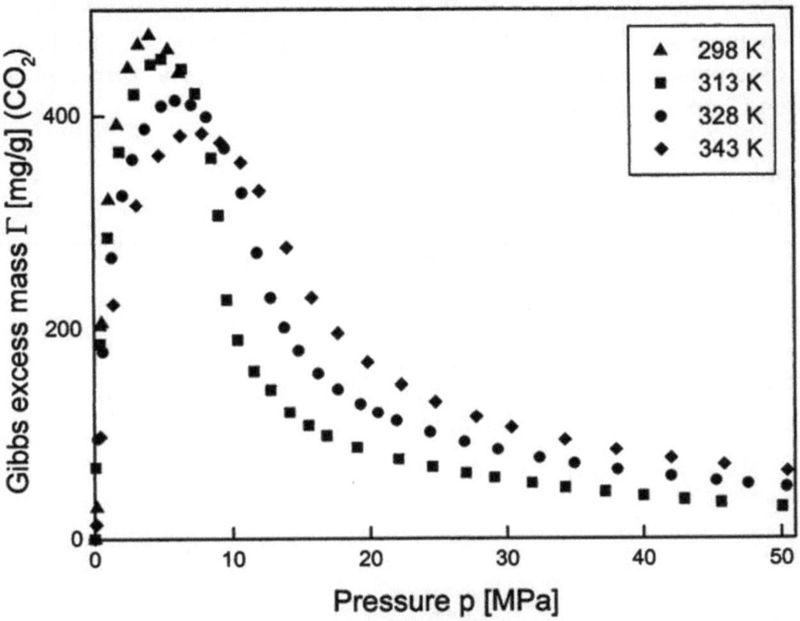

Figure 3.20. Gibbs excess mass ($m^a_{GE} = \Omega_G + \rho^f V^s_{He}$) of carbon dioxide adsorbed on activated carbon NORIT R1 EXTRA at (298 K – 343 K) for pressures up to 50 MPa, [3.26].

A similar effect can be recognized in Figures 3.18, 3.20 for the (helium approximation of the) Gibbs excess mass of N_2 and CO_2 adsorbed in NORIT R1. In the low pressure region it increases considerably with increasing pressure to reach a maximum and then falls off but still maintains positive numerical values. This decrease of the Gibbs excess mass with increasing pressure in the high pressure region poses a serious thermodynamic problem as then the stability condition $(\partial m^a / \partial p)_T \geq 0$, which is a consequence of the Second Law of Thermodynamics, does not hold, cp. Chaps. 1, 7. As according to experience adsorbate phases also at high pressures are thermodynamically stable phases and up to now indications of high pressure induced phase changes could be observed [3.39, 3.51]. Thus we prefer not to use the concept of the Gibbs excess mass (m^a_{GE}) as a basis for developing thermodynamics of adsorbate phases. This rather should be done on the basis of the concept of absolute masses or amounts adsorbed (m^a, n^a), cp. Chaps. 1, 9, the Gibbs excess (m^a_{GE}) still being a reasonable approximation for the absolute mass adsorbed (m^a) in the low pressure region where the thermodynamic stability condition mentioned above still holds. Examples for the difference between (m^a_{GE}) and (m^a) already have been given in Chap. 1. A method to measure the absolute mass adsorbed (m^a) without introducing a hypothesis on the volume of either the porous solid nor the adsorbate phase is outlined in Chap. 8, cp. [3.40].

Example 5
Adsorption of CO_2 on MS5A

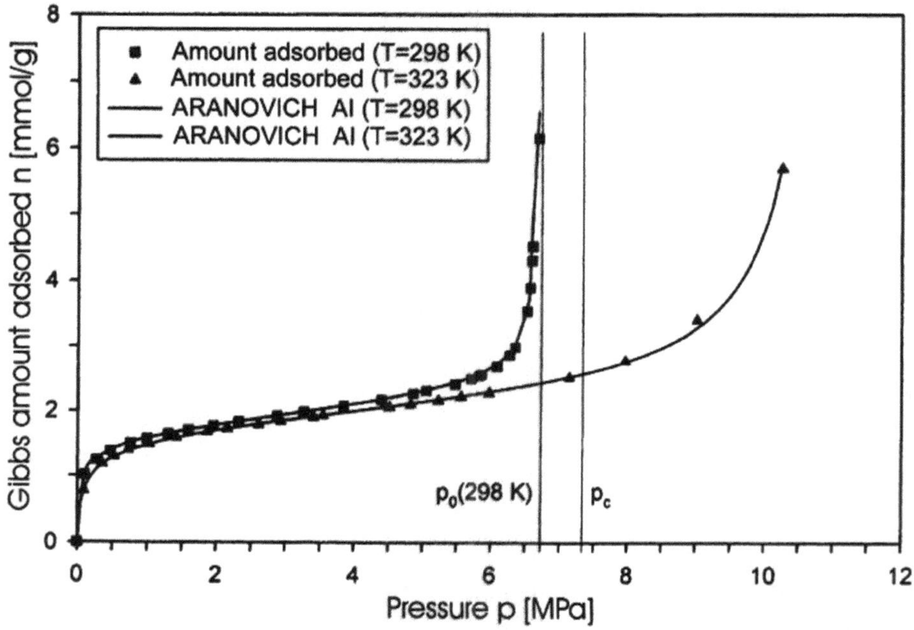

Figure 3.21. Adsorption of carbon dioxide (CO_2) on molecular sieve (UOP, 5A) at subcritical temperature (298 K) and supercritical temperature (323 K) for pressures up to 10 MPa.

As a last example for gravimetric measurements of pure gas adsorption equilibria we present data of sub- and supercritical carbon dioxide (CO_2) adsorbed on molecular sieve (MC) (UOP 5A) for pressures up to 10 MPa, [3.5], Fig. 3.18. Data were taken using a magnetic suspension balance, Fig. 3.4, Fig. 4.5. Relative uncertainties are (σ_n / n) ≤ 0.5 %. Temperatures were 298 K < T_{CCO2} = 304.2 K < 323 K. Data have been correlated using the adsorption isotherm of Aranovich, cp. Chap. 7. As can be seen from the subcritical isotherm (T = 298 K), in approaching the saturation pressure (p_{SCO2} (298 K) = 6.7 MPa) pore and bulk condensation occurs. However, also for supercritical adsorption (T = 323 K) the amount of CO_2 adsorbed seems to increase without limits as the pressure increases well above its critical value (p_{CCO2} = 7.35 MPa). This indicates occurrence of a liquid-like adsorbate phase in the pores of the molecular sieve or, equivalently, a shift of the critical parameters (pressure, temperature) of the pore fluid compared to its values of the bulk fluid phase [3.26].

3. THERMOGRAVIMETRY

Thermogravimetry, i. e. weighing material samples at elevated and possibly changing temperatures today is the standard method to investigate several kinds of thermally induced processes, especially desorption processes of gases or vapors, i. e. volatile compounds or chemical decomposition processes of solid samples [3.1, 3.8, 3.9]. By use of modern magnetic suspension balances equipped with electric heating systems, a range of pressure up to 50 MPa and temperatures up to 1250 K becomes feasible for experiments – at least on principle [3.6]. At elevated pressures measurement of the density of the gas surrounding the sample fixed to the balance causes some problems. These can be solved using a sinker also fixed to the balance and performing buoyancy measurements between the weightings of the material sample. Details of this procedure are given in Chap. 4, Sect. 3.1.

Also at temperatures above 700 K thermal decoupling of the sample and the permanent magnet inside the (hot) reaction vessel is mandatory. Convection flows initiated by temperature gradients have to be suppressed or at least reduced by appropriate flow guide tubes.

A list of commercial suppliers of thermogravimetric instruments is given at the end of this Section, cp. Tab. 3.3, [2.3].

In Figure 3.22 a schematic diagram of a magnetic suspension balance is given including a high temperature electric furnace which thermally is decoupled from the magnetic suspension by a spacing tube (height: ca. 30 cm) including tubes to reduce gas convection.

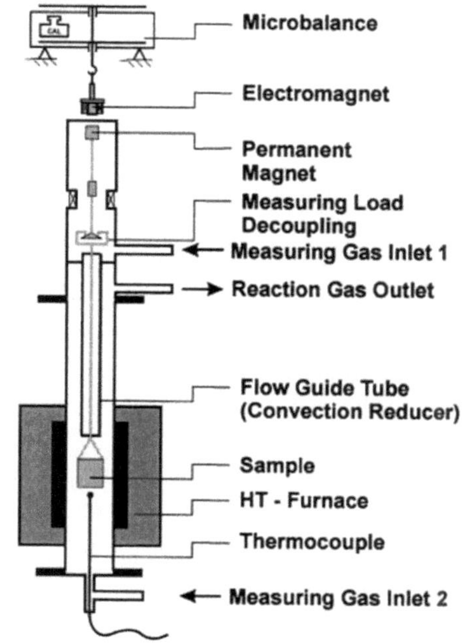

Fig. 3.22.
Schematics of a magnetic suspension balance (Rubotherm, Bochum, Germany) including an electric furnace for thermogravimetric measurements.
© Rubotherm GmbH, Bochum, Germany

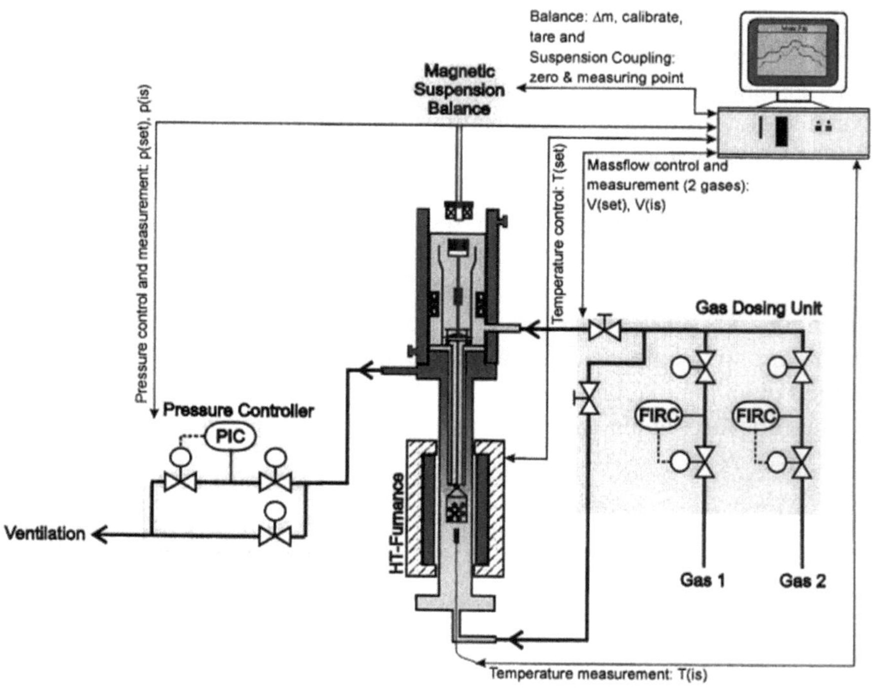

Figure 3.23. Schematics of an instrument for thermogravimetric analysis of substances including volatile components (TG-Analysis). The instrument is equipped with a gas dosing system and pressure control devices. © Rubotherm GmbH.

In Figure 3.23 the basic flow diagram of an instrument for thermo-gravimetric measurements with an automated gas dosing system and pressure control device is sketched. Application oriented details and hints for the measurement procedure are given in [3.5, 3.6].

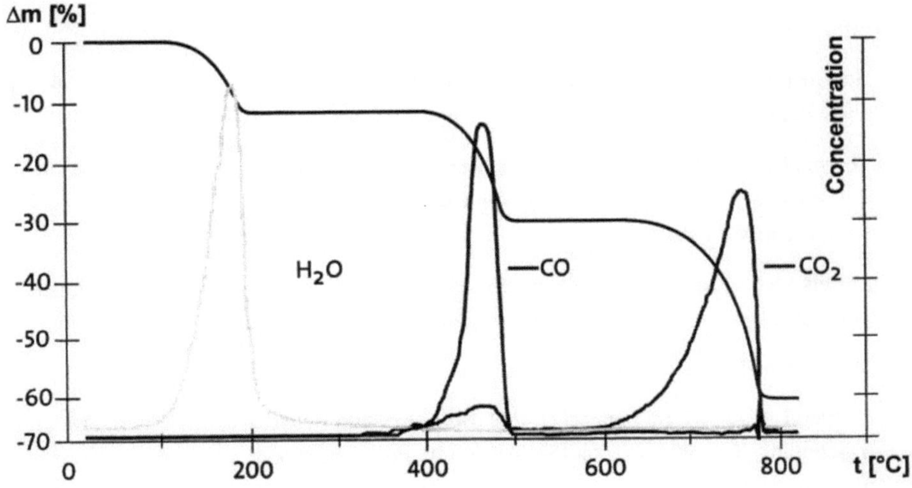

Figure 3.24. Thermogram of the decomposition process of Ca-oxalate in a nitrogen atmosphere at p = 0.1 MPa in the temperature range (20 °C-800 °C).

As an example for thermogravimetric measurements the drying and decomposition curve of Ca-oxalate in the temperature range (20 °C-800 °C) in a N_2-atmosphere at p = 0.1 MPa is sketched in Fig. 3.24. The reduction in mass due to desorption of water (H_2O) and thermal decomposition leading to evolution of CO and CO_2 from the substance is presented as function of temperature. Also concentration peaks of these components (H_2O, CO, CO_2) in the circulation gas as function of temperature are indicated. Data have been provided by Rubotherm GmbH, Bochum, and Bayer AG, Leverkusen, which is gratefully acknowledged [3.6].

A list of suppliers of commercial thermogravimetric instruments is given in Table 3.3 below.

Table 3.3. Manufacturers of thermogravimetric instruments*)

Manufacturer	Balance
Bähr-Thermoanalyse GmbH, Altendorfstr. 12, Postfach 1105, D-32603 Hüllhorst, Germany, Tel.: 05744-9302-0, Fax: -90, info@baehr-thermo.de, www.baehr-thermo.de	Bähr
Beckman Instruments, Fullerton, CA 82834, USA, www.beckmancoulter.com	Beckman
Thermo Cahn, 5225 Verona Road, Madison, WI 53711 USA Tel.: +1-608-276-633, Fax –273-6827, www.cahn.com www.thermocahn.com	Cahn
Linseis Meßgeräte GmbH, Viellitzer Str. 43, D-95100 Selb, Germany, Tel.: 09287-880-0, Fax: -70867, linseis@t-online.de, www.linseis.de	Linseis
MC^2 Thermal Systems, 667 Pinewood Ave., Troy, NY 12180, USA	
Mettler-Toledo AG, Im Langacher, P.O. Box, CH-8606 Greifensee, Switzerland, Tel.: 0041-1-944-45-45, Fax: -10, info.ch@mt.com, www.mt.com	Mettler
Netzsch Gerätebau GmbH, Wittelsbacher Str. 42, D-95100 Selb, Germany, Tel.: 09287-881-0, Fax: -44, www.ngb.netzsch.com	Sartorius
Perkin-Elmer Instruments, 710 Bridgeport Ave, Shelton CT 06484-4794, USA, Tel.: 01-203-925-4600, Fax: -4654, www.perkinelmer.com, info@perkinelmer.com	Perkin-Elmer
Rheometric Scientific, Inc. One Possumtown Road, Piscataway NJ 08854, USA, Tel.: 01-732-560-8550, Fax: 7451, www.rheometricscientific.com, www.rheosci.com	Rheometric Scientific
Rubotherm, Universitätsstr. 142, D-44799 Bochum, Germany, Tel.: +49-234-70996-0, Fax: -22, www.rubotherm.com	Rubotherm + Mettler or Satorius
SETARAM, 7 rue de l'Oratoire, F-69300 Caluire, Frankreich +33 (0)4 72 10 25 25 fax: +33 (0)4 78 28 63 55, www.setaram.com	Setaram
Shimadzu Corporation, 3 Kanda-Nishikicho 1-chome, Chiyoda-ku, Tokyo 101, Japan, www.shimadzu.com	Shimadzu
TA Instruments, 109 Lukens Drive, New Castle, DE 19720-2795, USA, Tel. :01-302-427-400-0, Fax : -1, www.tainstruments.com	

*) This table in no way claims to be exclusive or complete.

4. GRAVIMETRIC MEASUREMENT OF MULTICOMPONENT GAS ADSORPTION EQUILIBRIA (N>1)

4.1 Experimental

Coadsorption equilibria of gas mixtures of known initial concentrations on porous sorbent materials can be measured fairly easily using a microbalance combined with a gas chromatograph or mass spectrometer to analyze the sorptive's gas concentration at adsorption equilibrium. A schematic of such an instrument is given in Figure 3.25. It basically consists of a microbalance, gas supply system, a sample loop including a gas circulator to maintain equal and constant sorptive gas concentrations within

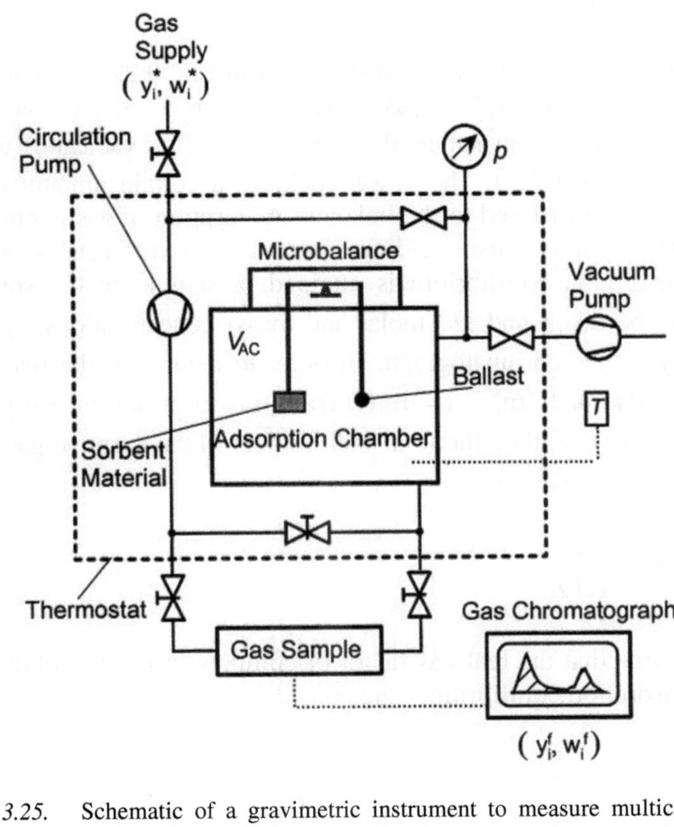

Figure 3.25. Schematic of a gravimetric instrument to measure multicomponent gas adsorption equilibria. Concentrations of the sorptive gas mixture supplied are assumed to be known. After equilibration, concentrations of the sorptive gas components are determined in a gas chromatograph or mass spectrometer [3.16, 3.20, 3.41].

the system, a thermostat and thermometers and manometers. The sample vessel should be connected to a gas chromatograph allowing one to analyze the concentrations of the sorptive gas in equilibrium. Helium at about 400 K-500 K is recommended as carrier gas. Materials of vessels, tubes, and valves should be chosen according to recommendations already given in Chap. 2, Section 2.1. Further information on design and operation of such an instrument can be found in [3.16, 3.20, 3.41].

An instrument for gravimetric measurements of multicomponent gas adsorption equilibria has been designed and operated during 1993-2003 at the IFT, University of Siegen. It was part of a multipurpose instrument for different types of gas coadsorption measurements. A photo of it is included in Chap. 4, cp. Fig. 4.2.

4.2 Theory

We consider the microbalance installation in Fig. 3.22 and assume that a certain mass (m^*) of sorptive gas mixture at known mass concentrations (w_i^*, $i = 1...N$) has been supplied to the (originally evacuated) adsorption vessel with volume (V_{AC}). The vessel includes a certain amount of sorbent material of mass (m^s) fixed to the balance. Adsorption of gas is enhanced by the circulation pump thus avoiding concentration differences within the vessel. If adsorption equilibrium is attained, a sample of the sorptive gas mixture may be taken and its (molar and mass) concentrations (y_i^f, w_i^f) be determined in a gas chromatograph. In order to determine the masses of all components adsorbed (m_i^a, $i = 1...N$) from this information, we proceed as follows: First we calculate the total mol number of the sorptive gas (n^f) from its equation of state (EOS)

$$n^f = \sum_{i=1}^{N} n_i^f = \frac{pV^f}{RTZ} \qquad 3.42$$

Here we assume that the real gas factor or compressibility (Z) of the sorptive gas in the adsorption equilibrium state considered,

$$Z = Z(p, T, y_1^f ... y_N^f) \qquad 3.43$$

is a known function of pressure p, temperature T, and molar concentration y_i^f, $i = 1...N$. The volume V^f of the sorptive gas inside the adsorption vessel and connecting tubes and valves is also assumed to be known from the relation

$$V^f = V_{AC} - V^{as} \qquad 3.44$$

Here the volume of the adsorption chamber (V_{AC}) etc. has to be determined – for example – from helium or nitrogen expansion experiments using a sample mass of well known volume (ballast or tare), cp. Chap. 2. For the volume of the sorbent mass and the adsorbed phase (V^{as}) one has again to introduce a model assumption. We here restrict the discussion to the Gibbs excess mass of all components adsorbed. Hence we choose the so-called helium approximation for V^{as}:

$$V^{as} = V^s_{He}, \qquad 3.45$$

V^s_{He} being the volume of the sorbent measured by exposing it to helium at ambient temperature and low pressure (1 atm) and assuming helium neither to be adsorbed nor absorbed by the sorbent sample.

From the mol number n^f and the concentrations (y^f_i) the total mass (m^f) of sorptive gas can be calculated by

$$m^f = \left(\sum_{i=1}^{N} y^f_i M_i\right) \cdot n^f, \qquad 3.46$$

M_i indicating the molar mass of component i = 1…N. Now the total mass of sorptive gas (m^*) originally supplied to the sorption vessel can be calculated from the mass balance

$$m^* = m^f + m^a_{GE} \qquad 3.47$$

as the total mass adsorbed (m^a_{GE}) can be calculated from the reduced mass (Ω_G) measured at the microbalance, cp. eq. (3.5, 3.14),

$$\Omega_G = m^a_{GE} - \rho^f V^s_{He} \qquad 3.48$$

and the density ρ^f of the sorptive gas given by (3.42), which can be rewritten as

$$\rho^f = \frac{p}{RTZ}\sum_i^N y_i M_i \tag{3.49}$$

Hence we have from (3.42-3.49)

$$m^* = \frac{pV_{AC}}{RTZ}\sum_i^N y_i M_i + \Omega_G \tag{3.50}$$

As the Gibbs excess mass adsorbed of component i is

$$m^a_{iGE} = w^*_i m^* - w^f_i m^f \tag{3.51}$$

we finally get in view of (3.50) and (3.42, 3.46):

$$m^a_{iGE} = \left(\frac{pV_{AC} - V^s_{He}}{RTZ} M^f + \Omega_G\right) w^*_i - \frac{p(V_{AC} - V^s_{He})}{RTZ} M^f_{wi}, \; i = 1 \ldots N \tag{3.52}$$

with the molar mass of the sorptive gas in the adsorption equilibrium state $M^f \equiv \sum_{i=1}^{N} y^f_i M_i$.

The mass concentrations (w^f_i) needed in eq. (3.52) and the molar concentrations (y^f_i) of the sorptive gas measured at the gas chromatograph are interrelated by the equations

$$w^f_i = y_i M_i / \left(\sum_k^N y_k M_k\right), \quad y_i = (w_i / M_i) / \sum_k^N (w_k / M_k), \; i = 1 \ldots N \tag{3.53}$$

From the Gibbs excess masses (3.52), the absolute masses adsorbed $(m^a_i, i = 1 \ldots N)$ can be calculated using eqs. (2.31), (2.32) already given in Chap. 2.

Also the *selectivities* (S_{ik}) of any two components i, k of the gas adsorption system can be calculated from eq. (2.33), Sect. 4.2 of Chap. 2 as adsorbate's Gibbs excess molar concentrations

$$x_i^a = (m_{iGE}^a / M_i) / \sum_{k=1}^{N} (m_{kGE}^a / M_k), \quad i = 1...N \tag{3.54}$$

can be calculated from eqs. (3.52). Examples will be given in Sect. 4.4 of this chapter.

4.3 Uncertainties or Errors of Measurement

Uncertainties of Gibbs excess masses (3.52) as represented by their mean square deviations (MSD) (σ^2_{miGE}) can be calculated by applying the Gauss Law of propagation of uncertainty of error to eqs. (3.51) and (3.52) respectively:

$$\sigma^2_{miGE} = \sigma^2_{i*} + \sigma^2_{if}, \tag{3.55}$$

$$\sigma^2_{i*} = m^{*2}\sigma^2_{wi*} + w_i^{*2}\sigma^2_{mf} \tag{3.55a}$$

$$\sigma^2_{if} = m^{f2}\sigma^2_{wif} + w_i^{f2}\sigma^2_{mf} \tag{3.55b}$$

in view of (3.50) and (3.42, 3.46) we have

$$\sigma^2_{m*} = \left[\left(\frac{\sigma_p}{p}\right)^2 + \left(\frac{\sigma_{VAC}}{V_{AC}}\right)^2 + \left(\frac{\sigma_T}{T}\right)^2 + \left(\frac{\sigma_Z}{Z}\right)^2 + \sum_{i=1}^{N} M_i^2 \sigma^2_{yif}\right] \bullet \left(\frac{pV_{AC}M^f}{RTZ}\right)^2 + \sigma^2_{\Omega} \tag{3.56a}$$

$$\sigma^2_{mf} = \left[\left(\frac{\sigma_p}{p}\right)^2 + \frac{\sigma^2_{VAC} + \sigma^2_{VHe}}{(V_{AC} - V^s_{He})^2} + \left(\frac{\sigma_T}{T}\right)^2 + \left(\frac{\sigma_Z}{Z}\right)^2 + \sum_{i=1}^{N} M_i^2 \sigma^2_{yif}\right] \bullet \left(\frac{p(V_{AC} - V^s_{He})}{RTZ} M^f\right)^2 \tag{3.56b}$$

To provide an overview of all parameters to be measured and uncertainties to be considered we list them in a table:

w_i^*	w_i^f	y_i^f	p	V_{AC}	V_{He}^s	T	Z	Ω
σ_{wi*}	σ_{wif}	σ_{yif}	σ_p	σ_{VAC}	σ_{VHe}	σ_T	σ_Z	σ_Ω

As can be recognized from eqs. (3.55a, b, 3.56a, b) accurate measurements of gas concentrations (y_i^*, y_i^f) or, equivalently (w_i^*, w_i^f) are very important in order to get small dispersions of adsorbate's mass m_{iGE}^a. Due to experience it also can be said that accurate measurements of the system's pressure (p) and temperature (T) are essential, whereas the influence of the real gas factor (Z) and its uncertainly often is rather small. Given MSDs of gas concentrations $\sigma_{wi*} \simeq \sigma_{yif} < 10^{-2}$, of microbalance measurement $(\sigma_\Omega/\Omega)<10^{-3}$ and of all the other quantities to be measured at the orders of magnitude mentioned in Sects. 2.3, 4.3 of Chap. 2, one can expect relative uncertainties of adsorbed masses $(\sigma_{iGE} / m_{iGE}^a)$ for binary systems (N = 2) of about 2 % and for ternary systems (N = 3) of ca. 5 %. Often the less adsorbed component poses a real problem as it is very difficult to determine small changes of gas concentrations with satisfying accuracy [3.20, 3.22, 3.27]. An example for this experimental problem is given in the next section.

4.4 Examples

Coadsorption equilibria of gas mixtures on porous solids normally are measured using the volumetric/manometric method described in Chap. 2, Sect. 4, cp. also [3.42-3.44]. Despite the fact that purely gravimetric-chromatographic measurements of coadsorption equilibria today are scarce, it should be emphasized that this method still has the advantage that, contrary to the manometric procedure, the approach to equilibrium can be accurately observed from the microbalance signal. Hence it can be checked whether or not the adsorption system investigated is in equilibrium or still in a transient non-equilibrium state. For this reason several gravimetric-chromatographic measurements of binary (N = 2) and ternary (N = 3) coadsorption equilibria of non-corrosive gases on both activated carbons and zeolites have been performed during the last 10 years at IFT, University of Siegen. In the following we first present adsorption equilibria of a ternary gas mixture on activated carbon. Data were taken by gravimetric measurements of the total mass adsorbed and analyzing the sorptive gas phase, cp. Sect. 4.2. This method is cumbersome but reliable from an experimenter's point of view. To avoid lengthy and therefore expensive gas analyzing procedures it has been suggested in the literature [3.45], to complement gravimetric measurements with a thermodynamic model allowing one to calculate from the total mass adsorbed and thermodynamic equilibrium conditions, i. e. equality of chemical potentials of the sorptive and the sorbate phase, the mass of all components adsorbed (van Ness model). Though this method sometimes seems to work very well, it cannot be recommended as coadsorption equilibria are fairly complex phenomena which only for sorptive gas mixtures with similar molecules approximately can be calculated by this or related methods (IAST), [3.38].

As a supplement we add pure and binary mixture data taken for adsorption equilibria in a *liquid*-solid system. This is to be understood as an example demonstrating that microbalances can also provide information on liquid adsorption processes, a field which in view of protein adsorption phenomena for separation certainly has potential for future development.

Example 1

Adsorption of ternary gas mixtures (CH_4, CO_2, N_2) on activated carbon (AC) NORIT R1, [3.22, 3.27]. Corresponding data of binary mixtures are presented in Chap. 4, Sect. 2.4.

Data have been measured gravimetrically (Sartorius) aided by gas phase chromatography (Perkin Elmer). For AC's activation procedure and some of its microscopic properties we refer to Sect. 2.3 of this chapter. Time for equilibration depended considerably on the methane concentration of the sorptive gas and ranged between 1 h for high and 8 h for low concentrations. In Figure 3.26 coadsorption equilibria data of the amounts of CH_4, $CH_4 + CO_2$ and $CH_4 + CO_2 + N_2$ being adsorbed on AC NORIT R1 at 298 K for sorptive gas concentrations $y_{CH4} = 48$ %, $y_{CO2} = 8$ %, $y_{N2} = 44$ % in the pressure range $p < 6$ MPa are sketched. Uncertainties of data are about 2-3 the size of the graphical symbols chosen. The full lines indicate adsorption isotherms calculated by fitting the data to a 2-site generalized AI of Langmuir type, cp. Chap. 7 and [3.27].

Data show that the AC has a good selectivity for CO_2: although its gas concentration is only $y_{CO2} = 8$ %, its concentration in the adsorbate is about $x_{CO2} = 25$ %. Hence this AC can be considered as a sorbent for purification of low energy gas consisting of (CH_4, CO_2, N_2) from CO_2.

For practical purposes so-called Gibbs ternary diagrams are very useful to present ternary adsorption equilibria. An example for this is given in the next Figure 3.27. It again refers to (CH_4, CO_2, N_2) gas mixtures being adsorbed at four different concentrations on AC NORIT R1 at $T = 298$ K and a total gas pressure $p = 1$ MPa, [3.27].

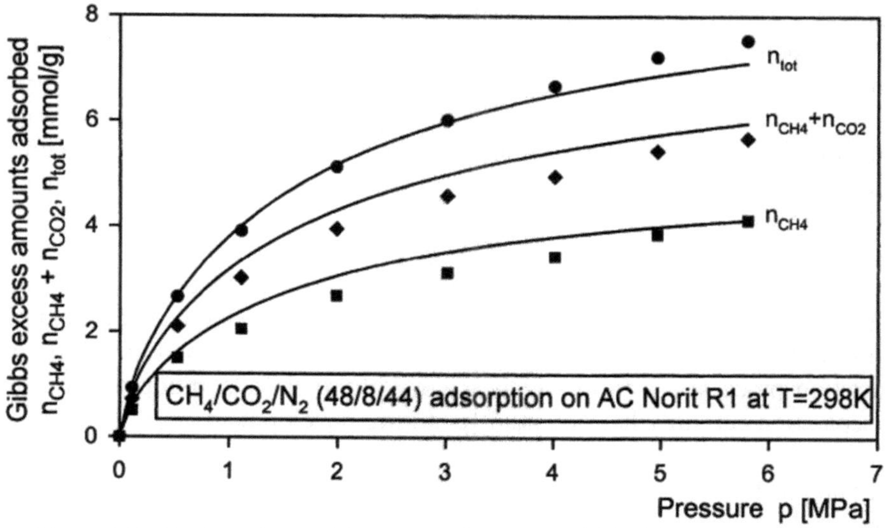

Figure 3.26. Coadsorption equilibria data of the system (CH_4, CO_2, N_2) on AC Norit R1 for sorptive gas concentrations y_{CH4} = 48 %, y_{CO2} = 8 %, y_{N2} = 44 %, p < 6 MPa, [3.27].

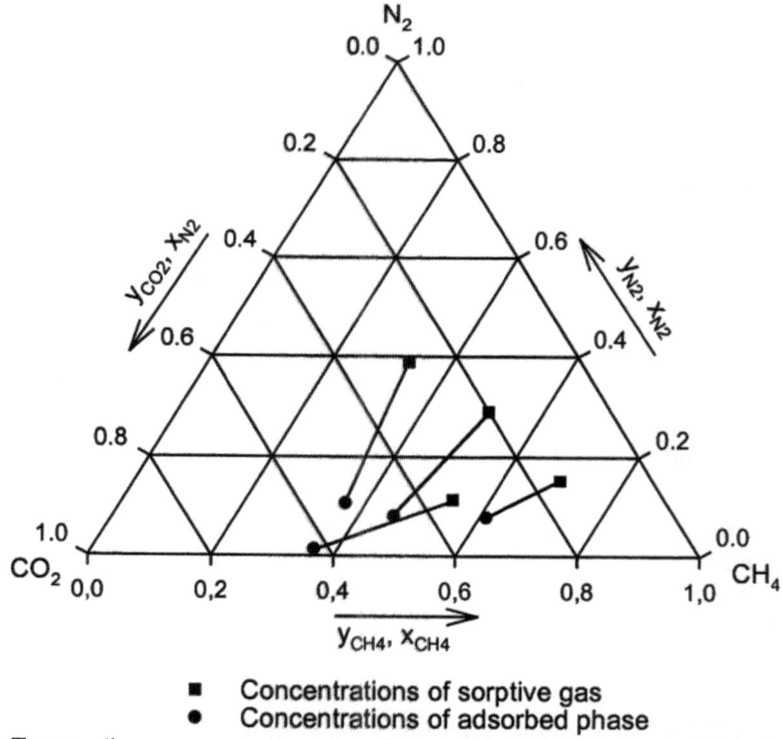

Figure 3.27. Ternary diagram representing adsorption equilibria of gas mixtures (CH_4, CO_2, N_2) (■) with adsorbed phases (•) on AC NORIT R1 at T = 298 K and p = 1 MPa, [3.27].

From this figure it clearly can be seen that the AC is selective for CO_2 as both the N_2 as well as the CH_4 concentration in the adsorbed phase are considerably reduced compared to their values in the sorptive gas phase. Information of this kind is very valuable for the selection of adsorbents in gas purification and separation processes [3.37].

Example 2

Adsorption of dye from an aqueous solution onto activated carbon.

The adsorption of dye Levafix ® Brilliant Red E-4BA from an aqueous solution onto activated carbon F300 (Chemviron Carbon) at 24 °C and ambient pressure has been investigated using a magnetic suspension balance (MSB) (Rubotherm GmbH, Bochum, Germany) [3.46, 3.47]. The experimental setup is shown in Figure 3.28. The dark region within the adsorption vessel of the MSB indicates the volume filled with liquid solution. More details of design and materials are given in [3.47] and literature cited there. Results of gravimetric measurements are given in the next Figure 3.29. The lower curve refers to the sorption process of pure water into the activated carbon. There are experimental indications that the density of the water in the pores is about the same as the density of ice at 0 °C (0.92 g / cm^3), i. e. lower than the density of water in its liquid state. Indeed, it is known that pore liquids can exhibit three dimensional structures [3.48] and that adsorption in micropores often is accompanied by kind of freezing process which also can be simulated by statistical molecular models [3.49]. The upper curve shows the increase of mass of an activated carbon sample due to the combined adsorption of water and dye. Seemingly both processes are fairly slow and did not reach equilibrium after a period of obeservation of 30 hours and more. Gravimetric measurements were complemented by traditional beaker experiments allowing thus to calculate separately the masses of water and dye adsorbed in the activated carbon. The procedure is similar to that described in the next Chapter 4 (Volumetric-Gravimetric Measurements). Details are given [3.47].

Though these measurements certainly are of preliminary character and need more detailed discussion, it should be emphasized that they demonstrate the possibility to explore adsorption processes of complex molecules like dye Levafix ® Brilliant Red E-4BA (whose chemical formulae is classified) on microporous sorbents. This certainly is of interest not only for textile engineering but also for waste water management and for environmental protection.

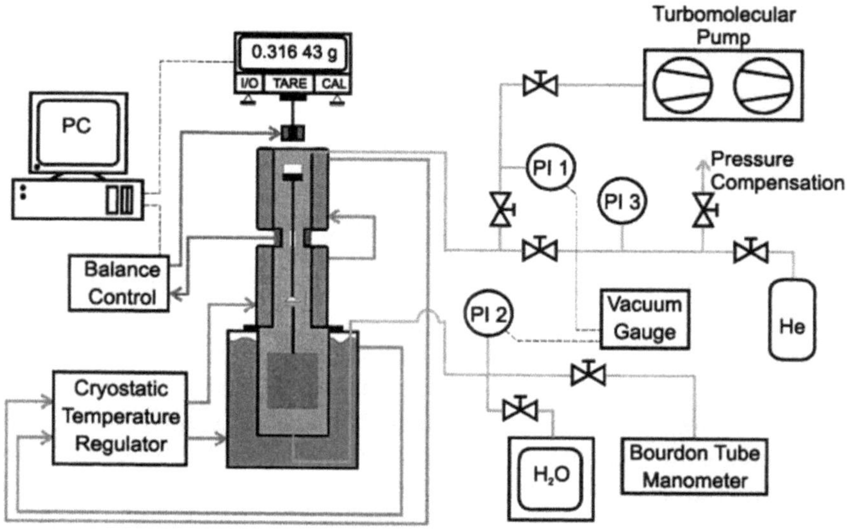

Figure 3.28. Experimental installation of an instrument for gravimetric measurements of adsorption processes from the liquid phase using a magnetic suspension balance. The dashed region within the adsorption vessel below the balance indicates the volume filled with liquid [3.46, 3.47]. © Rubotherm GmbH, Bochum, Germany.

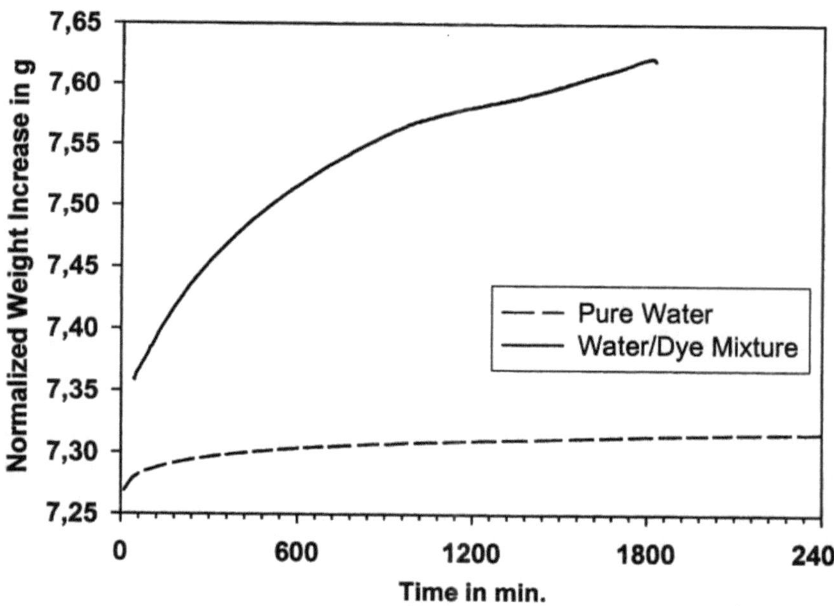

Figure 3.29. Gravimetric measurements of the adsorption processes of pure liquid water (lower curve) and of dye Levafix ® Brilliant Red E-4BA and water from aqueous solution (upper curve) in activated carbon F 300 (Chemviron Carbon), [3.47]. © Rubotherm GmbH, Bochum, Germany.

5. PROS AND CONS OF GRAVIMETRY

This section is devoted to the advantages and disadvantages of gravimetric measurement of gas adsorption equilibria. It reflects experiences we made in our experimental work in this field during the last 15 years using both two beam and magnetic suspension balances.

5.1 Advantages

1. Accuracy
   All microbalances which are commercially available today, cp. Tab. 3.1, [3.7-3.9] normally exhibit high reproducibility, sensitivity, and accuracy (up to 10^{-8}) of measurements. Hence they allow one to determine adsorbed gas masses on porous solids much more accurately than manometric measurements. The gain in accuracy often is about one order of magnitude; at very low pressures it may be two orders of magnitude or even more.

2. Amount of sorbent material
   For highly sensitive microbalances (Thermo Cahn, Hiden, Mettler-Toledo, Rubotherm, Setaram, TA Instruments, VTI) only tiny amounts of sorbent materials are needed to measure gas adsorption equilibria. This is advantageous for investigations in newly developed sorbent materials where often only small amounts are available. However, results may be misleading as the sorbent sample used does not represent a "statistically averaged" sample of a technical sorbent and thus may exhibit considerable statistical deviations.

3. Approach to equilibrium
   Microbalances with alphanumerical display and electronic data recording systems allow one to observe the approach to equilibrium for gas adsorption processes in porous sorbent samples. Typical relaxation times can be one or several seconds, minutes, hours, and – sometimes – even days, cp. helium adsorption data Sect. 2 of Chap. 1. Hence gravimetric measurements do allow one to check whether an adsorption system actually has reached its equilibrium state, i. e. these measurements deliver in principle also information concerning the kinetics of the adsorption process, represented for example by (phenomenological) diffusion coefficients, cp. Sect. 2.3 and Sect. 4.4 and [3.27, 3.48).

4. Wall sorption
   In gravimetric adsorption measurements of pure gases adsorption of the gas on walls of tubes and vessels does not pose a problem as no mass balances of the gas are necessary. However, in multicomponent adsorption measurements it may influence the sorptive gas concentration to a certain extent, especially at low gas pressures and temperatures ($T < 77K$), and if– for example in a binary gas mixture – one component is strongly, the other only weakly adsorbed. Electropolishing of inner surfaces of all tubes and vessels, preferably made of stainless steel, can reduce this problem considerably.

5. Extreme pressures
   Contrary to manometry/volumetry, very high and very low pressures of the sorptive gas do not pose a serious problem in gravimetric adsorption measurements. This is a due to the fact that in gravimetry, the adsorbed mass is determined by its weight, i. e. a quantity which in principle is physically independent of the gas pressure.

   Actually, it is possible to use a microbalance equipped with a dense sinker of accurately known mass and volume as a manometer: by measuring its weight in a gas, one can calculate from the buoyancy term the density of the gas and from it and the gas temperature via the thermal equation of state of the gas, its pressure. This method already proved to be useful for low pressures ($p < 10$ Pa) [3.1, 3.23-3.25]. However, it always should be checked whether radiation equilibrium between the (normally gold coated) sinker and the (black) walls of the gas vessel is maintained. If not, creeping gas flows may occur which may cause serious uncertainties in the balance's measurements.

   It also is possible to use two balances in parallel, one bearing a sinker and acting as a gás manometer, the other including the sorbent sample and serving as sorption measuring instrument.

   However, it should be mentioned that even gravimetric sorption measurements may become uncertain at very low pressures due to gas adsorption at the vessel or basket including the sorbent or other parts of the microbalance. Therefore it is recommended always to perform calibration measurements, i. e. measurements with the empty balance bearing no sorbent material but being exposed to the same sorptive gas.

6. Kinetics
   Modern microbalances do allow one to take weight data every tenth of a second. If the kinetics of a pure gas sorption process is slow compared to this time, it can be easily recorded. Examples have been given in the foregoing Sects. 2.3, 4.4. Here we only want to mention that curves depicting the mass of a sorbent / sorbate sample as function of time do not always show a simple exponential approach to an equilibrium state but may be much more complicated. An example of practical importance is chemisorption of SO_2 gas on activated carbon (AC). Due to catalytic properties of the AC, SO_2 can be converted via (SO_4) to sulfuric acid (H_2SO_4) which at near ambient conditions periodically falls down in droplets from the carbon sample leading thus to saw-tooth like curves in the balance's recordings.

   Also kinetic curves of ad- and desorption processes can give hints to hysteresis phenomena as they may occur in mesoporous sorbent materials.
   Mixture adsorption processes normally cannot be detected by simple gravimetric / manometric measurements. The reason for this is that during such a process the sorptive gas mixture has to be circulated in order to avoid local concentration gradients. As the gas flow inside the adsorption vessel causes dynamic forces acting on the sorbent sample, the balance recording is changed. According to our experience the influence of these dynamic forces, is often irreproducible due to the complex geometry of piled sorbent pellets and can not easily be taken into account by calibration experiments with a non-adsorbing "dummy sorbent".

7. Activation
   In practice, activation of a sorbent material much more easily can be accomplished in gravimetric than in volumetric sorption instruments. The main reason for this is that in gravimetry the mass of the sample can be recorded even during the activation process whereas in volumetry / manometry it normally can not. Hence in gravimetry the initial state of a sorbens is fairly well known. Also presorption in a new sorbent sample and remnant sorption after a desorption experiment easily can be checked.

5.2 Disadvantages

1. Complexity
   Modern microbalances, especially the magnetic suspension balances are fairly complex systems which often are sensitive to electromagnetic or – to lesser extend – mechanical disturbances from outside. Oscillations of the basement caused by nearby subways or trucks may cause disturbances, though balances recently (2002) have become less sensitive to these. However, electromagnetic fields radiating from electric heating systems or other wires inside a lab can lead to serious problems in the performance of either the balance itself or the data recording system.

   Also fine grained sorbent materials, especially activated carbon fibers or powder may cause problems as they may change their position within a vessel mounted to the balance or simply are blown out of it due to the sorptive gas flow.

   We mention this here to indicate to the reader that microbalances are fairly sensitive and – by the way – expensive instruments and require some experience and – foremost – patience in handling to get accurate and reliable results.

2. Measuring Techniques
   Gravimetric measurements of gas adsorption equilibria are laborious as they require
   - detection of several thermophysical quantities like pressure, temperature, and sometimes concentrations of sorptive gas mixtures,
   - various calibration experiments,
   - careful handling of sorbent material, and
   - sophisticated software for balance controlling and data handling, cp. also Chap. 4.

   It also should be mentioned that measurement of sorbent material's temperature is not as easy as in volumetric experiments as – for example – a thermocouple only can be placed in the vicinity of the sample but not inside it.

   Automation of gravimetric sorption instruments is not an easy matter and for precision measurements permanent supervision of the adsorption / desorption process by an experienced co-worker is strongly recommended.

6. LIST OF SYMBOLS AND ABBREVIATIONS

A list of most symbols and abbreviations used in this Chapter is given.

B_α	Nm	moment of buoyancy force at sorbent site ($\alpha = s$) and ballast or tare site ($\alpha = k$) respectively
G_0	N	weight of the empty suspension inside the adsorption vessel in vacuum
G_0^K	N	weight of the suspension inside the adsorption vessel bearing a ballast or tare (m^K) in vacuum
G_f^K	N	weight of the suspension inside the adsorption vessel bearing a ballast or tare (m^K) in sorptive gas atmosphere (f)
G_0^s	N	weight of the suspension inside the adsorption vessel loaded with sorbent mass (m^s) in vacuum
G_f^{sa}	N	weight of magnetic suspension floating freely in the sorptive gas or fluid (f)
$g = 9.81$	m/s^2	gravity constant of the earth
l_α	m	length of beams of a microbalance at sorbent site ($\alpha = s$) or ballast or tare site ($\alpha = k$) of a microbalance
M_0	Nm	mechanical moment of electromagnetic forces equilibrating the empty balance, i. e. without sorbent mass and ballast or tare in vacuum
M_f	Nm	mechanical moment of electromagnetic forces equilibrating the empty balance in a sorptive gas atmosphere

$M^f = \sum_i^N y_i^f M_i$	g/mol	molar mass of a sorptive gas mixture
M_i	g/mol	molar mass of component i = 1...N of a sorptive gas mixture
M_α	Nm	moment of gravity forces of beam, wire, and basket at the sorbent site ($\alpha = s$) or ballast or tare site ($\alpha = k$) of a microbalance
M_0^0	Nm	mechanical moment of electromagnetic forces equilibrating the balance without sorbent mass and without ballast or tare in vacuum
M_f^0	Nm	mechanical moment of electromagnetic forces equilibrating the balance without a sorbent sample in sorptive gas atmosphere
M_0^s	Nm	mechanical moment of electromagnetic forces equilibrating the balance loaded with sorbent mass and ballast or tare in vacuum
M_f^{sa}	Nm	mechanical moment of electromagnetic forces at the induction coil of a two beam microbalance needed to equilibrate the balance loaded with sorbent and ballast or tare in a sorptive gas atmosphere
m^B	kg	mass of basket to keep a sorbent material
m^K	kg	mass of a dense ballast or tare or sinker to be added to a suspension balance for gas density measurements
m^a	kg	mass adsorbed on a certain mass (m^s) of sorbent
m^k	kg	mass of ballast or tare at a two beam balance

m^*	kg	mass of sorptive gas supplied to the adsorption chamber prior and during the adsorption process
m^s	kg	mass of sorbent material
m^a_{GE}	kg	Gibbs excess mass of an adsorbed phase
m^a_i	kg	mass of component i = 1...N being adsorbed on a certain mass of sorbent (m^s)
m^a_{iGE}	kg	Gibbs excess mass of component i of an adsorbed phase
m^{PM}	kg	mass of permanent magnet fixed to a magnetic suspension
n^f	mol	number of moles of a sorptive gas mixture in a gas-adsorption equilibrium state
p	Pa	pressure of sorptive gas
R = 8.314	J / Kmol	universal gas constant
T	K	absolute temperature
V_{AC}	m^3	volume of adsorption chamber
V^B	m^3	volume of basket to keep a sorbent material
V^K	m^3	volume of a dense ballast or tare or sinker to be added to a suspension balance for gas density measurements
V^{as}	m^3	volume of a sorbate – sorbent system which can not be penetrated by a sorptive gas
V^f	m^3	volume of a sorptive gas phase
V^k	m^3	volume of a ballast or tare used in a two beam microbalance

V^{PM}	m^3	volume of permanent magnet fixed to a magnetic suspension
V^s_{He}	m^3	void volume of a porous solid (sorbent) measured by helium expansion experiments
w^f_i	1	mass concentration of component i = 1 . . .N in a sorptive gas mixture in adsorption equilibrium
w^*_i	1	mass concentration of component i = 1 . . .N in a sorptive gas mixture supplied to an instrument prior to adsorption
y^f_i	1	molar concentration of component i = 1 . . .N in a sorptive gas mixture in adsorption equilibrium
$Z = (pV_m / RT)$	1	compressibility of a gas (real gas factor)
ρ^f	kg/m^3	density of a sorptive gas
σ^2_x	$[x^2]$	dispersion, variance or mean statistical deviation (MSD) of a measurable quantity (x)
σ^2_{mGE}	kg^2	dispersion, variance or mean statistical deviation (MSD) of the Gibbs excess mass of an adsorbate
$\Omega = m^a - \rho^f V^{as}$	μg	reduced mass of an adsorbate
Ω_G	μg	reduced mass of an adsorbed phase measured gravimetrically; also microbalance signal, eqs. (3.6, 3.10, 3.30)

REFERENCES

[3.1] **Kochsiek M., Gläser M., (Eds.)**
Comprehensive Mass Metrology, Wiley-VCH, Berlin, 2000, ISBN 3-527-29614-X, p. 366-369, p. 494-496.

[3.2] **Lösch H. W., Kleinrahm R., Wagner W.**
Neue Magnetschwebewaagen für gravimetrische Messungen in der Verfahrenstechnik, Jahrbuch 1994, „Verfahrenstechnik und Chemieingenieurwesen", p. 117-137, VDI-Verlag, Düsseldorf, 1994.

[3.3] **Lösch H. W.**
Entwicklung und Aufbau von neuen Magnetschwebewaagen zur berührungsfreien Messung vertikaler Kräfte, VDI-Fortschritt-Berichte, Reihe 3 Verfahrenstechnik, Nr. 138, Düsseldorf, VDI-Verlag, 1987.

[3.4] **Lösch H. W., Vietinghoff J. v., Wagner W., Kleinrahm R.**
Magnetic suspension balances for research and industry, Proceedings of the 25th Conf. on Vacuum Microbalance Techniques, Siegen, 1993, J. U. Keller and E. Robens (Eds.), p. 79, Brentwood: Multi-Science-Publishing, 1995, ISBN 0-906-522-10-2.

[3.5] **Lösch H. W.**
Renaissance of Gravimetry in Metrology Under Controlled Environments – The Magnetic Suspension Balance, World Markets Series, Business Briefing, Global Chemical Processing and Engineering Industry, London, Febr. 1999, p. 104-109, www.wmrc.com

[3.6] **Rubotherm Präzisionsmesstechnik GmbH**
Suspension Balances, International Application Notes, available from Rubotherm GmbH, Universitätsstr. 142, 44799 Bochum, Germany, www.rubotherm.de, 2001.

[3.7] **Robens E., Keller J. U.**
Gassorptionsmessung und Thermogravimetrie, LABO, Magazin für Labortechnik + Life Sciences, p. 56, March 2002.

[3.8] **Keller J. U., Robens E.**
A Note on Sorption Measuring Instruments, J. of Thermal Analysis and Calorimetry, 71 (2003), 37-45.

[3.9] **Keller J. U., Robens E., Hohenesche C. du Fresne von**
Thermogravimetric and Sorption Measurement Techniques / Instruments, Studies in Surface Sciences and Catalysis Vol. 144, F. Rodriguez-Reinoso, B. Mc Enany, J. Rouquerol, K. Unger, Eds., p. 387-394, Elsevier Science B. V., N. Y., 2003.

[3.10] **Dabrowski A.**
Adsorption – from theory to practice, Adv. in Colloid and Interface Sci., 93 (2002), 135-224.

[3.11] **Robens E., et al., Eds.,**
Conferences on Vacuum Microbalance Techniques (VMT), 25th Conference, Siegen, September 1993, Proceedings published by Multi-Science Publishers, Brentwood, UK, 1994, ISBN 0-906-522-10-2.

[3.12] **Rodriguez-Reinoso F., Mc Enaney B., Rouquerol J., Unger K.**
Characterization of Porous Solids VI, Proceedings of the 6th Int. Symposium on the Characterization of Porous Solids, (COPS-VI), Alicante, May 2002, Studies in Surface Science and Catalysis, Vol. 144 Elsevier, Amsterdam etc., 2002.

[3.13] **Do D. D.**
Adsorption Analysis: Equilibria and Kinetics, Imperial College Press, London, 1998.

[3.14] **Kaneko K., Kanoh H., Hanzawa Y.**
Fundamentals of Adsorption 7, Proceedings of FoA 7 Conference, Nagasaki, May 2001, IK International, Chiba, Japan 2002.

[3.15] **Rudzinski W., Panczyk T.**
Phenomenological Kinetics of Real Gas-Adsorption-Systems: Isothermal Adsorption, J. Non.-Equilib. Thermodyn., 27 (2002), p. 149-204.

[3.16] **Seelbach M.**
Experimentelle und analytische Untersuchungen von Adsorptionsprozessen technischer Gase an mikroporösen Sorbentien, PH- D Thesis, Institute of Fluid- and Thermodynamics, University of Siegen, Siegen, Germany, 2003.

[3.17] **Robens E., Massen C. H., Poulis J. A., Staszcuk P.**
Fast measurements of adsorption on porous material using Jäntti's method, Adsorption Sci. & Techn., 17 (1999), No. 10, p. 801-804.

[3.18] **Toth J., Editor**
Adsorption, Theory, Modeling, and Analysis, Surfactant Science Series Vol. 107, M. Dekker, New York etc., 2002.

[3.19] **Webb P. A., Orr C.**
Analytical Methods in Fine Particle Technology, Micromeritics Instrument Corp., Norcross, GA, USA, 1997.

[3.20] **Tomalla M.**
„Experimentelle Untersuchung der Koadsorptionsgleichgewichte von CH_4/N_2 und CO_2/CH_4-Gasgemischen an Aktivkohle bei T = 298 K im Bereich p = 0 - 12 MPa" PHD-Thesis, Institute of Fluid- and Thermodynamics, University of Siegen, 1994.

[3.21] **Staudt R., Bohn S., Dreisbach F., Keller J. U.**
Gravimetric and Volumetric Measurements of Helium Adsorption Equilibria on Different Porous Solids, p. 261-266, Proceedings of IV Conference on Porous Solids, COPS IV, Bath UK, 1996, B. Mc .Enany et al., Eds., The Royal Society of Chemistry, Special Publ. No. 213, London, 1997.

[3.22] **Staudt R.**
„Analytische und experimentelle Untersuchungen von Adsorptionsgleichgewichten von reinen Gasen und Gasgemischen an Aktivkohlen und Zeolithen.“,
PHD-Thesis, IFT, University of Siegen, Siegen, 1994.

[3.23] **Robens E., Sandstede G., Gast Th.**
Apparatus for the measurement, automatic control and program control of the gas pressure in a receiver. United States Patent 3,572,364, 23.3.1971.
Sartorius-Werke GmbH, Göttingen.

[3.24] **Sandstede G., Robens E.**
System for automatic control of gas pressure.
United States Patent 3,516,429. 23.6.1970. Sartorius Werke GmbH.

[3.25] **Robens E., Sandstede G.**
Anordnungen zur präzisen Druckmessung und -regelung
im Bereich von 0,1 bis 760 Torr. Vakuum-Technik 16 (1967) 125-130.

[3.26] **Keller J. U., Abdel-Ghani T.,Dreisbach F., Tomalla M.**
Measurement of Adsorption Equilibria of Pure and Mixed Corrosive Gases:
The Magnetic Suspension Balance,
Proceedings 5th Int. Conference Fundamentals of Adsorption, Asilomar, May 1995, Kluwer Academic Publishers, p. 259-268, Boston, Mass., 1996.

[3.27] **Dreisbach F.**
„Untersuchung von Adsorptionsgleichgewichten methanhaltiger Gasgemische an Aktivkohle als Grundlage zur Auslegung technischer Adsorptionsanlagen”
PHD-Thesis, IFT University of Siegen, Siegen, 1998, cp. also Fortschritt-Berichte VDI, Reihe 3, Verfahrenstechnik, No. 547, VDI-Verlag, Düsseldorf, 1998.

[3.28] **Dreisbach F., Staudt R., Keller J. U.**
High Pressure Adsorption Data of Methane, Nitrogen, Carbon Dioxide and their Binary and Ternary Mixtures on Activated Carbon, Adsorption, 5 (1999), 215-227.

[3.29] **Horvath G., Kawazoe K.**
Method for the Calculation of Effective Pore Size Distribution in Molecular Sieve Carbon, J. of. Chem. Engng. of Japan, 16 (1983), 470-475.

[3.30] **Gumma S., Talu O.**
Gibbs Dividing Surface and Helium Adsorption, Adsorption, 9 (2003), p. 17-28.

[3.31] **Rouquerol F., Rouquerol J., Sing K.S.W.**
Adsorption by Powders and Porous Solids, Academic Press, San Diego, USA, 1999.

[3.32] **Sing K.S.W. et al.**
Reporting Physisorption Data for Gas/Solid Systems with Special Reference to the Determination of Surface Area and Porosity, IUPAC Recommendations 1984, Pure & Appl. Chem., 57 (1985), 603-619.

[3.33] **Sing K.S.W. et al. (Eds.)**
IUPAC Recommendations 1994 for Reporting Physisorption Data,
Pure & Appl. Chem., 66 (1994), 1739.

[3.34] **Rouquerol J., Sing K.S.W. et al.**
Guidelines for the Characterization of Porous Solids, article in: Characterization of Porous Solids III, J. Rouquerol et al. (Eds.), Elsevier, Amsterdam, 1994.

[3.35] **Iossifova N.**
Coadsorption Equilibria for Gas Separation and Purification Processes, PH-D Thesis, IFT, University of Siegen, in preparation, 2003.

[3.36] **Talu O.**
Needs, status, techniques, and problems with binary gas adsorption experiments, Adv. Colloid Interface Sciences, 76-77 (1998), p. 227-269.

[3.37] **Yang R.**
Gas Separation by Adsorption Processes, Imperial College Press, London, 1997.

[3.38] **Myers A. L., Prausnitz J. M.**
Thermodynamics of Mixed Gas Adsorption, AIChE-Journal, 11 (1965), p. 121-127.

[3.39] **Dreisbach F., Lösch H. W., Harting P.**
Highest Pressure Adsorption Equilibria Data: Measurement with Magnetic Suspension Balance and Analysis with a New Adsorbent/Adsorbate-Volume, Adsorption 8 (2002), p. 95-109.

[3.40] **Keller J. U., Zimmermann W., Schein E.**
Determination of Absolute Gas Adsorption Isotherms by Combined Calorimetric and Dielectric Measurements, Adsorption, 9 (2003) p. 177-188.

[3.41] **Iossifova N.**
Untersuchungen von Gemischgleichgewichten bei adsorptiven Gastrenn- und Reinigungsverfahren, Fortschritt-Berichte VDI, Reihe 3, Verfahrenstechnik, in preparation, VDI-Verlag, Düsseldorf.

[3.42] **Münstermann U.**
Adsorption von CO_2 aus Reingas und aus Luft am Molekularsieb-Einzelkorn und im Festbett. Berücksichtigung von Wärmeeffekten und H_2O-Präadsorption, PH-Dissertation, Technical University of Munich, 1984.

[3.43] **Busweiler U.**
Nichtisotherme Ad- und Desorptionskinetik an Einzelkörnern Technischer Adsorbentien am Beispiel der Wasserdampfadsorption an Silicagel und Molekularsieb, PH-Dissertation, Technical University of Darmstadt, Darmstadt (D17), 1984.

[3.44] **Reich R., Ziegler W. T., Rogers K. A.**
Adsorption of Methane, Ethane, and Ethylene Gases and their Binary and Ternary Mixtures and Carbon Dioxide on Activated Carbon at 212 K – 301 K and Pressures up to 35 atm, Ind. Eng. Chem. Processes Des. Dev., 19 (1980) 336-344.

[3.45] **Beutekamp S.**
Adsorptionsgleichgewichte der reinen Gase CO_2, CH_4, N_2 und deren binärer Gemische an verschiedenartigen porösen Stoffen, PH-Dissertation University of Leipzig, Leipzig, 2002, Fortschritt-Berichte VDI, Reihe 3, Verfahrenstechnik, Nr. 736, VDI-Verlag, Düsseldorf, 2002.

[3.46] **Ulbig P., Wenz P.**
Determination of the average density of water in porous activated carbon by a magnetic suspension balance, Chem. Technik (Leipzig), 52 (2000), p. 138–144.

[3.47] **Ulbig P.**
Adsorption of reactive dye Levafix 8R9 Brilliant Red E-4BA onto activated carbon monitored by a magnetic suspension balance, Chem. Technik (Leipzig), 000,000.

[3.48] **Kast W.**
Adsorption aus der Gasphase, Ingenieurwissenschaftliche Grundlagen und technische Verfahren, Verlag Chemie, Wertheim, Germany, 1988.

[3.49] **Radakrishnan R., Gubbins K. E., Sliwenska-Bartowiak M., Kaneko K.**
Understanding Freezing Behaviour in Porous Materials, p. 341–345,Fundamentals of Adsorption 7, K. Kaneko et al., Eds., IK International, Chiba Japan, 2002.

[3.50] **Radakrishnan R., Gubbins K. E., Sliwenska-Bartowiak M., Kaneko K.**
Understanding Freezing Behaviour in Pores, p. 234–238, Adsorption Science & Technology, D. D. Do, Ed., World Scientific, Singapore, 2000.

[3.51] **Herbst A.**
Exzessadsorption reiner Gase im Druckbereich bis 50 MPa, PH-D Thesis, University of Leipzig, 2003, Fortschrittberichte VDI, Reihe 3, Verfahrenstechnik, in preparation, VDI-Verlag Düsseldorf, 2003.

[3.52] **Janssen M. J. G., van Oorschot C. W. M.**
The Characterization of Zeolites by Gas Adsorption, article in: Zeolites: Facts, Figures, Future, P. A. Jacobs, R. A. van Santen, Eds., Elsevier, London, 1989.

[3.53] **Dreisbach F., Lösch H. W.**
Magnetic Suspension Balance for Simultaneous Measurement of a Sample and the Density of the Measuring Fluid, J. of Thermal Analysis and Calorimetry, 62 (2000), p. 515–521.

[3.54] **Dreisbach F., Seif R., Lösch H. W., Keller J. U.**
Gravimetric Measurement of Adsorption Equilibria of Gas Mixture CO/H_2 with a Magnetic Suspension Balance, Proceedings of FoA7 Conference, Nagasaki, May 2001, p. 255–262, Fundamentals of Adsorption 7, K. Kaneko, Y. Hanzawa, Eds., IK International, Chiba, Japan 2001.

[3.55] **Keller J. U., Dreisbach F., Rave, H., Staudt R., Tomalla M.**
Measurement of Gas Mixture Adsorption Equilibria of Natural Gas Compounds on Microporous Sorbens,
Adsorption, 5 (1999), 199–214.

This page intentionally left blank

Chapter 4

VOLUMETRIC – GRAVIMETRIC MEASUREMENTS

Abstract Combined volumetric and gravimetric measurements allow one to determine the coadsorption equilibria of binary gas mixtures without sorptive gas analysis, i. e. without using a gas chromatograph or mass spectrometer. The experimental setup, a basic theory and several examples of this method are presented. Two modifications of it, namely densimetric – gravimetric and densimetric - volumetric measurements are outlined. These especially are suited to do quick but still accurate measurements of binary coadsorption equilibria for industrial process control and / or design. These methods also can be used to measure adsorption of gases and vapors on walls of vessels, tubes or surfaces of any other solid materials. List of symbols. References.

1. INTRODUCTION

Volumetric / manometric measurements, Chap. 2, and gravimetric measurements, Chap. 3, can be performed simultaneously on the same gas adsorption system in a single instrument. For pure gas adsorption this will not lead to basically new information on the system as both methods lead to the same result, i. e. the reduced mass of the adsorbate phase, cp. Eqs. (2.4) and (3.5). However, for binary gas mixture adsorptives these measurements allow one to determine the masses of both components of the adsorbate without analyzing the (remnant) adsorptive gas mixture, i. e. without needing a gas chromatograph or a mass spectrometer. Measurements of this type seem to have been performed first in the authors' group in 1989 and published in 1990, cp. [4.1-4.3], [2.20], [3.20, 3.22] for CH_4 / N_2 and CH_4 / CO gas mixtures at ambient temperature up to pressures of 12 MPa. In fact this method can be used for any binary gas adsorptive with non-isomeric components, i. e. components with different molecular masses. Meanwhile this method has been commercialized by BEL – Japan, Osaka, with this company offering a fully automated instrument equipped with a magnetic suspension balance, Rubotherm, Bochum for volumetric-gravimetric measurements, cp. Sect. 2.4.

In this chapter we will present experimental information (Sect. 2.1), the theory of measurement (Sect. 2.2), and uncertainties (Sect. 2.3), and several examples (Sect. 2.4) of this method. Two modified versions of the measurement procedure which may be called densimetric-gravimetric and densimetric-volumetric / manometric methods (which especially seems to be suited for online industrial coadsorption measurements) are also outlined (Sect. 3). These methods also may be used to measure adsorption of gases and / or vapors on surfaces of arbitrary solid materials as for example the inner walls of vessels, tubes, valves etc. of the experimental device(s) used (Sect.3.6). Advantages and disadvantages of the methods proposed are discussed in Sect. 4. A list of symbols used is given in Sect. 5, followed by references to journal articles and books cited.

2. VOLUMETRIC – GRAVIMETRIC MEASUREMENTS OF BINARY COADSORPTION EQUILIBRIA

2.1 Experimental

A multipurpose instrument allowing multicomponent coadsorption measurements by the volumetric/manometric-chromatographic method, Chap. 2, as well as by gravimetric-chromatographic measurements, Chap. 3, has been designed and built for the first time in the working group of the authors in 1989, [2.5], [4.2]. It mainly consisted in a gas storage vessel (V_{SV}), an adsorption chamber (V_{AC}), a two beam microbalance outside the adsorption vessel (Sartorius 4104, Göttingen), a gas circulation loop including a pump (Brey, GK 24-02N, Memmingen), a gas chromatograph, thermostats, gas supply, vacuum pump, thermocouples and manometers to take measurements of temperatures and pressures. A schematic diagram of this installation is given in Figure 4.1, followed by a photo, cp. also [2.5], p. 207. More information on the instrument, especially specifications of vessels, tubes, valves etc. can be found in the literature [2.6, 2.20, 4.2–4.4].

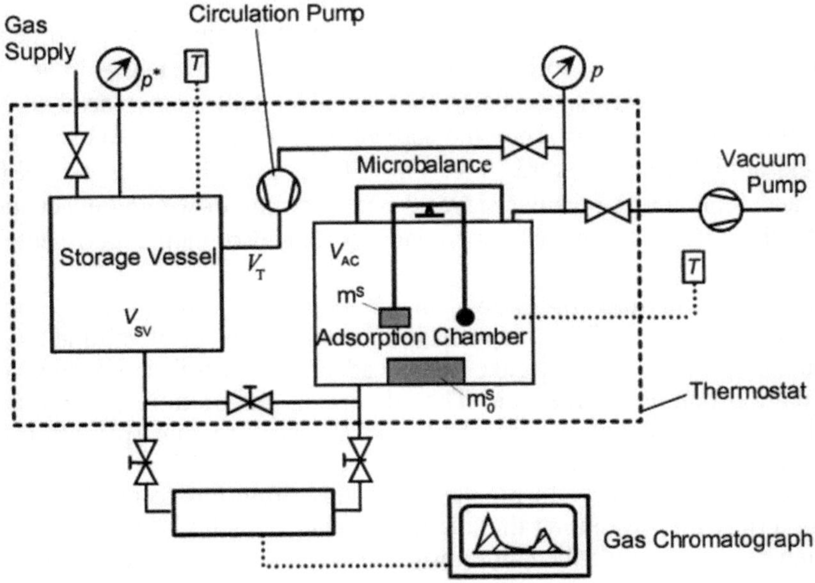

Figure 4.1. Experimental setup of a multipurpose instrument for coadsorption measurements of multicomponent gas mixtures on porous solids. The instrument allows volumetric-chromatographic, gravimetric-chromatographic and also combined volumetric-gravimetric measurements for binary coadsorption equilibria.

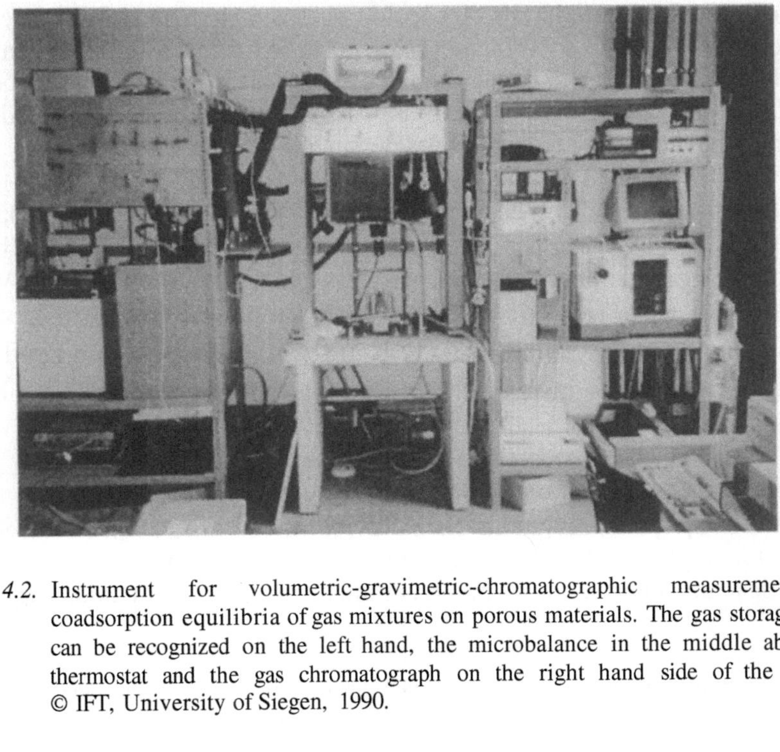

Figure 4.2. Instrument for volumetric-gravimetric-chromatographic measurements of coadsorption equilibria of gas mixtures on porous materials. The gas storage vessel can be recognized on the left hand, the microbalance in the middle above the thermostat and the gas chromatograph on the right hand side of the picture. © IFT, University of Siegen, 1990.

To measure binary coadsorption equilibria by the volumetric-gravimetric method one proceeds as follows: A sorbent sample of 1 g – 3 g and appropriate counterweights, typically lead or silver balls, are placed to the buckets of the microbalance. Then the sorbent is "activated" by exposing it to helium gas at higher temperatures, i. e. 433 K for activated carbons, 673 K for zeolites and inorganic molecular sieves.*) After cooling down and evacuation (< 10 Pa) the adsorption chamber is prepared for an adsorption experiment.

Next a gas mixture of known molar concentrations $(y_i^*, \ i = 1,2)$, pressure p^*, temperature T, is prepared in the storage vessel of volume (V_{SV}). Then the valves between the storage and the adsorption vessels are opened and the circulation pump is turned on. The adsorption gas mixture is circulated until adsorption equilibrium is reached, i. e. the gas pressure (p) within the system and the signal from the microbalance (Ω) remain "constant" after stopping the gas circulator. In practice this state often has to be defined in technical terms, i. e. by stating that the microbalance reading does not change more than a certain prescribed amount $(\Delta\Omega)$ within a certain period of time (Δt). Such a state may be called "technical equilibrium" which for $\Delta t \rightarrow \infty$, i. e. $\Delta t > 1$ hour and $\Delta\Omega \rightarrow 0$, i. e. $\Delta\Omega < 10^{-6} \, |\Omega|$ tends to physical or exact thermodynamic equilibrium. We experienced very quick adsorption processes where equilibrium was reached within a few minutes, but also very slow ones which took days and even weeks as for example the penetration process of helium into porous materials, Chap. 1. Hence it is a matter of practice to choose proper values of $(\Delta\Omega)$ and (Δt) according to the nature and behavior of the adsorption system and the accuracy of the data needed. From recorded data $(p^*, \ p, \ T, \ y_i^*, \ i = 1.\ 2, \ \Omega)$ the masses of both components adsorbed (m_1^a, m_2^a) can be calculated as will be outlined in the next section.

If the mass of the sorbent material (m^s) attached to the microbalance is very small, it may happen that the change of pressure caused by gas adsorption is tiny compared to that caused by pure gas expansion from volume (V_{SV}) to $(V_{SV} + V_{AC})$. In this case it is helpful to place a considerable amount of physico-chemically identical sorbent material (m_0^s) on the bottom of the adsorption chamber to increase the mass of gas adsorbed and also the change of pressure which is due to gas adsorption, cp. Fig. 4.1.

The experiment itself may be continued by adding more gas to the storage vessel and expanding it again to the adsorption vessel to realize a new state of adsorption equilibrium at a higher pressure. As uncertainties of measured quantities are additive, it is not recommended to perform this step up (or

*) Two beam microbalances have to be cooled during this process as their thermal range of operation normally is restricted to temperatures below 350 K.

likewise step down) procedure more often than 3 – 4 times, cp. remarks in Chaps. 2 and 3.

The instrument Figs. 4.1, 4.2 is subject to several severe restrictions which we would like to discuss in brief. First, all two beam balances commercially available today [2.3, 4.5], are designed such that the sorptive gas contacts the inner region of the balance including parts like electromagnetic coils which are highly sensitive to corrosion. Hence these balances only can be used for non-corrosive sorptive gas mixtures including components like (He, Ar, N_2, CH_4, (CO)). Of course a permanent flow of inert gas could be used to cover and thus to protect the microbalance region. But this technique, though in former days having often been used in traditional thermogravimetric experiments, has not proved to lead to reproducible and accurate results as it seems to be difficult to attain stable equilibrium conditions in systems including a permanent gas flow contacting a microbalance [2.3, 2.4, 4.5]. Also traditional microbalances normally can operate in fairly narrow ranges of temperature and pressure – usually about ambient conditions – only. Hence it would be desirable to extend these ranges of temperature and pressure and also to allow corrosive gases like (NH_3, H_2S, SO_2, CO_2 etc.). An instrument which can do this is provided by the magnetic suspension microbalance already mentioned in Chap. 3 (2.2).*) We will give more details about this in Sect. 2.4 and Sect. 3 of this chapter.

2.2 Theory

The masses of a two component adsorbate (m_1^a, m_2^a) can be determined from the results of a volumetric-gravimetric experiment, namely the gas pressure (p) in the system, the temperature (T), and the microbalance recording (Ω) as follows: We first denote the mass balance equations for each component

$$m_i^* = m_i^f + m_i^a, \qquad i = 1, 2, \qquad 4.1$$

the thermal equation of state (EOS) of the sorptive gas mixture with masses $(m_i^f, \quad i = 1,2)$

$$\frac{m_1^f}{M_1} + \frac{m_2^f}{M_2} = \frac{pV^f}{RTZ} \qquad 4.2$$

*) Manufacturer: RUBOTHERM Praezisionsmesstechnik GmbH, Universitätsstr. 142, D-44799 Bochum, Germany, www.rubotherm.de

with the real gas factor

$$Z = Z\,(p, T, m_1^f, m_2^f) \tag{4.3}$$

the microbalance equation

$$\Omega = m_1^a + m_2^a - \rho^f V^{as} \tag{4.4}$$

with the density of the sorptive gas given by

$$\rho^f = \frac{m_1^f + m_2^f}{V^f}, \tag{4.5}$$

and its volume being

$$V^f = V_{SV} + V_{AC} - V^{as} + V_T \tag{4.6}$$

In Eq. (4.1) the masses (m_i^*) of the gas components initially prepared in the storage vessel can be calculated from the EOS as

$$m_i^* = y_i^* M_i \frac{p^* V_{SV}}{RTZ^*} \quad , \qquad i = 1, 2 \tag{4.7}$$

Here y_i^* is the (known) molar concentration of component i = 1, 2 of the gas, M_i the molar mass of component i = 1,2, p^* the pressure in the storage vessel prior to the adsorption experiment, V_{SV} its volume, R is the universal gas constant and

$$Z^* = Z(p^*, T, y_1^*, y_2^*) \tag{4.8}$$

is the compressibility factor of the sorptive gas in the storage vessel. The adsorption chamber (V_{AC}) is assumed to be evacuated at the beginning of the experiment.

The microbalance recording (Ω) is an experimental quantity comprising the numerical results of four different measurements performed on the microbalance. These are

(1) Calibration measurement of the empty balance parameters as for example the so-called "asymmetry of beams" in a two beam balance or the mass m_0 of the permanent magnet and auxiliary equipment inside a magnetic suspension balance: $\Omega_0 = m_0$.
(2) Measurement of the loaded balance in vacuum to determine the mass (m^s) of sorbent material: $\Omega^s = m_0 + m^s$.
(3) Measurement of the empty balance in an inert gas to determine the buoyancy ($\rho^f V_0$) of the inner parts like beams, buckets etc. of the balance: $\Omega_f = m_0 - \rho^f V_0$.
(4) Measurement of the loaded balance in the sorptive gas atmosphere:

$$\Omega^{sf} = m_0 + m^s + m^a - \rho^f (V_0 + V^{as}) \quad 4.9$$

$$m^a = m_1^a + m_2^a \quad 4.10$$

From these four measurements Ω as defined in Eq. (4.4) can be calculated as, cp. Chap. 3:

$$\Omega = \Omega^{sf} - \Omega^f - (\Omega^s - \Omega_0) \quad 4.11$$

In physical terms Ω gives the total mass adsorbed, i. e. the sum of the masses of both components ($m_1^a + m_2^a$) minus the buoyancy correction term caused by the volume (V^{as}) of both the sorbent (s) and the adsorbate phase (a), cp. Eq. (4.4).

In Eq. (4.6) the volume of tubes and of the dead space of the gas circulator (V_T) has been added. This quantity is assumed to be known from inert gas calibration experiments. It normally is small compared to the volumes of the vessels (V_{SV}, V_{AC}) but in principle should be taken into account.

Equations (4.1-4.4) provide four algebraic relations from which the four unknown masses (m_i^a, m_i^f, i = 1, 2) can be determined. Indeed, by adding Eqs. (4.1) and introducing the microbalance equation (4.4) we get the relation

$$m_1^f + m_2^f = \frac{1}{1 + (V^{as} / V^f) M^s} (m^* - \Omega M^s) \quad 4.12$$

with

$$M^s \equiv 1 + \frac{m_0^s}{m^s} \tag{4.13}$$

and

$$m^* \equiv m_1^* + m_2^* \tag{4.14}$$

Now equations (4.2, 4.9) can be solved for m_1^f, m_2^f.. The result is

$$m_i^f = \frac{M_i}{M_{i+1} - M_i}\left(\frac{\Omega M^s - m^*}{1 + (V^{as}/V^f)M^s} + \frac{pV^f M_{i+1}}{RTZ(p,T,m_1^f,m_2^f)}\right) \tag{4.15}$$

Here we have used the definition

$$M_{i+1} = \begin{cases} M_2 \ldots i = 1 \\ M_1 \ldots i = 2 \end{cases}$$

The volume V^f is determined by Eq. (4.6) and the helium approximation, i.e. $V^{as} \cong V_{He}^s$, cp. (2.7). As immediately can be seen, solutions (4.12) become singular if $M_1 = M_2$. Indeed in this case equations (4.2, 4.12) become linearly dependent and do not allow one to determine the masses of the adsorptive gas m_1^f, m_2^f. Hence the volumetric – gravimetric method cannot be used for isomeric gas components. It should be noted that equations (4.15) can only be solved iteratively by starting with the ideal gas approximation for the compressibility, i. e. choosing Z = 1. The results of this calculation can be used to determine an approximate value of Z according to Eq. (4.3) via an appropriate thermal EOS for the gas mixture. Repeating this procedure one normally will get after a few steps (≤ 5) stable and accurate numerical values of the adsorptive masses (m_1^f, m_2^f) from which the Gibbs excess masses of the sorbate's components (m_{1GE}^a, m_{2GE}^a) immediately can be calculated via the mass balance equations (4.1).

$$m_{iGE}^a = m_i^* - m_i^f, \qquad i = 1,2 \tag{4.16}$$

$$= y_i^* M_i \frac{p^* V_{SV}}{RTZ^*} - \frac{M_i}{M_{i+1} - M_i}\left(\frac{\Omega M^s - m^*}{1 + (V_{He}^s/V^f)M^s} + \frac{pV^f M_{i+1}}{RTZ(p,T,y_i^f)}\right) \tag{4.17}$$

Similarly one may proceed in order to get the absolute masses adsorbed (m_1^a, m_2^a), in this case using for (V^{as}) instead of the helium approximation (V_{He}^s) (2.7) the model equation (2.9).

One may argue that combined volumetric – gravimetric measurements of pure gas adsorption equilibria could provide a means to measure the mass adsorbed (m^a) and the void volume of the sorbent and the sorbate phase (V^{as}) simultaneously. Unfortunately this does not hold true. Instead, specializing equations (4.1–4.6) to the case of pure gases ($m_1 \rightarrow m$, $m_2 \equiv 0$) and combining them one gets

$$\Omega = m^* - \rho^f (V_{SV} + V_{AC} + V_T) \qquad 4.18$$

As all quantities in this relation either can directly be measured or are the results of measurements, it provides a consistency condition for the volumetric – gravimetric experiment performed, but does not allow one to determine either (m^a) or (V^{as}) without introducing an additional hypothesis, cp. Eq. (2.7).

If the two sorptive gas components (1,2) are mixed with a carrier gas (0) which is not adsorbed on the sorbent material considered, equations (4.15) have to be modified. This situation may occur for example in purification processes of air or natural gas including polar components which are strongly adsorbed on zeolitic sorbent materials compared to non-polar components like (N_2, O_2, CH_4, etc.). Then the basic equations (4.2 – 4.5) should be substituted by

$$(4.2): \quad \frac{m_0^*}{M_0} + \frac{m_1^f}{M_1} + \frac{m_2^f}{M_2} = \frac{pV^f}{RTZ} \quad , \qquad 4.19$$

$$(4.3): \quad Z = Z(p, T, m_1^f, m_2^f, m_0^f) \; , \qquad 4.20$$

$$(4.4): \quad \Omega = m_1^a + m_2^a - \rho^f V^{as} \quad , \qquad 4.21$$

$$(4.5): \quad \rho^f = \frac{1}{V^f}(m_1^f + m_2^f + m_0^f) \; . \qquad 4.22$$

The mass of inert gas (m_0^*) originally provided to the storage vessel (V_{SV}) of the instrument, Fig. 4.1 can be calculated from the EOS of the gas mixture with components (1, 2, 0) cp. Eq. (4.7),

$$m_0^* = y_0^* M_0 \frac{p^* V_{SV}}{RTZ^*}, \qquad 4.23$$

$$Z^* = Z(p^*, T, y_1^*, y_2^*, y_0^*) \qquad 4.24$$

Here y_0^* is the initial molar concentration of the inert component in the gas mixture. It is subject to the condition $(y_1^* + y_1^* + y_0^* = 1)$. Neglecting adsorption of the inert component, i. e. assuming its mass in the equilibrium sorptive gas mixture (m_0^f) to be the same as in the initial state prior to adsorption, $m_0^f = m_0^*$, equations (4.1), (4.19-4.22) again provide 4 algebraic equations from which the sorptive gas masses (m_1^f, m_2^f) and adsorbate's masses (m_1^a, m_2^a) can be determined. The result is

$$m_{iGE}^a = m_i^* - m_i^f, \qquad i = 1, 2 \qquad 4.25$$

with $(m_i^*, i = 1,2)$ as given by (4.7), (4.24) and

$$m_i^f = \frac{M_i}{M_{i+1} - M_i}\left[\left(\frac{pV^f}{RTZ} - \frac{m_0^*}{M_0}\right)M_{i+1} + \right.$$
$$\left. + \frac{M^s\left(\Omega + \frac{V_{He}^s}{V^f} m_0^f\right) - m^*}{1 + \frac{V_{He}^s}{V^f} M^s}\right], \qquad i = 1, 2 \qquad 4.26$$

(4.6): $\quad V^f = V_{SV} + V_{AC} + V_T - V_{He}^s$

(4.13): $\quad M^s \equiv 1 + \frac{m_0^s}{m^s}$

(4.14): $\quad m^* \equiv m_1^* + m_2^*$

Also the convention ($M_{i+1} = M_2$ for i = 1, $M_{i+1} = M_1$ for i = 2) should be taken into account.

Equations (4.25, 4.26) require again an iteration procedure for numerical solution as they include the compressibility factor

$$Z = Z(p, T, m_1^f, m_2^f, m_0^*)$$

depending on the unknown masses of sorptive components (m_1^f, m_2^f). However, for low initial concentrations of the sorptive components, i. e. $m_i^* << m_0^*$, i = 1,2, Z may be approximated by the real gas factor of the pure carrier gas, i. e.

$$Z \cong Z_0(T, \rho_0 = m_0^* / V^f) \quad 4.27$$

Coadsorption measurements of sorptive gases including inert components are presently done at the author's institution. Results will be published in a forthcoming paper [4.7].

2.3 Uncertainties or Errors of Measurement

As already outlined in Chap. 2, Sects. 2.3 and 4.3 uncertainties or errors of measurements are important quantities for any kind of experimental work and should be provided for any quantity measured. Consequently we here denote the dispersions or mean square deviations (MSD) (σ^2_{miGE}), i = 1, 2 of the Gibbs excess masses (m^a_{iGE}) of a two component adsorbate which have been measured by the volumetric-gravimetric method, i. e. calculated by Eqs. (4.1)

$$m^a_{iGE} = m_i^* - m_i^f \quad , \quad i = 1, 2 \quad 4.28$$

with m_i^f given by (4.15) including the helium approximation (2.7), i. e. $V^{as} \cong V^s_{He}$. Applying the Gauss law of propagation of error or uncertainty we get from (4.28) and (4.15) the expression

$$\begin{aligned} \sigma^2_{miGE} = {} & \sigma^2_{mi*} + \left(\frac{M_i}{M_{i+1} - M_i}\right)^2 \bullet \left[\left(\frac{M^s}{1 + (V^s_{He}/V^f)M^s}\right)^2 \sigma^2_\Omega + \right. \\ & + \left(\frac{\Omega}{1 + (V^s_{He}/V^f)M^s} - \frac{\Omega M^s - m^*}{(1 + (V^s_{He}/V^f)M^s)^2} \bullet \frac{V^s_{He}}{V^f}\right)^2 \sigma^2_{MS} \\ & + \frac{\sigma^2_{m*}}{(1 + (V^s_{He}/V^f)M^s)^2} + \frac{(\Omega M^s - m^*)^2}{(1 + (V^s_{He}/V^f)M^s)^4} \sigma^2_{VHeVf} \\ & + \left(\frac{V^f M_{i+1}}{RTZ}\right)^2 \sigma^2_p + \left(\frac{p M_{i+1}}{RTZ}\right)^2 \sigma^2_{Vf} \end{aligned}$$

$$+ \left(\frac{pV^f M_{i+1}}{RT^2 Z}\right)^2 \sigma_T^2 + \left(\frac{pV^f M_{i+1}}{RTZ^2}\right)^2 \sigma_Z^2 \Bigg] \quad , \qquad i = 1, 2 \tag{4.29}$$

Here we have used the abbreviations

$$\sigma_{MS}^2 = \frac{\sigma_{m0s}^2}{m^{s2}} + \left(\frac{m_0^s}{m^{s2}}\right)^2 \sigma_{ms}^2 \tag{4.30}$$

$$\sigma_{m*}^2 = \sigma_{m1*}^2 + m_{m2*}^s \tag{4.31}$$

$$V^f = V_0 - V_{He}^s$$
$$V_0 = V_{SV} + V_{AC} + V_T \tag{4.32}$$

$$\sigma_{VHeVf}^2 = \left(\frac{1}{V_0 - V_{He}^s} + \frac{V_{He}^s}{(V_0 - V_{He}^s)^2}\right)^2 \sigma_{VHe}^2 + \left(\frac{V_{He}^s}{(V_0 - V_{He}^s)^2}\right)^2 \sigma_{V0}^2, \tag{4.33}$$

$$\sigma_{V0}^2 = \sigma_{VSV}^2 + \sigma_{VAC}^2 + \sigma_{VT}^2 \tag{4.34}$$

$$\sigma_{Vf}^2 = \sigma_{V0}^2 + \sigma_{VHe}^2 \tag{4.35}$$

$$\sigma_Z^2 = \left(\frac{\partial Z}{\partial T}\right)_p^2 \sigma_T^2 + \left(\frac{\partial Z}{\partial p}\right)_T^2 \sigma_p^2 + O(\sigma_{yi}^2 \, , i = 1,2) \tag{4.36*)}$$

The symbol "O" indicates residual terms depending on the dispersions of the gas concentrations σ_{yi}.
For sake of simplicity we list the various measurable quantities needed to calculate (m_{iGE}^a) via (4.25), (4.12) and their dispersions below:

m_i^*	Ω	m_0^s	m^s	V_{He}^s	V_0	V_{SV}
σ_{mi*}	σ_Ω	σ_{m0s}	σ_{ms}	σ_{VHe}	σ_{V0}	σ_{VSV}

4.37

V_{AC}	V_T	V^f	Z	T	p	y_i^f
σ_{VAC}	σ_{VT}	σ_{Vf}	σ_Z	σ_T	σ_p	σ_{yi}

*) The symbol "O" indicates "order of magnitude" of this term.

As immediately can be seen from (4.29), dispersions (σ_{miGE}) become large if the difference between the molar masses of the adsorbed components (M_{i+1} – M_i) becomes small. Hence this method is not recommended for example to investigate cosorption phenomena of isotopes. Even for mixtures of nitrogen and oxygen (air) it may fail as (σ_{miGE}) may be of the order of magnitude of (m_{iGE}) itself. According to our experience the most important dispersions within the set mentioned in (4,37) are $(\sigma_\Omega, \sigma_p, \sigma_T)$, i. e. the balance measurements and measurements of the equilibration pressure (p) and thermostat's temperature (T) are most important to get accurate values of (m_i^{GE}), i. e. dispersions $(\sigma_{miGE} \approx 1\% \, m_i^{GE})$. Also it should be noted that for supercritical sorptive gas mixtures far away from saturation in (4.26) the ideal gas approximation may be used, i. e. one may assume the compressibility factor $Z = 1$ and hence $\sigma_Z = 0$.

2.4 Examples

In this section we want to present data for binary coadsorption equilibria of gas mixtures on activated carbon (NORIT R 1) which have been taken by combined volumetric-gravimetric measurements. We start with a set of data which have been measured with the instrument depicted in Figs. 4.1, 4.2. Then we describe in brief a new type of volumetric-gravimetric instrument including a magnetic suspension balance allowing also measurements with corrosive sorptive gases. Equilibria as well as kinetic data taken at this instrument will be presented. Finally we hint at a commercialized version of this instrument offered by BEL – Japan company.

A set of data describing the coadsorption equilibria of methane (CH_4) - nitrogen (N_2) gas mixtures at T = 298 K in the pressure range p = 0-12 MPa on activated carbon NORIT R 1 is given in Table 4.1 below [4.11]. It includes consecutively the sorptive gas pressure (p), the molar fraction of CH_4 in the sorptive gas (y_{CH4}), the density of the sorptive gas (ρ), the number of moles of (CH_4) and (N_2) adsorbed per unit mass of sorbent (n_{CH4}) and (n_{N2}) respectively and the total mol number of gases adsorbed per unit mass of sorbent ($n_{CH4} + n_{N2}$). As can be seen from the table, the molar fraction of CH_4 in the sorptive gas is changing from about 18.1 % at p = 0.11 MPa to 23.7 % at p = 7.93 MPa. This is due to the experimental technique applied: As coadsorption equilibria basically are unknown, the amount and concentrations of the sorptive gas mixture to be prepared in the storage vessel of the instrument, Fig. 4.1, are unknown and have to be chosen either according to approximate calculations from pure component adsorption data using for example the IAST-formalism [4.10], or by experience. Hence the graphical representation of data of Tab. 4.1 in Fig. 4.3 is not exactly an adsorption isotherm referring to a constant sorptive gas concentration, but in the three-dimensional (p, y_{CH4}, n)-space a three-dimensional curve of slightly spiral character within the concentration interval 18.1 % $\leq y_{CH4} \leq$ 23.7 %. If data at an exactly prescribed sorptive gas concentration are needed, it is recommended to aim at 2 or (better) 3 equilibria states in the vicinity of the wanted state, i. e. with neighboring gas concentrations, and use linear interpolation.

The volumetric-gravimetric data presented in Tab. 4.1, Fig. 4.3 have been checked by doing gravimetric-chromatographic measurements simultaneously, i. e. by analyzing the sorptive gas phase by a gas chromatograph [4.11]. Coincidence always was satisfying, i. e. results of both methods were identical within error bounds of data which were about 1 % for mole numbers and up to 2 % for sorptive gas concentrations.

To overcome several restrictions of the volumetric-gravimetric instrument of Figs. 4.1, 4.2 which already have been mentioned at the end of Section 2.1 of this chapter, another instrument for volumetric, gravimetric, and chromatographic measurements of multicomponent gas adsorption equilibria has been designed at the authors' institution in the years (1993–1998). A schematic diagram of this instrument is given in Figure 4.4 below followed by a snapshot, Fig. 4.5. The instrument mainly consists on two vessels for gas storage, a circulation pump, microbalance, necessary valves and tubing and supply systems for gases and vacuum. Details about design, materials, fittings, etc. can be found in Ref. [2.5], [2.6], [4.3], [4.4], and [4.8].

The main advantage of this instrument compared to its predecessor, Figs. 4.1, 4.2, was the use of a magnetic suspension decoupling the gas-filled adsorption chamber from the microbalance. This allows measurements with corrosive gases (H_2S, SO_2, NH_3) etc., in ranges of pressure ($p < 100$ MPa) and temperature ($T < 1500$ K) which are considerably extended compared to those where traditional two beam balances can be used. The magnetic suspension already has been described in more detail in Chap. 3. It is manufactured by RUBOTHERM-Präzisionsmesstechnik GmbH, Bochum, Germany and can be used for many different types of high precision weighing experiments [4.9].

Table 4. 1. Data for coadsorption equilibria of CH_4/N_2 gas mixtures at T=298K on activated carbon NORIT R1, [4.11]. p...sorptive gas pressure, y_{CH4}...molar concentration of CH_4 in the gas phase, ρ...density of the sorptive gas, n_{CH4}, n_{N2}...number of moles of CH_4, N_2 adsorbed per unit mass of sorbent.

p/MPa	y_{CH4}	ρ/kg/m³	n_{CH4}/mmol/g	n_{N2}/mmol/g	$n_{CH4}+n_{N2}$/mmol/g
0.110	0.1809	1.15	0.3300	0.3099	0.6399
0.284	0.1987	2.94	0.6856	0.6529	1.3385
0.828	0.2123	8.53	1.4369	1.2815	2.7184
1.387	0.2188	14.29	1.8894	1.7074	3.5968
1.906	0.2228	19.64	2.1600	2.0080	4.1680
2.804	0.2272	28.93	2.5881	2.3389	4.9270
3.976	0.2307	41.10	2.9344	2.6548	5.5892
5.031	0.2327	52.09	3.2331	2.8133	6.0465
6.032	0.2337	62.54	3.5031	2.8958	6.3989
7.929	0.2372	82.26	3.7131	3.0964	6.8095

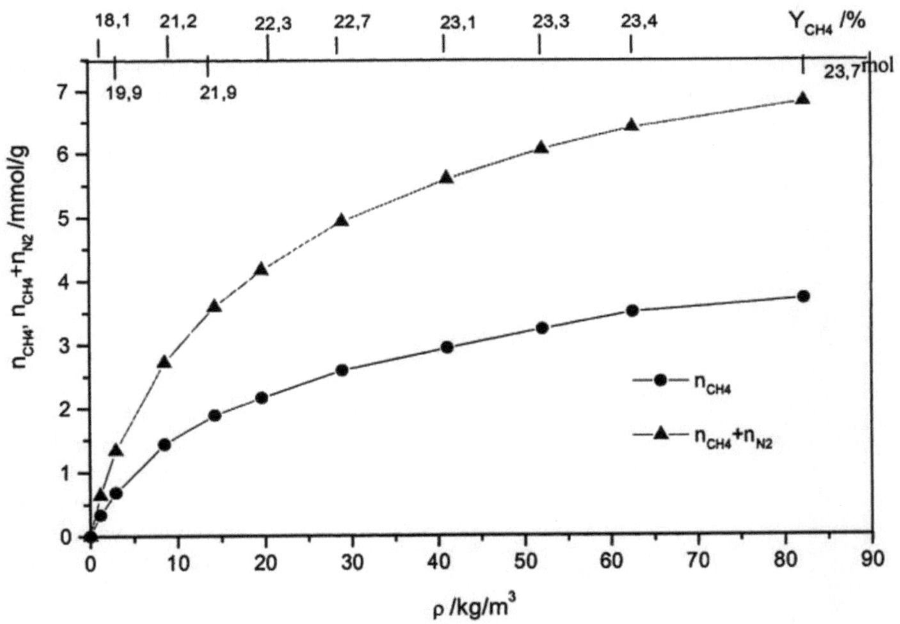

Figure 4.3. Coadsorption equilibria of CH_4 / N_2 gas mixtures at T = 298 K on AC Norit R1 for sorptive gas concentrations ranging from y_{CH4} = 18.1 % at p = 0.11 MPa to y_{CH4} = 23.7 at p = 7.93 MPa. Uncertainties of the data are approximately three times of the diameter of the graphical symbols used [4.11].

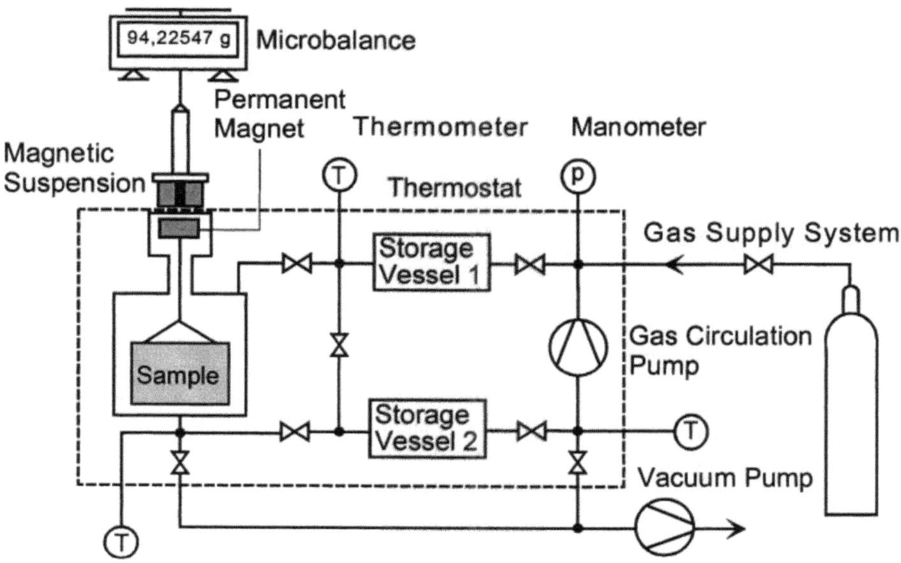

Figure 4.4. Schematic diagram of an installation for volumetric-gravimetric measurements of binary coadsorption equilibria without using a gas chromatograph. The installation includes a magnetic suspension balance (RUBOTHERM, Bochum, Germany) allowing also measurements with corrosive gases in a large range of temperature (T < 1500 K) and pressure (p < 100 MPa).

Figure 4.5. Instrument for volumetric-gravimetric measurements of binary coadsorption equilibria including a magnetic suspension balance. It can be recognized in the upper left section of the installation.
© IFT University of Siegen, Lab PB-A 0126, 1995.

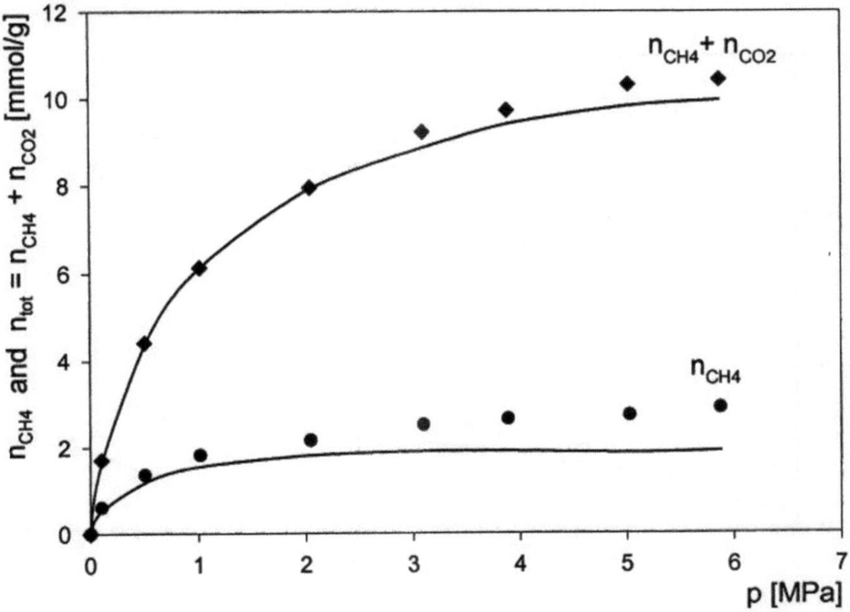

Figure 4.6 a. Coadsorption equilibria of a binary gas mixture CH_4 / CO_2 with y_{CH4} = 53 %, y_{CO2} = 47 % on AC Norit R1 at 298 K. Date were taken by volumetric-gravimetric measurements and checked selectively by gravimetric-chromatographic measurements [4.4].
Total amount adsorbed ($n_{CH4} + n_{CO2}$): (◆).
Partial amount of CH_4 adsorbed (n_{CH4}) (●).
Data are given in [mmol/gAC] The lines show predicted coadsorption equilibria based on pure component data and the IAST formalism [4.10].

In Figure 4.6a data of coadsorption equilibria of a gas mixture of y_{CH4} = 53 %mol of methane (CH_4) and y_{CO2} = 47 %mol carbon dioxide (CO_2) on activated carbon (AC) NORIT R1 at 298 K for pressures up to 6 MPa are shown [4.4]. Data have been checked selectively by simultaneous gravimetric-chromatographic measurements. Uncertainties of data are about twice the size of the graphical symbols. Data have been correlated by using the simple Ideal Adsorbed Solution Theory (IAST) of A. Myers and J. M. Prausnitz [4.10]. Selectivities of the carbon for this (nearly equimolar) gas mixture at – for example – p = 2 MPa are $S_{CH4\text{-}CO2} = 0.37$, $S_{CO2\text{-}CH4} = 2.7$, cp. (2.33).

Similarly, in Figure 4.6 b coadsorption equilibria data of a gas mixture CH_4 / N_2 with y_{CH4} = 9 %, y_{N2} = 91 % on AC Norit R 1 at 298 K for pressures up to 6 MPa are pictured [4.4]. The data show that despite the low concentration ratio of $y_{CH4} : y_{N2} = 1 : 10$ in the sorptive phase, the sorbate shows a ratio of concentrations of about x_{CH4} x_{N2} = 1 : 2 leading to a selectivity of $S_{CH4\text{-}N2} \cong 5$.

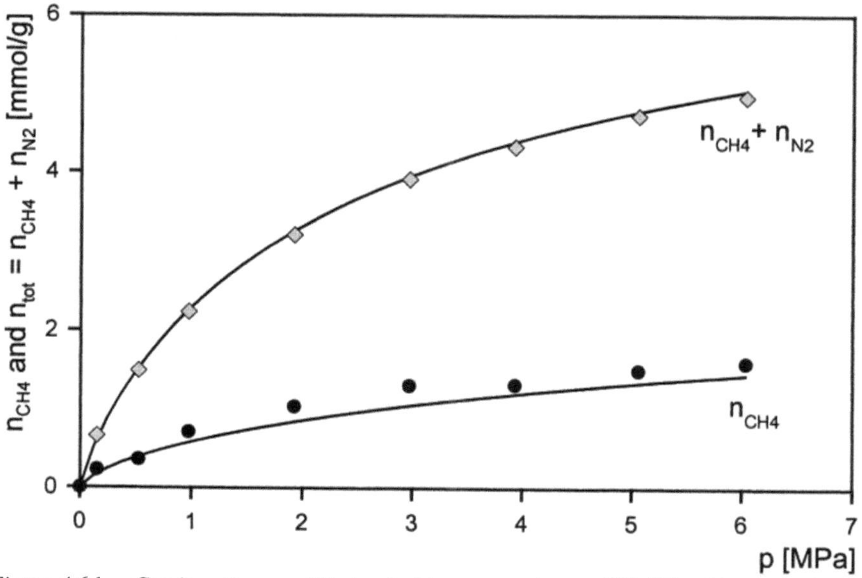

Figure 4.6 b. Coadsorption equilibria of a binary gas mixture CH_4 / N_2 with $y_{CH4} = 9$ % on AC Norit R 1 at 298 K. Data were taken by volumetric – gravimetric measurements [4.4].
Total amount adsorbed ($n_{CH4} + n_{N2}$): (◈).
Partial amount of CH_4 adsorbed (n_{CH4}): (●).
Data are given in [mmol/gAC]. The lines show the correlation of the data by a 2-site Langmuir isotherm, Chap. 10.

In Figure 4.6 c coadsorption equilibria data of gas mixtures CO_2 / N_2 with $y_{CO2} \cong 20$ %, $y_{N2} \cong 80$ % on AC Norit R 1 at 298 K for pressures up to 6 MPa are presented [4.4]. Selectivities are about $S_{CO2-N2} \cong 12$ showing that at ambient, but still subcritical temperature (298 K), CO_2 is much more (and stronger) adsorbed on the highly microporous AC chosen than N_2.

In Figure 4.7 all coadsorption data taken for the gas mixture CO_2 / N_2 at (approximate) sorptive concentrations $y_{CO2} = 20$ %, 45 %, 86 % at 298 K on AC Norit R 1 are presented in a 3 D-diagram (p, y_{CO2}, n), [4.4]. This provides an overview of the data and clearly shows that at all concentrations (CO_2) is much more adsorbed than (N_2). Though we do not have a clear molecular picture of the nature of the co-adsorbed phase, we argue that adsorption of (CO_2) predominantly is taking place at the active groups of the AC, i. e. at the edges of its slit-like micropores thus clogging the space between the graphite-like carbon layers and preventing (N_2) molecules from entering. The reader interested in this kind of problems is referred to the literature [2.2, 4.12, 4.13, 4.14]. The coadsorption instrument Figs. 4.4, 4.5, also allows one to get information about the kinetics of binary coadsorption processes.

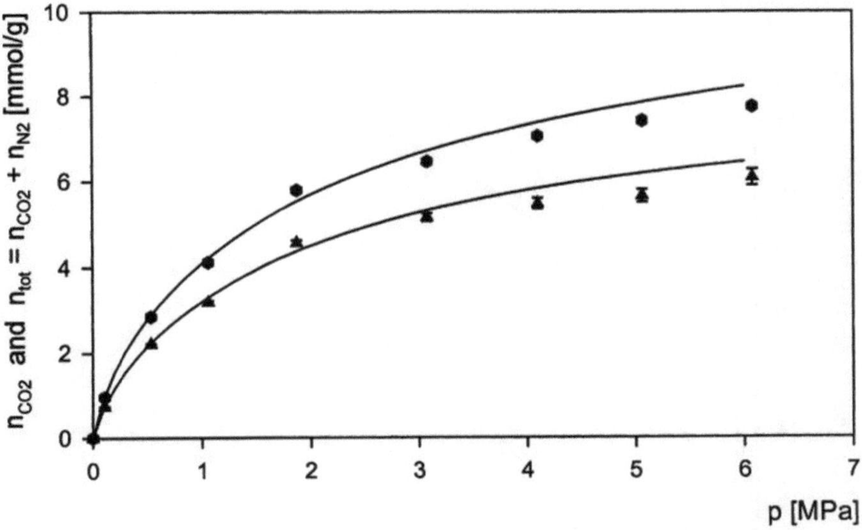

Figure 4.6 c. Coadsorption equilibria of a binary gas mixture CO_2 / N_2 with $y_{CO2} \cong 20$ %, $y_{N2} \cong 80$ % on AC Norit R 1 at 298 K. Data were taken by volumetric – gravimetric measurements [4.4].
Total amount adsorbed ($n_{CO2} + n_{N2}$): (●).
Partial amount of CO_2 adsorbed (n_{CO2}): (▲).
Data are given in [mmol/gAC]. The lines show the correlation of the data using a virial adsorption isotherm, cp. Chap. 10.

This is due to the fact that the basic data of volumetric-gravimetric measurements, i. e. the suspension balance recording (Ω) and the sorptive gas pressure (p) at least can be taken every second to give via equations (4.1 – 4.4) and an efficient data acquisition and handling system the number of moles of both components being adsorbed as a sequence of time.

This of course would not be possible with measurement procedures including chromatographic analyses. An example of the kinetics of a binary coadsorption process is given in Figure 4.8 below [4.4]. It refers to the adsorption of CH_4 and CO_2 on AC Norit R 1 at 298 K, cp. also Fig. 4.6 a. Here the sorbent sample first has been exposed to pure CH_4, equilibrium being attained after 4 hs. Then CO_2 is added to the sorptive gas (t = 0) and data are taken and recorded every second for the next hour.

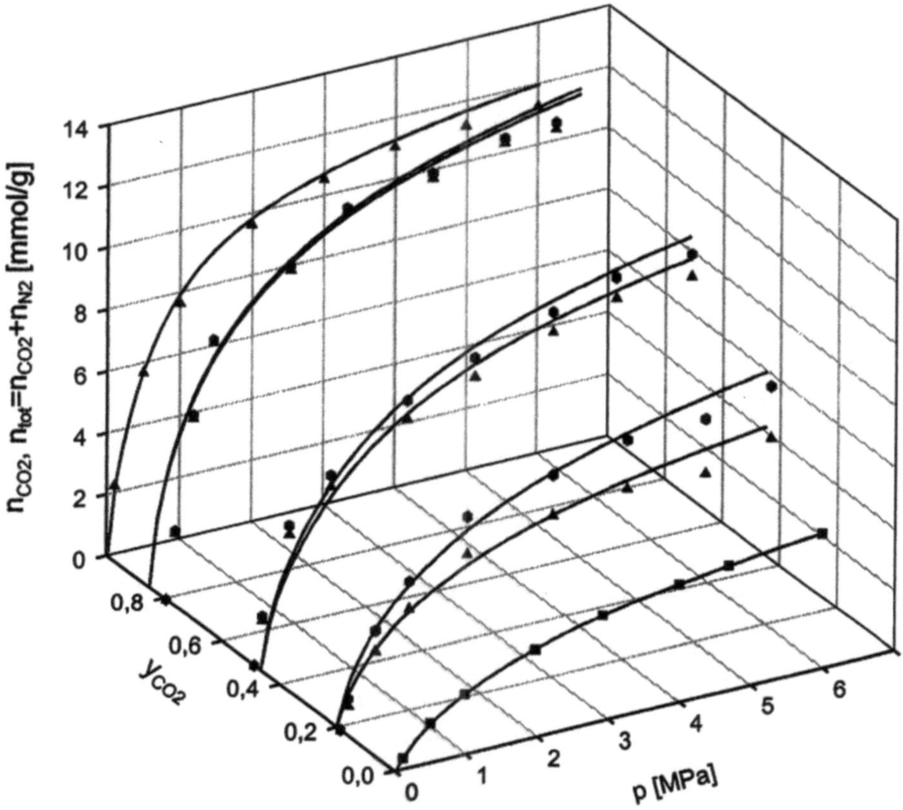

Figure 4.7. 3D diagram of coadsorption data (p, y_{CO2}, ●...n_{CO2} + n_{N2}, ▲...n_{CO2}) of the gas mixture CO_2 / N_2 at 298 K on AC Norit R 1, [4.4]. Data have been measured by the volumetric - gravimetric method using the instrument shown in Figs. 4.4, 4.5. Sorptive gas concentrations are $y_{CO2} \cong$ 20 %, 45 %, 86 % and change along an isotherm up to 3 %. Exact numerical values are given in [4.4]. Data have been correlated by using a virial adsorption isotherm including 6 parameters. Correlations are shown by lines (——). Relative deviations between calculated and measured data often are below 5 % but sometimes may exceed 10 %. (Reprint by permission granted by VDI, Düsseldorf.)

The resulting Gibbs excess mole numbers per gram AC of both components CH_4, CO_2 adsorbed ($n_{CH4\ ex}$, $n_{CO2\ ex}$) as well as the total mole number (n_{ex} = $n_{CH4\ ex}$ + $n_{CO2\ ex}$) are shown as functions of time in Figure 4.8. As can be seen, $n_{CH4\ ex}$ decreases considerably during the first 3 minutes to increase again and finally approach its equilibrium value. This decrease of $n_{CH4\ ex}$ is due to the "aggressive" adsorption of CO_2 being reflected in the steep ascent of $n_{CO2\ ex}$ during the first 5 minutes. Both effects result in a slight decrease of the total amount adsorbed (n_{ex}) during an initial period of about 3 minutes to be followed by a slow increase which asymptotically is leveling of to approach the equilibrium value at about $n_{ex\,eq}$ = 1.95 mmol/g.

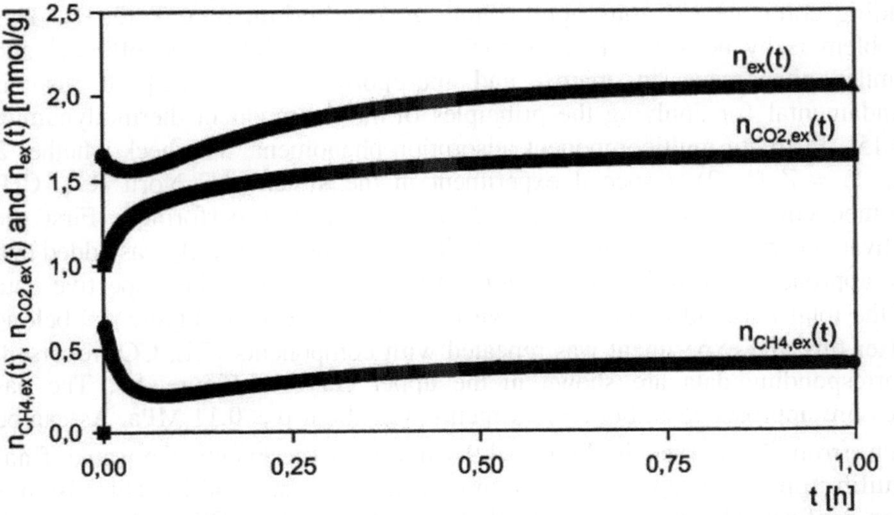

Figure 4.8. Time sequences of Gibbs excess mole numbers ($n_{CH4\,ex}$, $n_{CO2\,ex}$) and the total mole number ($n_{ex} = n_{CH4\,ex} + n_{CO2\,ex}$) of gases CH_4, CO_2 being adsorbed per gram AC Norit R 1 at 298 K. Asymptotic ($t \to \infty$) values of sorptive gas pressure is p = 0.113 MPa, with methane concentration y_{CH4} = 40 %. The sorbent sample initially is saturated with CH_4 at n_{CH4} (0) = 0.65 mmol/g. Than CO_2 gas is added and the coadsorption process started. The temperature measured below the sorbent sample in the sorptive gas decreased during the sorption process by 4 K from its initial value (298 K) but approached it again after approximately 1 h. (Reprint by permission granted by VDI, Düsseldorf, Germany.)

In coadsorption (N = 2) or multicomponent (N $\geqq$ 3) adsorption experiments it is decisive to make sure that in equilibrium sorptive's gas concentration (y_i, i = 1, 2...) are everywhere the same within the system. If this is not granted, the state of the adsorbate may be changed due to diffusion of sorptive's components causing "creeping adsorption" which may last for many hours and even days. Hence it is always recommended to provide a gas circulator to coadsorption instruments, cp. Figs. 4.1, 4.4, and to circulate the sorptive gas during the adsorption process. However, the circulator should be shut down immediately prior to measurement to avoid dynamic effects of the gas flow at the balance reading.

In engineering gas adsorption processes, for example in air separation processes [4.14], it may happen that a sorbent first is exposed to pure component 1 of a sorptive gas and after equilibration component 2 is added thus starting a coadsorption process approaching a certain equilibrium state Z

(1, 2). Then the question can be raised whether the state of equilibrium attained will be the same after reversing the order of components 1 and 2, that is, first to realize adsorption equilibrium with pure component 2 and then adding component 1 thus approaching an equilibrium state Z (2, 1). This problem is by no way trivial in view of so-called "ink bottle effects" and similar phenomena in micro- and mesoporous solids [2.2]. It also is fundamental for applying the principles of multicomponent thermodynamics [4.15], [4.16] to multicomponent adsorption phenomena. To check whether Z (2, 1) = Z (1, 2) a special experiment in the system AC Norit R 1, CH_4 (component 1), CO_2 (component 2) at 298 K was performed. First the activated carbon was exposed to pure CH_4 for 20 hs. Then CO_2 was added and the approach to equilibrium monitored for another 20 hs. The respective data of the total mass adsorbed are shown in the lower curve of Figure 4.9 below. After this, the experiment was repeated with components CH_4, CO_2 reversed. Corresponding data are shown in the upper curve of Figure 4.9. The gas pressure approached in both experiments was about $p \simeq 0.11$ MPa. As can be seen from the curves in Figure 4.9, in both experiments the same final equilibrium state is approached asymptotically, i. e. after 40 hs. This also has been confirmed by analysis of adsorbates' partial loads of CH_4 and CO_2 which turned out to be the same in both experiments [4.4].

The instrument sketched in Figure 4.4 above also can be used to investigate changes of adsorption capacities of sorbent materials due to pre-adsorption of gases or vapors. An example for this phenomenon is given below. Figure 4.10 shows data of Gibbs excess amounts of carbon dioxide (CO_2) adsorbed on dry molecular sieve (MS) NAX13 (Linde, UOP) at 323 K and the corresponding data for this sorbent material including 43 mg/g MS and 56 mg/g MS pre-sorbed water (H_2O) respectively. As can be seen from the data, the adsorption capacity of the molecular sieve for CO_2 is reduced by 40 % due to the pre-adsorbed water, the amount of which is only slightly changed during the adsorption process of CO_2 [4.15].

The instrument for combined volumetric – gravimetric measurements of binary coadsorption measurements, Figs. 4.4, 4.5 has been redesigned and automated by BEL-Japan Inc., Osaka, Japan. It is now available commercially from BEL-Japan Inc. or Rubotherm GmbH. A schematic diagram of this instrument including a magnetic suspension balance is shown in Fig. 4.11 a, followed by a snapshot taken at lab PBA-0126/1 of IFT, Fig. 4.11 b. This instrument is used today by several institutions in Japan, Germany and the USA. Examples of measurements of coadsorption equilibria of not only gases but also (subcritical) vapors are given in the literature [4.16 a, b].

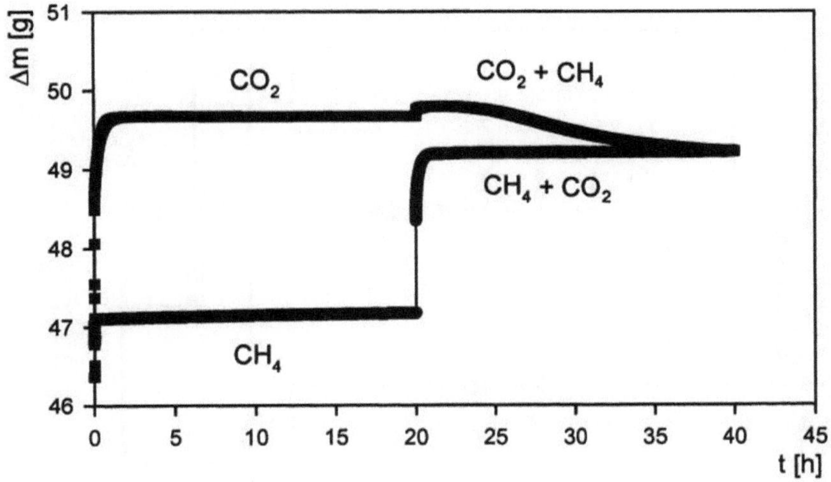

Figure 4.9. Total mass of gas adsorbed on AC Norit R 1 at 298 K in two complementary two-step experiments. First the AC is exposed to pure CH_4 and after 20 hs CO_2 is added. Data are presented by the lower curve ($CH_4 + CO_2$). In a second experiment the AC is first exposed to pure CO_2 and after 20 hs CH_4 is added. Respective data are shown as upper curve in the Figure. Equilibrium pressures in both experiments were p = 0.11 MPa, [4.4].

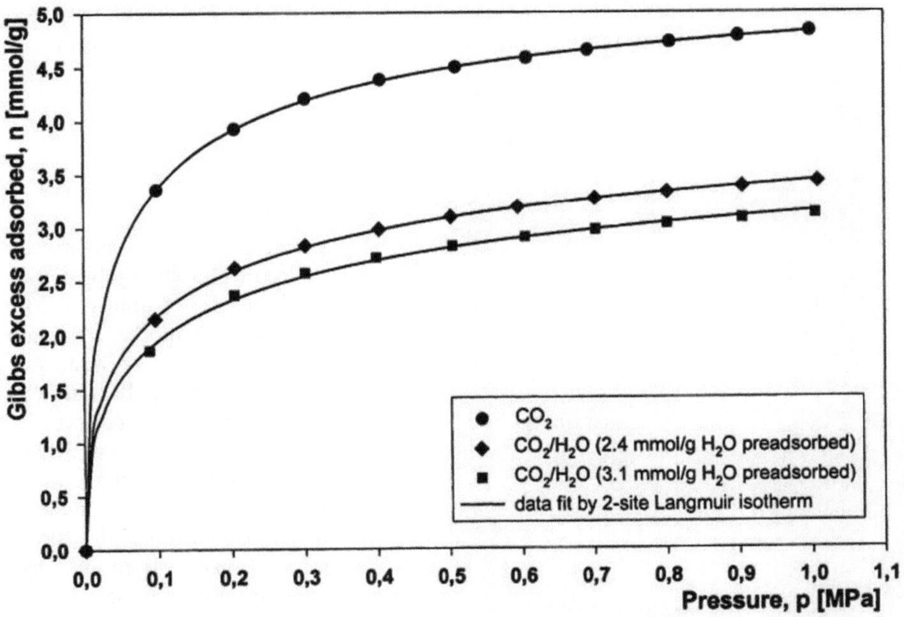

Figure 4.10. Adsorption isotherms of carbon dioxide (CO_2) on molecular sieve NAX13 (Linde, UOP) in dry state (upper curve) and in states preloaded with water (lower curves) at 323 K, [4.15].

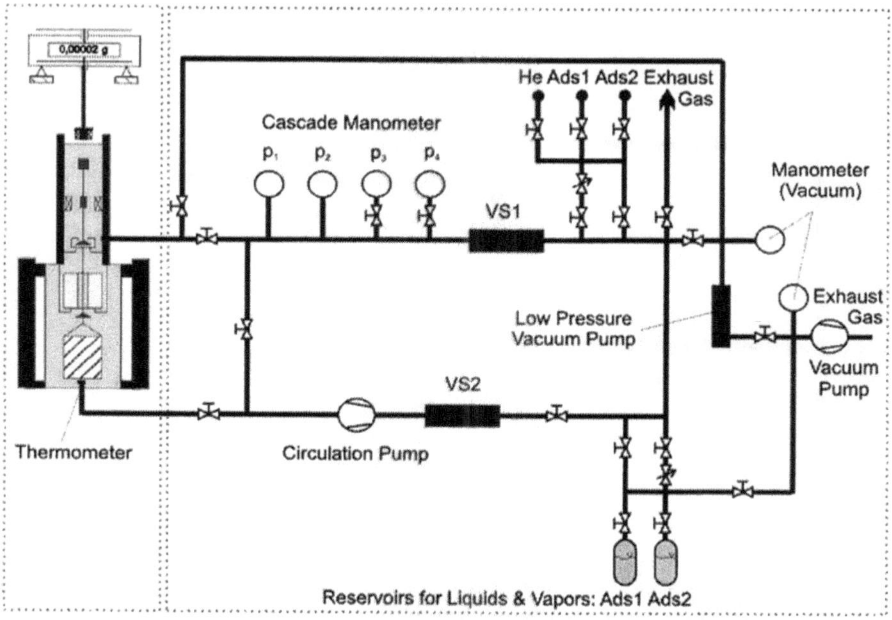

Figure 4.11a. Schematic diagram of an instrument for automated binary coadsorption equilibria measurements using the volumetric-gravimetric method. The installation has been designed by BEL-Japan Corp. and includes a magnetic suspension balance, RUBOTHERM, Bochum. (Reprint by permission of BEL-Japan.)

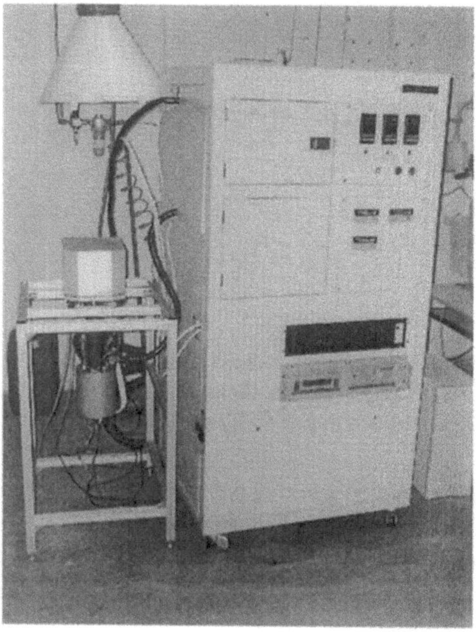

Figure 4.11 b.
Instrument for automated combined volumetric-gravimetric measurements of binary co-adsorption equilibria of gas mixtures with non-isometric components. The RUBOTHERM magnetic suspension balance is on the left side. The large white closet includes the volumetric / manometric part of the instrument and is manufactured by BEL-Japan, Osaka. (Reprint by permission of BEL-Japan.)

3. DENSIMETRIC – GRAVIMETRIC MEASUREMENTS OF BINARY COADSORPTION EQUILIBRIA

3.1 Experimental

The magnetic suspension balance in Figure 4.4 can be operated in two different positions. In the first one – the so-called zero point position – only the permanent magnet at top and the stem below of it are floating whereas the basket including the sorbent material is mechanically decoupled from it so that its weight is transmitted via a load decoupling system to a mechanical support and from there to the instrument. This position is used to tare and calibrate the microbalance outside the suspension. In the second position the permanent magnet and the stem are lifted somewhat so that the basket including the sorbent material is decoupled from its support and also is floating. This so-called "measuring position (1)" of the suspension is used to determine the weight of the sorbent material reduced by its buoyancy in a fluid i. e. gaseous or liquid environment. Now, in 1995, it has been proposed by the authors to add to these two positions of the suspension a third one by lifting the permanent magnet, stem, and primary load somewhat further and by this to make a second load coupled similarly as the first one to the floating stem to float and thus allowing one to measure the weight of the sum of both loads in this "measuring position (2)". This would allow consecutive weighing of two masses in the same gaseous environment. Choosing as masses two different (or differently activated) sorbent materials, this could double the measuring capacity of the balance or reduce the time needed for gravimetric measurements by 50 %. Choosing as masses a sorbent material and a dense sinker of exactly known volume this would allow simultaneous measurements of the reduced mass adsorbed in the sorbent and the density of the sorptive gas (ρ^f). A simplified schematic diagram of such a suspension is sketched in Figure 4.12. Working positions (0), (1), (2) are marked from left to right.

Following a suggestion of the authors a magnetic suspension allowing 3 working positions was built by RUBOTHERM GmbH during (1996-1997) and installed in the volumetric-gravimetric coadsorption instrument Figs. 4.4, 4.5 in the lab PB-A 0126 of the Institute of Fluid- and Thermodynamics at the University of Siegen early in 1998. First measurements of weights of sorbent/sorbate samples and simultaneously of the density of the sorptive gas were performed on 11 February 1998, cp. snapshot Fig. 4.13 below. Data are presented in Sect. 3.3.

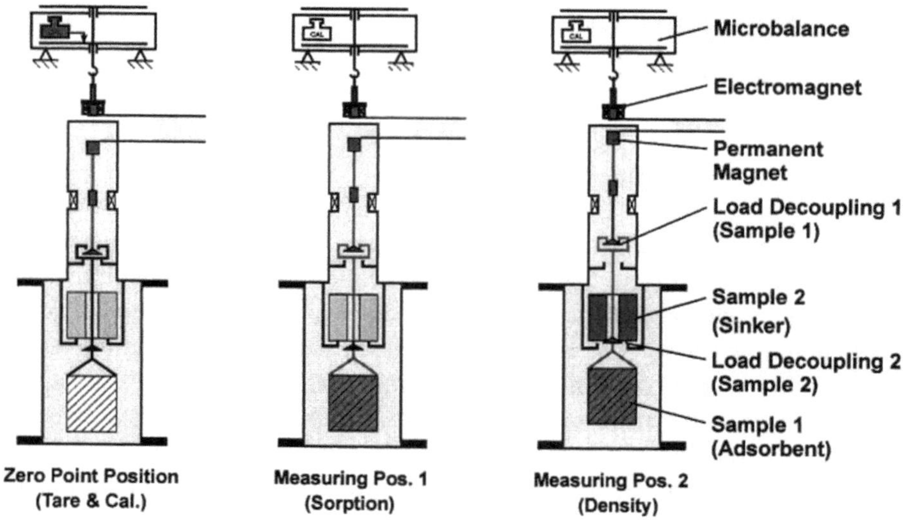

Figure 4.12. Schematic representation of the working positions of a dual mode suspension balance. In position (0) only the permanent magnet and the stem below of it are floating whereas both the sorbent in the basket and the sinker above of it are resting on their respective mechanical supports. In measuring position (1) the basket including the sorbent material is lifted. In measuring position (2) both the baskets filled with sorbent and the sinker above of it are lifted and the total weight being recorded at the microbalance on top of the installation. (Reprint by permission of RUBOTHERM GmbH, Bochum, Germany.)

An interesting consequence of simultaneous weighing and sorptive gas density measurements is that in the case of binary gas mixtures with non-isomeric components, coadsorption equilibria can be determined from the data. This procedure is much faster than volumetric – gravimetric measurements. A basic scheme of the experimental installation is given in Figure 4.14. It mainly consists of an adsorption chamber (AC) of volume (V_{AC}) including a 3 position magnetic suspension balance (RUBOTHERM GmbH, Bochum, Germany), supply lines for sorptive gases and vacuum, thermocouples and manometers to take temperatures and pressures inside the chamber.

To measure binary coadsorption equilibria by the densimetric–gravimet-ric method one should proceed as follows: After preparing a sorbent sample, i. e. activating it, cp. Sect. 2.1 and weighing it in vacuum to determine its mass (m^s), the adsorption chamber (AC) is filled with a gas mixture of known concentrations (y_1^*, y_2^*) and the circulation pump is turned on.

Figure 4.13. Simultaneous measurements of weights of porous sorbent materials exerted to a sorptive gas and density of the gas on site were performed for the first time on 11 February 1998 in lab PB-A 0126 at the Institute of Fluid- and Thermodynamics (IFT) of the University of Siegen. Members of the working group have been R. Seif (RUBOTHERM), C. Lösch (RUBOTHERM), F. Dreisbach (IFT), and M. Seelbach (IFT) (from left to right). (Reprint by permission of RUBOTHERM GmbH.)

After thermodynamic equilibrium is reached, i. e. the pressure (p) and the microbalance signal (Ω) are (at least technically) constant, the circulation pump can be turned off and data (p, T, ρ^f, Ω) be registered. These together with the known concentrations (y_1^*, y_2^*) of the sorptive gas mixture originally supplied to the system, provide the basis from which the masses (m_1^a, m_2^a) of both gas components adsorbed on (m^s) can be calculated as will be described in the next section. Naturally, a set of similar step up experiments can be performed with increasing sorptive gas pressures in the (AC). However, it should be taken into account that in such a procedure, uncertainties of results are adding up. Hence it is not recommended to perform more than – say – five sequential pressure step experiments to keep uncertainties of adsorbed masses and sorptive gas concentrations below or at 10 %, cp. also Sect. 3.4 of this chapter.

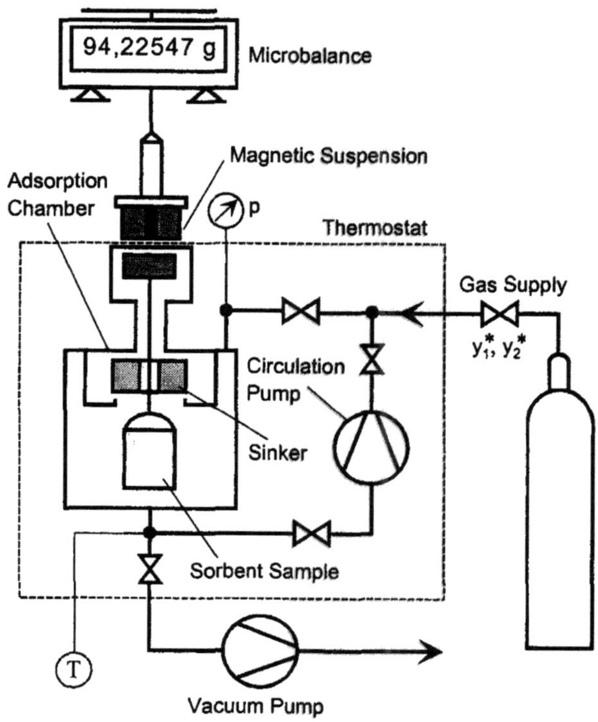

Figure 4.14. Schematic diagram of an instrument for densimetric – gravimetric measurements of binary coadsorption equilibria without using a gas chromatograph. The sorptive gas supplied to the system is assumed to be a binary mixture with well known initial molar concentrations (y_1^*, y_2^*).

3.2 Theory

In this section we will outline the algebraic procedure by which data for a binary coadsorption equilibrium state, namely the masses $(m_i^a, i=1,2)$ adsorbed on a known mass (m^s) of sorbent material and the concentrations of the sorptive gas (y_1^f, y_2^f) in equilibrium with the sorbed phase can be calculated from measured data (p, T, Ω, ρ^f) of a densimetric – gravimetric experiment described in Sect. 3.1. Here we assume that the adsorption chamber initially has been evacuated. (Step up experiments can be handled in a fairly similar manner and do not require additional information or algebraic procedures.)

The microbalance in working position (1), Fig. 4.12 delivers the sum of masses adsorbed minus the buoyancy of the sorbent sample, namely the quantity (Ω), cp. Eq. (4.4)

$$\Omega = m_1^a + m_2^a - \rho^f V^{as} \tag{4.40}$$

From measurements in working position (2) (Fig. 4.12), we get the numerical value of the sorptive gas density

$$\rho^f = \frac{m_1^f + m_2^f}{V^f} \tag{4.41}$$

with

$$V^f = V^* - V^{as} \tag{4.42}$$

Here V^* is the net volume of the adsorption chamber, i. e. its geometric volume minus the volume of all fixed parts of the magnetic suspension and of the other elements which permanently are included in the chamber as for example wires of thermocouples, the basket for sorbent sample etc. Also (V^{as}) indicates the volume of the sorbent / sorbate sample which here again is approximated by the helium volume, cp. Chaps. 2 and 3:

$$V^{as} = V_{He}^s \tag{4.43}$$

We also have the thermal equation of state (EOS) of the sorptive gas

$$\frac{m_1^f}{M_1} + \frac{m_2^f}{M_2} = \frac{pV^f}{RTZ(p,T,y_i^f)} \tag{4.44}$$

and the mass balances

$$m_i^* = m_i^f + m_i^a, \quad i = 1,2 \tag{4.45}$$

Here (m_i^*, i = 1, 2) indicate the masses of components i = 1, 2 supplied to the AC from the gas supply system and (m_i^f, i = 1, 2) the masses of components i = 1, 2 of the sorptive gas phase in equilibrium at pressure (p) and temperature (T). The quantity (Z) in (4.44) is the so-called compressibility or real gas factor of the sorptive gas. It is assumed to be a known function of pressure, temperature and the composition (y_i, i = 1, 2) of the gas mixture. Adding eqs. (4.45) and inserting (4.40) and (4.41), we get in view of (4.43) for the total gas mass (m^*) initially supplied to the adsorption chamber

$$\begin{aligned} m^* &\equiv m_1^* + m_2^* \\ &= m_1^f + m_2^f + m_1^a + m_2^a \\ &= \rho^f V^* + \Omega \end{aligned} \qquad 4.46$$

Hence this quantity can be calculated numerically. From it, masses (m_1^*, m_2^*) can be determined by the relations

$$m_i^* = w_i^* m^*, \quad i = 1,2 \qquad 4.47$$

with

$$w_i^* = \frac{y_i^* M_i}{y_1^* M_1 + y_2^* M_2}, \quad i = 1,2 \qquad 4.48$$

being the mass concentrations of the gas mixture originally supplied to the adsorption chamber.

Also eqs. (4.41) and (4.44) provide two equations for the sorptive gas (m_1^f, m_2^f) masses, an iterative solution of which is

$$m_1^f = \frac{M_1}{M_1 - M_2}(\rho^f - \frac{pM_2}{RTZ})V^f, \qquad 4.49\ (1)$$

$$m_2^f = \frac{M_2}{M_2 - M_1}(\rho^f - \frac{pM_1}{RTZ})V^f \qquad 4.49\ (2)$$

From these expressions the Gibbs excess masses adsorbed can be calculated via the mass balances (4.45) as

$$m_{1GE}^a = w_1^*(\rho^f V^* + \Omega) - \frac{M_1}{M_1 - M_2}(\rho^f - \frac{pM_2}{RTZ})(V^* - V_{He}^s) \qquad 4.50\ (1)$$

$$m_{2GE}^a = w_2^*(\rho^f V^* + \Omega) - \frac{M_2}{M_2 - M_1}(\rho^f - \frac{pM_1}{RTZ})(V^* - V_{He}^s) \qquad 4.50\ (2)$$

Numerical iteration in eqs. (4.49) should begin by choosing for the compressibility factor (Z) its ideal gas value, i. e. Z = 1. By inserting resulting values (m^f_{10}, m^f_{20}) in an EOS one gets $Z_0 = Z(p, T, y^f_{i0})$ and hence from eqs. (4.49) new values (m^f_{11}, m^f_{21}). Repeating this procedure i. e. calculating (m_{in+1}, $n = 0,1,2...$, $i = 1,2$) from eqs. (4.49) by using $Z_{n+1} = Z(p, T, y^f_{in})$, one will arrive in the limit of iterations ($n \to \infty$) at the true sorptive masses

$$m^f_i = \lim_{n\to\infty} m^f_{in}, \quad i = 1,2 \tag{4.51}$$

which via eqs. (4.50) will lead to the Gibbs excess masses adsorbed (m^a_{iGE}, $i = 1,2$). A similar procedure can be used to calculate the absolute masses adsorbed (m^a_i, $i = 1,2$), in this case using the model equation (2.9) for (V^{as}) in eq. (4.43) instead of the helium approximation (2.7).

In case the two sorptive gas components (1,2) are mixed with a carrier gas of molecular weight (M_o) which practically is not adsorbed on the sorbent material considered, densimetric – gravimetric measurements still can be used to determine binary coadsorption equilibria of the (non-isomeric) components (1, 2). However, the basic equations (4.41, 4.44, 4.47) have to be modified as follows:

Sorptive gas density:

$$(4.41): \quad \rho^f = \frac{1}{V^f}(m^f_1 + m^f_2 + m^*_0) \tag{4.52}$$

Thermal equation of state of sorptive gas mixture:

$$(4.44): \quad \frac{m^f_1}{M_1} + \frac{m^f_2}{M_2} + \frac{m^*_0}{M_0} = \frac{pV^f}{RTZ(p,T,y^f_1,y^f_2)} \tag{4.53}$$

Total mass of gas supplied the adsorption chamber:

$$\begin{aligned} m^* &= m^*_1 + m^*_2 + m^*_0 \\ &= m^f_1 + m^f_2 + m^f_0 + m^a_1 + m^a_2 \end{aligned} \tag{4.54}$$

$$(4.40)\,(4.52): \; m^* = \rho^f V^* + \Omega$$

Mass concentrations of gas originally supplied to the adsorption chamber:

$$w_i^* = \frac{y_i^* M_i}{y_1^* M_1 + y_2^* M_2 + y_0^* M_0} \qquad 4.55$$

As the molar concentrations $(y_i^*, i = 0,1,2)$ of the gas mixture originally supplied to the adsorption chamber are assumed to be known, the total masses of all components of the gas supplied can be calculated from (4.54, 4.55) by

$$m_i^* = w_i^* m^*, \; i = 0,1,2 \qquad 4.56$$

From these quantities and eqs. (4.52, 4.53) the Gibbs excess (m_{iGE}^a) of the adsorbed components (1, 2) can be calculated via eqs. (4.45) as

$$\begin{aligned} m_{iGE}^a &= m_i^* - m_i^f \quad i = 1,2 \\ &= (w_i^* + \frac{M_1 \, M_2 \, w_0^*}{M_0(M_i - M_{i+1})})(\rho^f V^* + \Omega) \\ &- \frac{M_i}{M_i - M_{i+1}}(\rho^f - \frac{pM_{i+1}}{RTZ})(V^* - V_{He}^s) \end{aligned} \qquad 4.57$$

with

$$M_{i+1} = \begin{cases} M_2 & i = 1 \\ M_1 & i = 2 \end{cases}$$

Equations (4.57) include the compressibility factor

$$Z \quad = Z(p, T, m_0^*, m_1^f, m_2^f) \qquad 4.58$$

and hence need an iterative procedure for solution similar to that described above, cp. eqs. (4.49, 4.51). For low initial concentrations of the sorptive components (1,2), i. e. $m_i^* << m_0^* \quad i = 1,2$, Z can be approximated by the real gas factor of the pure sorptive gas, i. e.

$$Z \quad \cong Z_0(T, \rho_0 = m_0^* / V^f) \qquad 4.59$$

thus avoiding any numerical iteration procedure.

3.3 Uncertainties or Errors of Measurement

For readers interested in applying the (fairly simple) densimetric-gravimetric method to measure binary coadsorption equilibria, we here denote the dispersion or mean square deviations (MSD) $(\sigma^2_{miGE}, i=1,2)$ of the Gibbs excess masses (m^a_{iGE}) of a two component adsorbate as given by eqs. (4.57), cp. also Sect. 2.3. Applying Gauss' law of propagation of error we get

$$\begin{aligned}\sigma^2_{miGE} &= (\rho^f V^* + \Omega)^2 \left[\sigma^2_{wi*} + \left(\frac{M_1 M_2}{M_0(M_i - M_{i+1})}\right)^2 \sigma^2_{w0*}\right] \\ &+ \left(w_i^* + \frac{M_1 M_2}{M_0(M_i - M_{i+1})} w_0^*\right)^2 \bullet \left[V^{*2}\sigma^2_{\rho f} + \rho^{f2}\sigma^2_{V*} + \sigma^2_{\Omega}\right] \\ &+ \left(\frac{M_i}{M_i - M_{i+1}}\right)^2 (V^* - V^s_{He})^2 \bullet \\ &\bullet \left[\sigma^2_{\rho f} + \left(\frac{M_{i+1}}{RTZ}\right)^2 \sigma^2_p + \left(\frac{pM_{i+1}}{RTZ}\right)^2 \sigma^2_T + \left(\frac{pM_{i+1}}{RTZ}\right)^2 \sigma^2_Z\right] + \\ &+ \left(\frac{M_i}{M_i - M_{i+1}}\right)^2 \left(\rho^f - \frac{pM_{i+1}}{RTZ}\right)^2 (\sigma^2_{V*} + \sigma^2_{VHe}), \; i = 1,2\end{aligned} \qquad 4.60$$

Here the abbreviation

$$\sigma^2_Z = \left(\frac{\partial Z}{\partial T}\right)^2_p \sigma^2_T + \left(\frac{\partial Z}{\partial p}\right)^2_T \sigma^2_p + O(\sigma^2_{yi}, i = 1,2,0) \qquad 4.61$$

has been used, the last term indicating expressions including uncertainties of the order (O) of sorptive gas concentrations (y_i, i = 1, 2). To elucidate the expression (4.60) we list a table of quantities to be measured and dispersions below:

w_i^*	w_0^*	ρ^f	V^*	Ω	p	T	Z	V^s_{He}
σ_{wi*}	σ_{w0*}	$\sigma_{\rho f}$	σ_{V*}	σ_{Ω}	σ_p	σ_T	σ_Z	σ_{VHe}

4.62

Similar to (4.29), the dispersions (σ^2_{miGE}) become large according to (4.60) if the difference between the molar masses of the adsorbed components $|M_i - M_{i+1}|$ is small compared to either M_i or M_{i+1}. Hence also this method is not recommended for measurements of coadsorption equilibria of gases or vapors with nearly equal molar masses like (N_2, O_2) or (CO_2, Ar) etc. For these mixtures always additional measurements of the sorptive gas concentrations by gas-chromatography or mass-spectrometry are recommended.

As already mentioned in Sect. 2.3 the dispersions of the balance measurement (σ_Ω), of pressure (σ_p) and of system's temperature (σ_T) are most important to get accurate values of adsorbed masses ($m^a_{iGE}, i = 1,2$), whereas the contributions of uncertainties in the compressibility factor (Z), often are rather small and hence may be neglected.

3.4 Example

Densimetric-gravimetric measurements of binary coadsorption equilibria have been performed at IFT using the instrument shown in Figs. 4.13, 4.14 in 2001-2002, [4.17]. The system chosen was carbon dioxide (CO_2), methane (CH_4), and activated carbon (AC) D 47/3 at T = 293 K for pressures up to 1.4 MPa. Gas mixtures were provided by mixing components of purities CO_2: 99.995 % (4.5), CH_4: 99.999 (5.0), delivered by Messer Griesheim, Düsseldorf. The AC is manufactured by German CarboTech company, Essen. It is used as nearly cylindrical pellets (length: 3 mm, diameter: 1 mm) having a bulk density of (470 ± 20) kg/m^3 and a BET-surface of (1000 ± 120) m^2/g. It mainly is used in industrial adsorption processes for purification of gas mixtures including either hydrogen or methane.

In Figure 4.15 an example of the kinetics of the coadsorption process of a gas mixture (CO_2/CH_4) on AC D47/3 is given. The equilibrium state realized after 4 hours was at p = 0.71 MPa, T = 293.25 K, and sorptive gas concentration y_{CO2} = 20 %mol, y_{CH4} = 80 %mol. The upper curve indicates the increase of the total absolute mass of CH_4 and CO_2 adsorbed. As can be seen from the figure, adsorption occurs rapidly in an initial period of about 5 min. accompanied by a sharp increase of sorptive's gas temperature (taken by a thermocouple immediately below the sorbent sample) of ca. 2.7 K and an almost instantaneous decrease of gas pressure from its initial value of about 1.1 MPa to 0.73 MPa. Then the uptake of mass slows down showing relaxational behavior of order (1) within the next 1.5 hours and reaching 99 % of its equilibrium value after 2 hours. At this time (135 min.) the gas circulator was shut down. This is reflected in all 3 curves by a tiny fluctuation, most clearly observable in the (upper) mass uptake curve. The relaxation in the mass uptake is accompanied by a similar relaxation of the sorptive gas temperature (lower curve) showing a small secondary decrease after circulator shut down of about (0.1K).

In Figure 4.16 coadsorption equilibria data of the above mentioned (CO_2, CH_4) gas mixture on AC D47/3 at T = 293 K for pressures up to 1.4 MPa are shown. Absolute masses of (CO_2) and of (CO_2 + CH_4) adsorbed per unit mass are presented as function of the sorptive gas pressure.

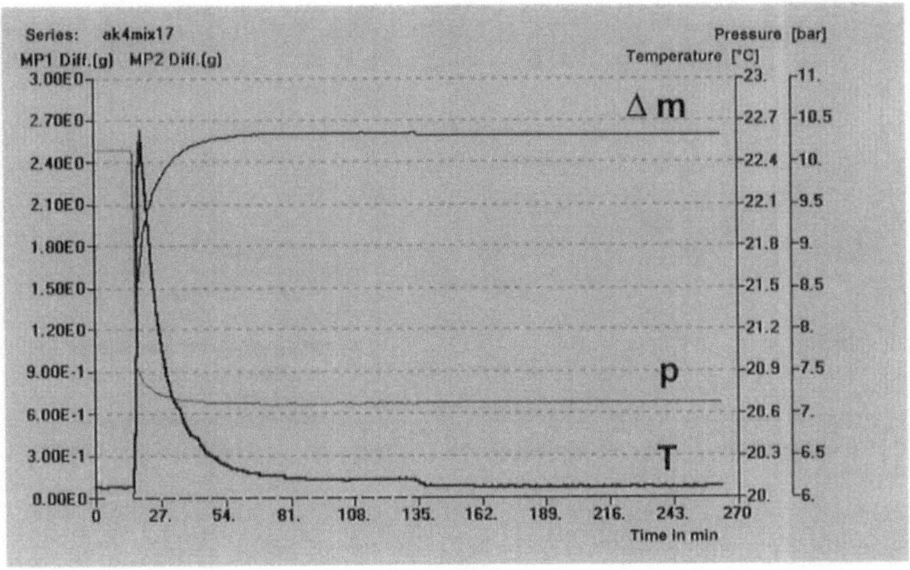

Figure 4.15. Kinetics of the adsorption process of a gas mixture (CO_2, CH_4) with equilibrium sorptive gas concentrations (y_{co2} = 20 %mol, y_{ch4} = 80 %mol). The upper curve (Δm) indicates the total absolute mass, cp. (2.31), of CH_4 and CO_2 adsorbed as function of time. Data have been taken from the microbalance recordings. The lower curves show the time variation of the pressure and the temperature of the sorptive gas mixture inside the adsorption vessel. As can be seen adsorption equilibrium is approached asymptotically and realized – in technical terms – after approximately 4 hours.

Data have been correlated by a generalized isotherm of the Langmuir type taking the fractal dimension of the internal surface of the AC into account, cp. Chap. 7. Uncertainties of data are about three times the size of the graphical symbols. The selectivity of the AC is $S_{CO2\text{-}CH4} \cong 4$, cp. (2.33).

A preliminary check of the quality of data and internal consistency easily can be done by sketching them in a diagram ln (p/n) – n. There, inconsistencies often are magnified, especially for low pressure data ($p \rightarrow 0$, $n \rightarrow 0$) and hence easily can be detected. In Figure 4.17 data of Fig. 4.16 are presented in this way and again correlated by a generalized Langmuir adsorption isotherm of fractal dimension, Chap. 7. Seemingly, data are consistent. However, it should be noticed that linear interpolation of high pressure data to intersections with the ln (p/n) axis would lead to numerical values of Henry-coefficients of the coadsorbate which are different from the values indicated by the correlation curves in Fig. 4.17.

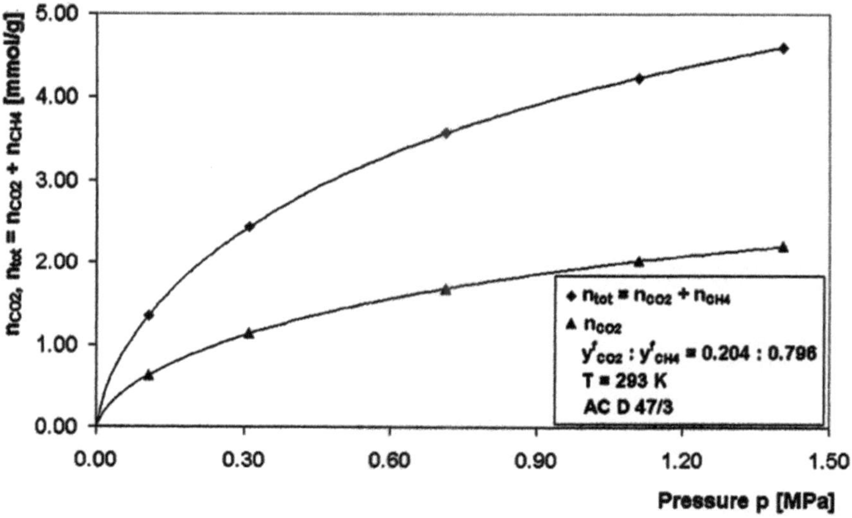

Figure 4.16. Coadsorption equilibria of a binary gas mixture CO_2/CH_4 with y_{co2} = 20.4 %mol, y_{ch4} = 79.6 %mol on AC D47/3 at T = 293 K. Data have been taken by densimetric gravimetric measurements [4.17].
Absolute mass adsorbed of both components ($n_{CO2} + n_{ch4}$): (♦).
Absolute value of partial amount of CO_2 adsorbed (n_{CO2}): (▲).
Data are given in [mmol/g AC] and correlated by a Langmuir isotherm of fractal dimension, $n_i = n_{i\infty} (bp_i)^{\alpha i} / (1+(bp_i)^{\alpha i})$, i = CO_2, CH_4, $\alpha(CO_2)$ = 0.706, $\alpha(CH_4)$ = 0.753, cp. Chap. 7.

As in Figs. 4.16, 4.17 we have chosen not to present Gibbs excess mole numbers, but absolute mole numbers of the co-adsorbate because we want to add information on the former to elucidate the difference between the two sets of data within the pressure range considered (p < 1.4.MPa). Table 4.2 includes numerical data of both the Gibbs excess mole numbers (n_{ex}) and the absolute mole numbers (n_{abs}) of components CO_2 and CH_4 adsorbed per unit mass of sorbent on AC D47/3 at T = 293 K. As can be seen, relative differences $|(n_{ex} - n_{abs})/n_{ex}|$ between corresponding values are very small at low pressures (0.1 %) but increase considerably for higher pressures (1.4 MPa) to approximately (2 %), [4.17].

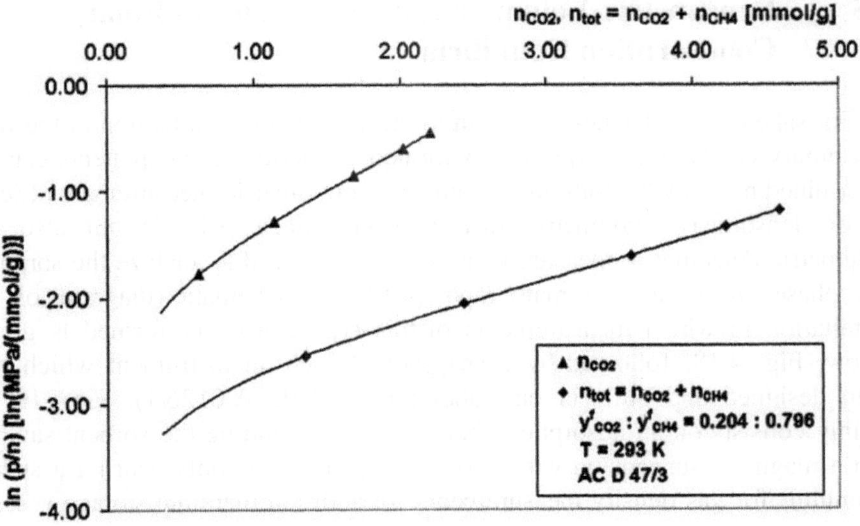

Figure 4.17. Coadsorption equilibria data for a binary gasmixture CO_2 / CH_4 with $y_{CO2} = 20.4$ %mol, $y_{CH4} = 79.6$ %mol on AC D47/3 at 293 K. The quantity ln (p/n) is correlated to the absolute amount adsorbed (n) for $n = n_{CO2}$....▲, and $n = n_{tot} = n_{CO2} + n_{CH4}$ ...◆. Lines refer to adsorption isotherms of the generalized Langmuir type for adsorbents with fractal dimensions cp. Chap. 7, [4.17].

Table 4.2. Gibbs excess masses and absolute masses of a CO_2 / CH_4 co-adsorbed phase on AC D47/3 at T = 293 K for pressures up to 1.4 MPa, [4.17]. For calculation of absolute masses adsorbed cp. also (2.7), (2.9), (2.31).

T = 293 K				
p [MPa]	CO_2		CH_4	
	n_{ex} [mmol/g]	n_{abs} [mmol/g]	n_{ex} [mmol/g]	n_{abs} [mmol/g]
0.105	0.621	0.621	0.718	0.720
0.31066	1.130	0.132	1.293	1.302
0.713	1.669	1.676	1.872	1.903
1.109	2.008	2.022	2.152	2.210
1.406	2.187	2.207	2.312	2.392

3.5 Densimetric-Volumetric Measurements of Binary Coadsorption Equilibria

For sake of completeness we want to draw the reader's attention to the fact that binary coadsorption equilibria with non-isomeric gas components can be determined not only by combined volumetric-gravimetric measurements (Sect. 2) or densometric-gravimetric measurements (Sects. 3.1-3.4) but also by volumetric-densimetric measurements without the need to analyze the sorptive gas phase in a gas chromatograph [4.18]. A schematic diagram of an installation in which measurements of this type can be performed is given below, Fig. 4.18, followed by a snapshot of such an instrument which has been designed and built in our laboratory (IFT PB A 0126/1), Fig.4.19. It mainly consists on an adsorption chamber (AC) including the sorbent sample and a magnetic suspension with two working positions only bearing a sinker (titanium) for gas density measurements. In addition there are storage vessels for the pure sorptive gas components and auxiliary equipment like a gas circulation pump, valves, manometers, and thermometers, a vacuum pump and a gas supply system. All vessels and tubes connecting them should be placed in a thermostat.

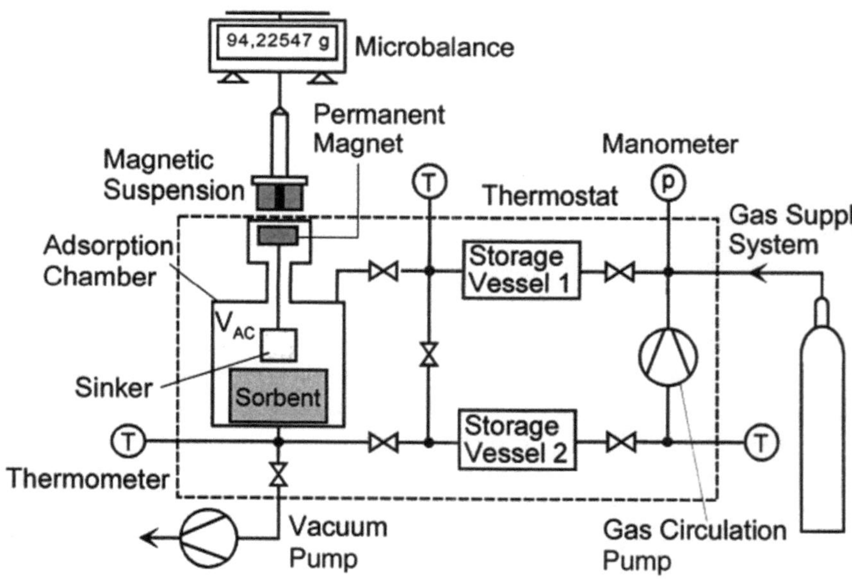

Figure 4.18. Instrument for volumetric-densimetric measurements of binary coadsorption equilibria of gas mixtures on porous solids without using a gas chromatograph. The sorptive gas prepared in the system is assumed to be a binary mixture with known initial molar concentrations (y_1^*, y_2^*). © IFT University of Siegen, 2002.

Figure 4.19. Instrument for volumetric-densimetric measurements of binary coadsorption equilibria of gas mixtures on porous solids without using a gas chromatograph. The magnetic suspension balance, cp. Fig. 4.18, near the center is complemented by a thermostat at left, an electronic unit for automated measurements and control in the rear, a monitor of a PC at right and a helium gas based instrument for leak detection in the foreground at right. © IFT University of Siegen, 2002.

The masses of both components of the sorptive gas being adsorbed ($m_i^a, i = 1,2$) can be determined from the mass balances, cp. (4.1), (4.45) as

$$m_i^a = m_i^* - m_i^f, \quad i = 1,2 \qquad 4.63$$

Here (m_i^*) indicates the total mass of sorptive gas component (i = 1, 2) supplied to the storage vessel. Usually this quantity is calculated from an EOS and measured data of pressure (p^*), temperature (T) and volume of the storage vessel (V_{SV}) including the volume of gas filled tubes and valves as

$$m_i^* = \frac{y_i^* p^* V_{SV}}{RTZ^*} M_i, \quad i = 1,2 \qquad 4.64$$

The quantity $Z^* = Z(p_i^*, T, y_1^*, y_2^*)$ is the compressibility factor of the gas mixture which today in fairly wide ranges of pressure and temperature is known for many industrial gas mixtures [4.19], [4.20]. The masses of the sorptive gas components in adsorption equilibrium $(m_i^f, i = 1,2)$ can be calculated from gas density measurements by the buoyancy of a sinker coupled to the magnetic suspension balance, i. e. the equation

$$\Omega = m_{si} - \rho^f V_{si} \tag{4.65}$$

Here (Ω) is a measured quantity coming from the microbalance readings, cp. Sect. 2.2, m_{si} and V_{si} indicate mass and volume of the sinker. As the density of the sorptive gas (ρ^f) now is a known quantity, the extensivity relation

$$m_1^f + m_2^f = \rho^f V^f \tag{4.66}$$

with

$$V^f = V^* - V^{as} \tag{4.67}$$

provides an algebraic equation for (m_1^f, m_2^f). It can be coupled with the EOS of the sorptive gas

$$\frac{m_1^f}{M_1} + \frac{m_2^f}{M_2} = \frac{pV^f}{RTZ(p,T,w_i^f)} \tag{4.68}$$

Solving (4.66-4.68) for (m_1^f, m_2^f) we get

$$m_i^f = \frac{M_i}{M_i - M_{i+1}}(\rho^f - \frac{pM_{i+1}}{RTZ(p,T,w_i)})V^f, \quad i = 1,2 \tag{4.69}$$

where one has the identity ($M_{i+1} = M_2$ for $i = 1$, $M_{i+1} = M_1$ for $i = 2$). Equations (4.69) still include the unknown mass concentrations $(w_i = (m_i^f / (m_1^f + m_2^f)), i = 1,2, w_1 + w_2 = 1)$ of the sorptive gas. These parameters can be calculated iteratively from the known gas density (ρ^f) via the EOS (4.68) which can be written as

$$w_i = \frac{M_1 M_2}{M_{i+1} - M_i}(\frac{p}{\rho^f RTZ(p,T,w_i)} - \frac{1}{M_{i+1}}), \quad i = 1,2 \tag{4.70}$$

Starting with the ideal gas value of the compressibility factor (Z=1), eq. (4.70) leads to certain mass fractions (w_{i1}, i = 1, 2) which inserted in an analytic expression for Z = Z(p, T, w_i) and eq. (4.70) can be iterated to converge (usually after ca. 5 steps) to the sorptive gas mass fractions (w_i, i = 1, 2) needed to calculate $(m_i^f, i=1,2)$ from eq. (4.69). Then from eqs. (4.63), (4.64), (4.67), (4.69) we have for the Gibbs excess masses adsorbed

$$m_{iGE}^a = \frac{y_i^* p^* V_{SV}}{RTZ^*} M_i - \frac{M_i}{M_i - M_{i+1}} (\rho^f - \frac{pM_{i+1}}{RTZ(p,T,w_i)}) \cdot (V^* - V_{He}^s), \quad i=1,2 \tag{4.71}$$

using again the convention ($M_3 \equiv M_1$).

For sake of completeness we also mention the dispersion or uncertainty of Gibbs excess masses m_{iGE}^a as given by eq. (4.71). Gauss' law of propagation of uncertainty leads to the expression

$$\begin{aligned} \sigma_{miGE}^2 &= (\frac{V_{SV} M_i}{RTZ^*})^2 \sigma_{pi*}^2 + (\frac{p_i^* M_i}{RTZ^*})^2 \sigma_{VSV}^2 \\ &+ (\frac{p_i^* V_{SV}}{RT^2 Z^*} M_i)^2 \sigma_T^2 + (\frac{p_i^* V_{SVi}}{RTZ^{*2}} M_i)^2 \sigma_{Z*}^2 \\ &+ (\frac{M_i}{M_i - M_{i+1}})^2 (V^* - V_{He}^s)^2 \Big[\sigma_{\rho f}^2 + (\frac{M_{i+1}}{RTZ})^2 \sigma_p^2 \\ &+ (\frac{pM_{i+1}}{RT^2 Z})^2 \sigma_T^2 + (\frac{pM_{i+1}}{RTZ^2})^2 \sigma_Z^2 \Big] \\ &+ (\frac{M_i}{M_i - M_{i+1}})^2 (\rho^f - \frac{pM_{i+1}}{RTZ})^2 (\sigma_{V*}^2 + \sigma_{VSHe}^2) \quad , \quad i=1,2; \end{aligned} \tag{4.72}$$

$$p_i^* = y_i^* p^*, \quad i=1,2 \tag{4.73}$$

$$\sigma_Z^2 = (\frac{\partial Z}{\partial T})^2 \sigma_T^2 + (\frac{\partial Z}{\partial p})^2 \sigma_p^2 + O(\sigma_{yi}^2, i=1,2) \tag{4.74}$$

Here the symbols $O(\sigma_{yi*}^2 \ldots)$ indicate terms of order (σ_{yi*}^2).

The dispersions used in eq. (4.72) are related to the respective measurable quantities as follows:

$p_i^{\bullet}$	V_{SV}	T	$Z^{\bullet}$	ρ^f	p	Z	$V^{\bullet}$	V_{He}^s
$\sigma_{pi\bullet}$	σ_{VSV}	σ_T	$\sigma_{Z\bullet}$	$\sigma_{\rho f}$	σ_p	σ_Z	$\sigma_{V\bullet}$	σ_{VSHe}

Numerical examples easily show that in order to get accurate masses of adsorbate ($m_{iGE}^a, i = 1,2$) it is essential to have precise measurements of the system pressure (σ_p) and temperature (σ_T) at hand. Also, as already has been mentioned in Sect. 3.3, the molar masses (M_1, M_2) of the sorptive gas components should be "fairly different", i. e. $|M_2 - M_1| \geq 0.25 \mathrm{Min}(M_1, M_2)$. If this condition does not hold, uncertainties (4.72) may well exceed 10 %. In this case again additional measurements of sorptive gas concentrations in the equilibrium state considered are necessary [4.17].

Densimetric-volumetric measurements have been performed recently at our Institute determining again coadsorption equilibria of (CO_2, CH_4) on AC D47/3 at T = 293 K for pressures up to 1.4 MPa, cp. Sect. 3.4. Results are identical within experimental uncertainties with those received by densimetric-gravimetric measurements, Sect. 3.4. Hence discussion of this method can be postponed to Sect. 4 where all experimental methods outlined in this chapter will be evaluated from both the experimental and the theoretical point of view.

3.6 Volumetric-Densimetric Measurements of Wall Adsorption

In this section we would like to address in brief the problem of wall adsorption. By this we understand the phenomenon that any gas or liquid contacting the inner surfaces of vessels, tubes, valves etc. of any device will be adsorbed to a certain, though normally only tiny, amount on them. This phenomenon first has observed in gas thermometers at very low temperatures i. e. low gas pressures [4.20]. Here it added considerably to the uncertainties of measurements. The same can be true for volumetric or volumetric-gravimetric adsorption measurements at low pressures or if only very small amounts of sorbent material – say 100 mg or less – are available, cp. Chap. 2. Therefore it sometimes may be necessary in setting up mass balances for adsorption measurements to take this phenomenon into account. A possibility to do this is to measure it. This on principle can be done by using the instrument sketched in Fig. 4.20 below. It mainly consists on three parts: a gas storage vessel (SV)*) equipped with a piston to change its volume (V_{SV}) and

*) For this volume changing vessel it is recommended to use a sapphire cell equipped with a smoothly movable piston. Detailed information about such a device is available at the

inner surface (A_{SV}), a microbalance – preferably a magnetic suspension balance (MSB) – to measure the density (ρ^f) of the sorptive gas and an adsorption chamber (AC) of volume (V_{AC}) and inner surface (A_{AC}) including a set of plates or tubes of volume (V) on the surface (A) of which adsorption should be measured. Naturally, the whole device is placed in a thermostat and systems for evacuation (VP) and sorptive gas supply (GS) are provided as well as manometers (p) and thermometers (T) to control pressure and temperature inside the vessels.

An experiment to measure adsorption on the surface (A) of the sample material is started by preparing a certain amount of sorptive gas at pressure (p^*), temperature (T) in the storage vessel of volume (V_{SV}) and the microbalance vessel where the density (ρ^{f*}) of the gas is determined by buoyancy measurements of a sinker (Si) of known volume.

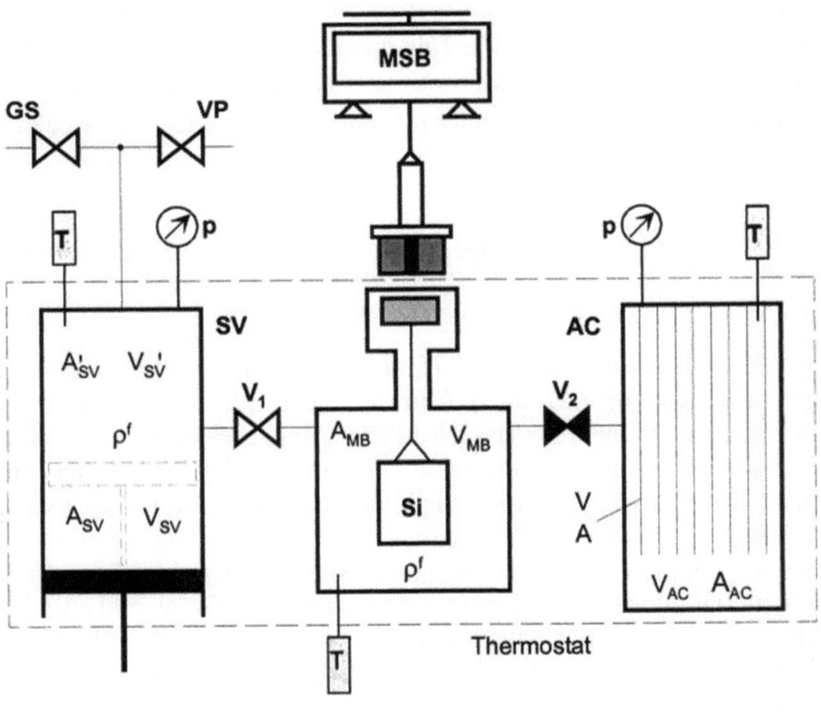

Figure 4.20. Schematic diagram of an instrument to measure wall adsorption by using the volumetric – densimetric method.
SV...gas storage vessel, MSB...magnetic suspension balance, AC...adsorption chamber, GS...gas supply system, VP...vacuum pump, T...thermometer, p...manometer, Si...sinker, V_1, V_2...valves.

Laboratory for Thermophysical Properties at the University of Oldenburg, Oldenburg, Germany.

The adsorption chamber is assumed to have been evacuated to as low a remnant gas pressure as possible. Upon opening the valve (V_2) the gas expands and adsorption occurs, both effects causing a decrease of the gas pressure to a certain value ($p < p^*$). After equilibrium is attained, the piston in the storage vessel is moved to reduce the volume of the vessel and thus increase the gas pressure till its initial value (p^*) is realized again. The values of the storage vessel's volume and internal surface are denoted by (V^1_{SV}) and (A^1_{SV}) respectively. The mass balance of the sorptive gas and the adsorbed phases is

$$\rho^* V_{SV} + \mu^*_W A_{SV} = \rho^* V^1_{SV} + \mu^*_W (A^1_{SV} + A_{AC}) + \rho^* (V_{AC} - V) + \mu_W A \qquad 4.75$$

Here $\mu^*_W = \mu^*_W(p^*, T)$ is the amount of mass adsorbed per unit area on the inner surface of the storage vessel (SV). This quantity normally can be assumed to be known from calibration measurements. Likewise (μ_W) denotes the mass adsorbed per unit area on the sample material. This quantity can be calculated from (4.75) as

$$\mu_W = \frac{1}{A}[\rho^* (V_{SV} - V^1_{SV} - V_{AC} + V) + \mu^*_W (A_{SV} - A^1_{SV} - A_{AC})] \qquad 4.76$$

Here we have assumed as in Eq. (4.75) that the storage vessel and adsorption chamber have been manufactured from the same material having the same surface properties so that the mass adsorbed per unit area on the inner surfaces of the vessels is the same for both vessels. According to Eq. (4.76) the surface adsorption $\mu_W = \mu_W(p^*, T)$ of the sample material is determined by two terms in parentheses ( ), the first of which describes the mass of gas being transferred from the gas phase to the adsorbed phase on the surface (A) of the sample material during the experiment. The second term in parentheses ( ) describes the change in the mass adsorbed on the internal surfaces of the vessels, i. e. areas (A_{SV}, A_{AC}) during the experiment. Numerical values of this term often are small compared to those of the first term. Hence it may be neglected, but in principle always should be considered as a correction term which possibly should be taken into account. It should be noted that in the balance equation (4.75) no terms related to the microbalance or its vessel occur. This is due to the kind of "isobaric" experiment chosen, namely the condition that the initial and the final gas pressure (p^*) within the system are the same and hence the respective terms cancel in the balance equation (4.75).

One may argue that for adsorption measurements of the type described above, gas density measurements and hence a microbalance are not needed in the device Fig. 4.20. This is true since measured values of the gas pressure (p^*) and the temperature (T) the density of the gas (ρ^*) can be calculated from its thermal equations of state (EOS). However, often neither pressure measurements nor EOS available today in the open literature are accurate enough to deliver precise and reliable numerical values of the gas density. Hence we recommend – according to our experience – the use of a microbalance to measure the gas density directly prior to and after the experiment.

Preliminary measurements of the wall adsorption (μ_W) show that this effect is becoming important at either very low pressures and in the vicinity of saturation pressures for subcritical adsorptives. Results and respective data will be published in a forthcoming paper.

4. PROS AND CONS OF VOLUMETRIC-GRAVIMETRIC MEASUREMENTS OF BINARY COADSORPTION EQUILIBRIA

In this section we will discuss the advantages and disadvantages of combined volumetric-gravimetric and related densimetric-gravimetric and densimetric-volumetric measurements of binary coadsorption equilibria (N = 2). Basically, volumetric-gravimetric measurements (VGMs) combine all advantages and disadvantages which occur in either volumetric or gravimetric measurement procedures. As these already have been discussed in Chaps. 2 and 3 we can make only a few remarks reflecting specific experiences we have had with VGMs on various binary coadsorption systems during the last 14 years.

4.1 Advantages

1. Automation

VGMs do not need adsorptive gas analysis by a gas chromatograph or mass spectrometer but only include measurements of pressure and temperature of the sorptive gas mixture and reading of a microbalance. Hence they easily can be automated which already has been done by BEL Japan company, cp. Sect. 2.4, Figs. 4.11 a, b.

2. Kinetics and Approach to Equilibrium

Use of a microbalance with a permanent recording program which – for example – is able to take measurements of the weight of the sorbent sample every second, allows one to get information concerning the kinetics of the coadsorption process and to observe approach to adsorption equilibrium, cp. example given in Fig. 4.15. However, it should be mentioned that for mixture gas adsorption measurements – either volumetrically or gravimetrically or by VGM – circulation of the sorptive gas mixture is always needed to avoid concentration differences within the gas phase. That is during adsorption a gas circulator should be turned on leading to certain, often small but basically not neglectable dynamic effects in the microbalance reading and sometimes also in the pressure transducer recordings, cp. remarks given at Fig. 4.15. The mass uptake curve(s) may serve as a basis to calculate "effective" or "equivalent" diffusion coefficients of the sorptive gas mixture within the sorbent material [4.4]. This information may prove useful in industrial adsorption processes.

In practice a method proposed originally by O. Jäntti [2.4, 4.21] some years ago and recently more elaborated by E. Robens et al. [4.22, 4.23] may prove useful in gravimetric measurements of adsorption kinetics. It allows one to curtail the microweighing time i. e. to avoid measurements of the approach to equilibrium by calculating from 3 data points taken at 3 consecutive times the asymptotic uptake of mass in the sorbent sample with fair accuracy.

3. Activation Procedure of the Sorbent Material

It is recommended to provide an electric heating system for the adsorption chamber of any instrument of VGMs. This will allow one to activate the sorbent material "on site" and thus avoid any pre-adsorption of – for example – ambient air's components, especially humidity, if the sample is activated outside the instrument and then built in the adsorption chamber.

4.2 Disadvantages

VGMs do have all the disadvantages of volumetric / manometric measurements which already have been discussed at the end of Chap. 2. These are in brief:

1. Large amount of sorbent material needed.
2. Wall sorption may cause serious errors.
3. Uncertainties of measurements add up in step-up adsorption (or step-down desorption) experiments.

Commercially available instruments for VGMs usually need several grams of sorbent material to perform reasonably accurate measurements. For coadsorption equilibria of binary mixtures of unsaturated gases (with or without a supercritical carrier gas), wall adsorption in instrument's tubes, valves, and vessel may cause a serious problem. Calibration experiments in the empty instrument, i. e. without a sorbent, may solve the problem but not always. In these cases additional measurements of sorptive gas' concentrations are indispensable.
In VGMs uncertainties of measurements also add up. Hence it is recommended not to do more than 3 steps in a step up (or step down) sorptive gas pressure experiment.

4. Isomeric Sorptive Gas Components

An intrinsic disadvantage of VGMs is that they fail for sorptive gas mixtures with isomeric components as for example (N_2, CO). In this case the resulting equations become linearly dependent and can not be solved algebraically. Hence chromatographic determination of sorptive gas concentrations is necessary. If these quantities are available, masses of the adsorbate's components can be calculated as outlined in Sect. 2.2.

In practice an instrument for volumetric-gravimetric coadsorption measurement requires a considerable amount of equipment and expertise to operate it properly. Hence the instrument will be quite expensive, especially if it is designed and equipped for automatic measurements.

4.3 Comparison of Densimetric-Gravimetric and Densimetric-Volumetric Binary Coadsorption Measurements

Prior to comparing densimetric-gravimetric measurements (DGMs) of binary coadsorption equilibria, cp. instrument Fig. 4.14, and densimetric-volumetric measurements (DVMs), cp. instrument Fig. 4.16, it seems to be useful to compare both methods with basic volumetric-gravimetric measurements (VGMs) discussed in Sects. 1, 2. For this it should be observed that the instruments sketched in Figs. 4.14, 4.16 are designed to use premixed sorptive gas mixtures. Consequently, adsorption equilibria will be attained at different sorptive gas concentrations $(y_i^f, i = 1,2)$, depending on the selectivity of the sorbent material used. If adsorption equilibria data at prescribed gas concentrations $(y_i^f, i = 1,2)$ are needed, instruments in Figs. 4.14, 4.16 have to be equipped with two separate gas storage vessels for pure components (1,2).

The masses of these have to be approximately determined from pure adsorption isotherms of both components (1, 2) and calculations of coadsorption equilibria including only pure components' parameters as for example the IAST-formalism [4.10, 4.4, 4.15, 4.17].

As both procedures DGMs and DVMs have their pros and cons it is appropriate to present them simultaneously in a table which would allow to compare both methods in a glance.

Table 4. 3. Comparison of experimental pros and cons of densimetric-gravimetric measurements (DGMs) and densimetric-volumetric measurements (DVMs) of binary gas adsorption equilibria without analyzing the sorptive phase.

Criterion	DGMs	DVMs
1. Equipment needed	Magnetic suspension balance (3 positions)	Spring balance (quartz), microbalance, magnetic suspension balance (2 positions)
2. Operation	sophisticated	fairly simple
3. Automation	sophisticated	fairly simple
4. Kinetics	can be observed	hardly observable
5. Activation of sorbent	inside the instrument i. e. controllable	outside the instrument i. e. often changed during transportation of sorbent to the instrument
6. Amount of sorbent needed	small (> 0.1 g)	large (5 g – 100 g)
7. Wall adsorption	neglectable	may cause serious errors or uncertainties
8. Uncertainties of measurements	add up in pressure step-up experiments	add up more rapidly in pressure step-up experiments
9. Thermostatization	easily achievable	achievable, but takes more time

In DGMs, approach to equilibrium can be accelerated by using Jäntti's method [2.4, 4.21], cp. Sect. 4.1. In both DGMs and DVMs, Knudsen pressure differences at low sorptive gas pressure may cause thermal gas flow which can interfere with balance's measurements. As a result from Tab. 4.3 it can be said that DGMs are recommended if coadsorption data of high precision and reliability are needed and if there is enough time for measurements available. However, if only data of moderate or low accuracy are needed, money for investment being scarce and time is pressing, DVMs may very well be appropriate as they quickly can provide data sufficient for basic decisions in – for example – designing an industrial adsorption process.

No commercial instruments for DGMs or DVMs are available today.

5. LIST OF SYMBOLS AND ABBREVIATIONS

A list of most symbols and abbreviations used in this Chapter is given.

A	m^2	surface area of a sample material
A_{AC}	m^2	internal surface of an adsorption chamber
A_{SV}	m^2	internal surface of a gas storage vessel prior to a wall adsorption experiment
A^1_{SV}	m^2	internal surface of a gas storage vessel after a wall adsorption experiment
DGMs		densimetric-gravimetric measurements (of binary coadsorption equilibria)
DVMs		densimetric-volumetric measurements (of binary coadsorption equilibria)
M_i	g/mol	molar mass of component (i = 1, 2) of a sorptive gas mixture
$m^a = m^a_1 + m^a_2$	kg	mass of an adsorbate on a certain mass (m^s) of sorbent

m_i^a	kg	mass of component (i = 1, 2) of an adsorbate on mass (m^s) of sorbent
m_{iGE}^a	kg	Gibbs excess mass of component (i = 1,2) of an adsorbate
m_i^f	kg	mass of component (i = 1, 2) of a gaseous sorptive
m_i^*	kg	mass of component (i = 1, 2) of a sorptive gas mixture prior to adsorption
m^s	kg	mass of sorbent material
m_{Si}	kg	mass of a sinker at the magnetic suspension balance
M_0	g/mol	molar mass of carrier gas, i. e. an inert gas component which is much less adsorbed than the sorptive gas components (1, 2)
m_0^s	kg	mass of a reference sample of sorbent material
m_0^*	kg	mass of inert gas carrying adsorptive gas components (1, 2)
n_i	mol	mol number of component (i = 1, 2) adsorbed on mass (m^s) of sorbent
p^*	Pa	pressure of sorptive gas prior to adsorption
$p_i = y_i^f p$	Pa	partial pressure of component (i = 1, 2) in a sorptive gas mixture
R = 8.314	J / Kmol	universal gas constant
T	K	absolute temperature
V_{AC}	m^3	volume of adsorption chamber

V^{as}	m^3	volume of a sorbate – sorbent system which can not be penetrated by a sorptive gas
V^f	m^3	volume of a sorptive gas phase
V^s_{He}	m^3	void volume of a porous solid (sorbent) measured by helium expansion experiments
V_m	cm^3 / mol	volume of 1 mol of a gas (molar volume)
V_{Si}	m^3	volume of a sinker at the magnetic suspension balance
V_{SV}	m^3	volume of storage vessel
V_{SVi}	m^3	volume of storage vessel i = 1, 2
V_T	m^3	volume of tubes, valves and dead volume of a gas circulator in an adsorption installation, cp. Fig. 4.1 and Eq. (4.6)
VGMs		volumetric-gravimetric measurements (of adsorption equilibria)
y^*_i	1	molar concentration of component (i = 1, 2) in a sorptive gas mixture supplied to an instrument prior to adsorption
$Z = (pV_m / RT)$	1	compressibility of a gas (real gas factor)
$Z^* = (p^* V^*_m / RT)$	1	compressibility factor of a sorptive gas (mixture) prior to adsorption
$\Delta\Omega$	μg	change of microbalance signal within a certain time interval (**Δt**)
μ^*_w	g/m^2	mass adsorbed per unit area on the internal surface of a vessel including a sorptive gas
μ_w	g/m^2	mass adsorbed per unit area on the surface of a sample material

ρ^f	kg/m^3	density of a sorptive gas
σ^2_{miGE}	kg^2	dispersion, variance or mean statistical deviation (MSD) of the Gibbs excess mass of component (i = 1, 2) of an adsorbate
σ^2_x	$[x^2]$	dispersion, variance or mean statistical deviation (MSD) of a measurable quantity (x)
$\Omega = m^a - \rho^f V^{as}$	µg	reduced mass of an adsorbate; also micro-balance signal during adsorption process measurements

REFERENCES

[4.1] **Keller J. U.**
Research into Coadsorption Phenomena of Natural Gases on AC's and MS's, Adsorption News, 1 (1990), No. 3, p. 4.

[4.2] **Keller J. U., Staudt R., Tomalla M.**
Volume-Gravimetric Measurements of Binary Gas Adsorption Equilibria, Ber. Bunsenges. Phys. Chem., 96 (1992), p. 28 - 32.

[4.3] **Staudt R.**
Analytische und experimentelle Untersuchungen von Adsorptionsgleichgewichten von reinen Gasen und Gasgemischen an Aktivkohlen und Zeolithen, PHD-Thesis, IFT, University of Siegen, Siegen, 1994.

[4.4] **Dreisbach F.**
"Untersuchung von Adsorptionsgleichgewichten methanhaltiger Gasgemische an Aktivkohle als Grundlage zur Auslegung technischer Adsorptionsanlagen" PHD-Thesis, IFT University of Siegen, Siegen, 1998, cp. also Fortschritt-Berichte VDI, Reihe 3, Verfahrenstechnik, No. 547, VDI-Verlag, Düsseldorf, 1998.

[4.5] **Robens E., Keller J. U.**
Gassorptionsmessung und Thermogravimetrie, LABO, Magazin für Labortechnik + Life Sciences, p. 56, March 2002.

[4.6] **Kochsiek M., Gläser M., (Eds.)**
Comprehensive Mass Metrology, Wiley-VCH, Berlin, 2000, ISBN 3-527-29614-X.

[4.7] **Schein E., He R., Keller J. U.**
Measurement of adsorption of gas mixtures including inert components on zeolite, Report of IFT – USI, Thermodynamics Series, 2003.

[4.8] **Abdel Ghani T., Tomalla M.**
Entwicklung, Aufbau und Inbetriebnahme einer volumengravimetrischen Koadsorptionsanlage mit Schwebekupplung, Diplomarbeit, IFT, Universität Siegen, 1994.

[4.9] **Dreisbach F., Lösch H. W.**
Suspension Balances, Int. Application Notes,
Rubotherm Präzisionsmesstechnik GmbH, Bochum, Germany, 2002.

[4.10] **Myers A. L., Prausnitz J. M.**
Thermodynamics of Mixed Gas Adsorption,
AIChE Journal 11 (1995), 121.

[4.11] **Tomalla M.**
Experimentelle Untersuchung der Koadsorptionsgleichgewichte von CH_4/N_2 und CH_4/CO-Gasgemischen an Aktivkohle (NORIT R1) bei T = 298 K im Bereich p = 0 - 12 MPa; PHD-Thesis, IFT, University of Siegen, Siegen, 1994.

[4.12] **Steele W.**
The Interaction of Gases with Solid Surfaces,
Pergamon, New York, 1974.

[4.13] **Adamson A. W., Gast A. P.**
Physical Chemistry of Surfaces,
J. Wiley & Sons, 6th Ed., New York, 1997.

[4.14] **Kaneko K., Kanoh H, Hanzawa Y., Eds.**
Fundamentals of Adsorption 7
Proceedings of FoA 7 Conference, Nagasaki, May 2001, IK International, Chiba, Japan 2002.

[4.15] **Seelbach M.**
Experimentelle und analytische Untersuchungen von Adsorptionsprozessen technischer Gase an mikroporösen Sorbentien,
PHD-Thesis, IFT, University of Siegen, in preparation, 2003.

[4.16 a] **Dreisbach F., Lösch H. W., Nakai K.**
Adsorption Measurement of Water / Ethanol Mixtures on Activated Carbon Fibre, Chem. Eng. Technol. 24 (2001) 10, p. 1001-1005.

[4.16b] **Dreisbach F., Lösch H. W., Nakai K.**
Messung von Wasser / Ethanol-Gemischisothermen an Aktivkohlefaser,
Chemie Ingenieur Technik (73) 3 (2001), p. 246-252.

[4.17] **Iossifova N.**
Untersuchungen von Gemischgleichgewichten bei adsorptiven Gastrenn- und Reinigungsverfahren,
PHD-Thesis, IFT, University of Siegen, Siegen, Germany, in preparation, 2003.

[4.18] **Keller J. U., Iossifova N., Schein E., Seelbach M.**
Densimetric-Gravimetric and Densimetric-Manometric Measurements of Binary Coadsorption Equilibria, Adsorption, Science & Technology, submitted, 2004.

[4.19] **Sengers J. V., Kayser R. F., Peters C. J., White Jr. H. F., Eds.**
Equations of State for Fluids and Fluid Mixtures, Part I
IUPAC Series in Experimental Thermodynamics, Vol. V,
Elsevier, New York etc., 2000.

[4.20] **Eder F. X.**
Arbeitsmethoden der Thermodynamik, Bd. 1: Temperaturmessung,
Springer, Berlin etc. 1981.

[4.21] **Jäntti O., Junttila J. Yrjänheikki E.**
On curtailing the microweighing time by an extrapolation method,
Suomen Kemistilehti, A 43 (1970), 214-218.

[4.22] **Robens E., Massen C. H., Poulis J. H., Staszczuk P.,**
Fast measurements of adsorption on porous materials using Jäntti's method,
Adsorption Sci. and Tech., 17 (1999) No 10, p. 801-804.

[4.23] **Massen C. H., Poulis J. H., Robens E.,**
Criticism on Jäntti's three point method on curtailing gas adsorption measurements,
Adsorption, 6 (2000), 229-232.

Chapter 5

OSCILLOMETRY

Abstract The physical and experimental technique of gas adsorption measurements by slow oscillations of a rotational pendulum or, likewise, the relaxational motion of a freely floating rotator are described. Combinations of the pendulum with either gravimetric or volumetric measurements are outlined. These especially are suited to measure the absorption or solubility of gases in non-rigid or swelling sorbent materials like polymers. Pros and cons of these methods are discussed in brief. List of symbols. References.

1. INTRODUCTION

The inertia of mass against acceleration provides another possibility to measure it, i. e. to compare it with a standardized sample mass. For practical measurements periodic motions, i. e. oscillations at high or low frequencies in linear or circular modes are used, cp. Ref. [3.1] *). For gas adsorption measurements it always should be taken into account that

- changes of mass due to adsorption are normally small, and
- the sorbent / sorbate is surrounded by a sorptive gas atmosphere causing additional damping to the sample's motion depending on the viscosity of the gas.

For these reasons we will be here restricted to consider only the slow rotational oscillations of sorbent masses in a sorptive gas atmosphere such that the masses moved geometrically are transformed into themselves and only initiate flows in the surrounding gas which are rotationally symmetric.

*) Today this method is used in practice to determine the mass of astronauts in zero gravity ($g = 0$) conditions like in "Spacelab". The astronaut is put on a sledge connected to a linear spring of known constant (k) and the frequency ($\omega = 2\pi / T$) or the time period of oscillations (T) is measured. Then the astronaut's mass (m) can be calculated as $m = k / \omega^2$, this quantity being an important indicator of his health status [5.9].

Hence the respective flow fields can be exactly calculated from the Navier Stokes equations , at least for low frequencies, i. e. laminar flows. This is not the case for linear oscillations at either low frequencies, i. e. the linear spring in gas system [3.1], or high frequencies, namely piezo-quartz systems bearing sorbent samples [5.10] or oscillating rod systems (SETARAM). Also at high frequencies the state of the sorbate phase may become "transient", i. e. deviate considerably from a thermodynamic equilibrium state.

The torsional pendulum of a dense disk moving slowly in a viscous medium has been developed by J. Kestin and co-workers about 1980 to measure simultaneously the density and the viscosity of the fluid [5.11]. This work was based on a couple of papers reporting on oscillating body viscometers using cups, spheres, cylinders, etc. published already in the period 1930-1960 [5.12]. Today pendulum viscosimeters are used to measure the viscosity of both gases, liquids, and even molten metals and salts at extreme temperatures (Ir(liquid melting state):2,800 K) and pressures (100 MPa), [5.13-5.15].

The rotational pendulum does not seem to have been used for gas adsorption measurements prior to 1993 when the respective theory was developed by one of the authors [5.1], and a pendulum was designed and built at the IFT, University of Siegen for this purpose [5.2, 5.7]. In this chapter we are going to describe in brief this method, i. e. give some experimental information, Sect. 2.1, present an outline of the theory, Sect. 2.2, and give examples for gas adsorption measurements on activated carbon, i. e. a nearly rigid sorbent material. In Section 3 we will discuss combined oscillometric – gravimetric gas sorption measurements. These allow one to determine the sorption of gases on non-rigid, i.e. swelling sorbent materials like polymers [5.2, 5.7, 5.8]. In these measurements the mass and the volume of a polymer sample in which gas has been dissolved can be determined independently from each other, these data allowing one together with data of temperature and pressure, to establish a thermal equation of state of the polymer – gas system. In Section 4 combined oscillometric-volumetric measurements of gas sorption equilibria are discussed. Pros and cons of these methods will discussed in Sect. 5 briefly, followed by a list of symbols, Sect. 6, and references.

2. MEASUREMENT OF PURE GAS ADSORPTION EQUILIBRIA (N=1) BY SLOW OSCILLATIONS OF A ROTATIONAL PENDULUM

2.1 Experimental

Pure gas adsorption equilibria can be measured by observing slow oscillations of a rotational pendulum bearing the sorbent material. An instrument of this type has been designed and built at the author's institute during (1995-1997), Fig. 5.1.*) It basically consists of a disk bearing a ring slit filled with homogeneously distributed sorbent material (pellets, powder etc.), fixed to a torsional wire and stabilized by a stem bearing a small mirror to reflect a laser beam, Fig. 5.2. The pendulum is placed in a vessel (adsorption chamber) which can be filled with sorptive gas. The adsorption chamber should be placed within a thermostat and equipped with manometer(s) and thermometer(s).

To observe the motion of the pendulum a laser beam is sent to the mirror, the reflected beam crossing during its motion in radial direction within a horizontal plane 2 diodes placed at known angles (α_1, α_2), Fig. 5.2. The diodes of a couple produce by photo effect electric signals of opposite signs, hence the time when the reflected beam is crossing the area between the two diodes can be detected very accurately, cp. signal sample given in Fig. 5.3. The series of time signals produced in this way by a slow damped oscillation of the pendulum in the gas provides the basis to fit an ideal damped harmonic motion, Eq. (5.7) to it and to calculate the respective angular frequency and logarithmic decrement by a Gaussian minimization procedure. To start the motion of the pendulum, the instrument is based on a ball bearing allowing to rotate it by an arbitrary angle, cp. Fig. 5.1. Choosing this at about 90 ° and keeping the instrument in this position for a while**), the pendulum inside the instrument starts to move due to the retarding moment of the torsional wire. Returning then the whole instrument to its original position will maintain motion of the pendulum, this procedure resulting now in an oscillation. This however will fade away slowly due to internal friction of the wire and the flow of gas surrounding the pendulum.

*) The instrument meanwhile has been transferred to the Institute of Thermodynamics, Dept. Chem. Engineering, University of Dortmund, Dortmund (2003).

**) The time interval to be chosen for this distorted position of the pendulum depends on its design parameters, especially its mass and may vary from a few seconds up to a minute.

Figure 5.1.
Rotational pendulum for measurements of gas adsorption equilibria by observing slow damped oscillations. Height of instrument: 1.5 m. More parameters of the instrument are given in Sect. 2.4.

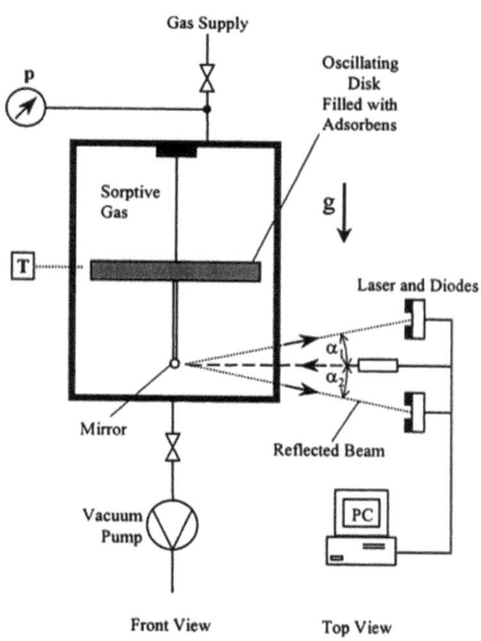

Figure 5.2. Experimental setup for oscillometric measurements of gas adsorption equilibria using a rotational pendulum [5.1, 5.2].

For the instrument shown in Figs. 5.1, 5.2 we could observe for various gases at different pressures rotational oscillations for up to 30 minutes having oscillation times of about (30-45) seconds and logarithmic decrements of about 10^{-4}, cp. Sect. 2.2. For choice of materials and design information, hints already given in the foregoing chapters should be observed. Details can also be found in [5.2, 5.7, 5.11-5.13]. Here we point out that

- oscillation experiments only should be started if one reasonably can expect that adsorption equilibrium is attained, i. e. the pendulum does not seem to be suited for adsorption kinetic measurements and
- the sorptive gas pressure should be at "intermediate level" as for very low pressures the theory presently available for the motion of the pendulum fails, cp. Sect. 2.2, and for high gas pressures the viscous damping of the motion of the pendulum is so severe that only few oscillations or mere creeping motions can be observed [5.2, 5.7]. In practice, measurements of adsorption equilibria of methane (CH_4) on activated carbon at 293 K with our instrument were limited to the pressure range 0.5 MPa – 10 MPa. We expect however, that these limits can be overcome or at least shifted considerably by proper, i. e. optimized, design of the pendulum.

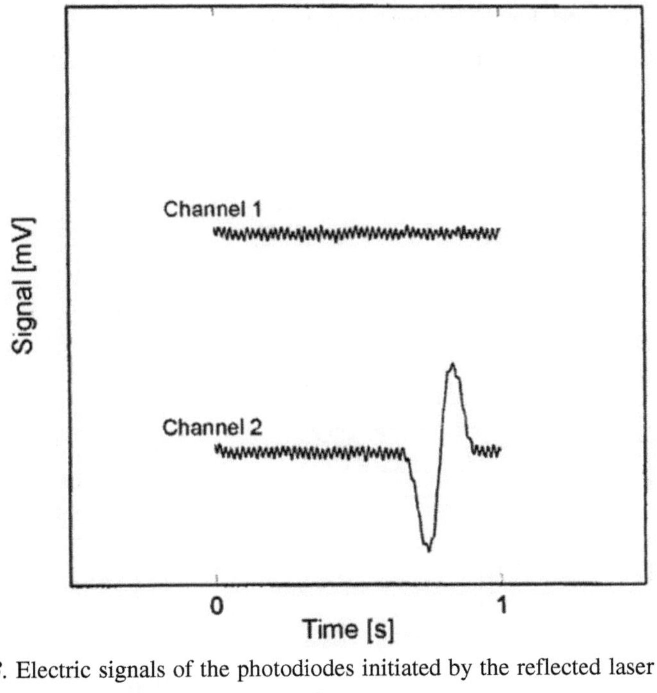

Figure 5.3. Electric signals of the photodiodes initiated by the reflected laser beam crossing diode channel 2.

2.2 Outline of Theory

In this section we present relations between the moment of inertia (J), of the rotational pendulum and measurable parameters of its slow, damped harmonic oscillations, namely the frequency (ω) and the logarithmic decrement (Δ). As the geometry of the pendulum normally is known, from these relations, set up for oscillations in both vacuum and in a sorptive gas atmosphere, the mass of the pendulum and hence the mass of an adsorbed phase included in it can be calculated – at least in principle.

To start we consider the scheme of a simple rotational pendulum as given in Figure 5.4. The pendulum consists of a disk of radius (R_a) bearing a ring slit (height: d, radii: $R_i < R_a$) filled with sorbent material of mass (m^s). (The ring slit is chosen instead of a compact disk in order to save sorbent material). The disk's center is connected to a torsional wire of length (L_W) and diameter ($2r_W$) having a retarding moment (D_0) and an internal damping coefficient (R_0). To stabilize the pendulum, i. e. to make sure its center of gravity is below the base plate, it is recommended to fix a cylindrical stem below it. This stem may bear a small mirror to reflect a laser beam for observing the oscillations of the pendulum, cp. Fig. 5.2. Above and below the pendulum thin plates should be mounted to cut off secondary flows which may be created in the fluid surrounding the pendulum during its oscillating motion.

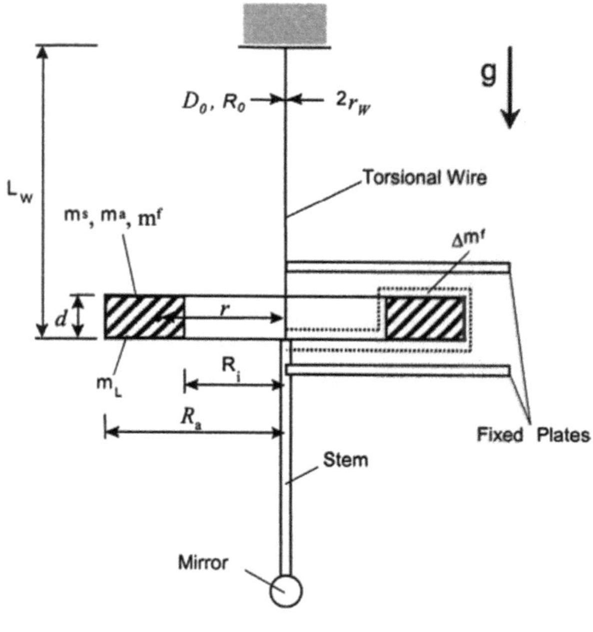

Figure 5.4.
Scheme of a rotational pendulum of mass (m_L) bearing a ring slit with radii (R_i, R_a) and height (d) which can be loaded with sorbent material of mass (m^s). The torsional wire of length (L_w) and diameter ($2r_w$) has a retarding moment (D_0) and internal damping coefficient (R_0).

2.2.1 The Motion of the Pendulum in Vacuum

The equation of motion of the rotational pendulum loaded with sorbent material (m^s) is [5.1, 5.2]:

$$J_0 \ddot{\alpha} + R_0 \dot{\alpha} + D_0 \alpha = 0 \qquad 5.1$$

Here $\alpha = \alpha(t)$ is the angular displacement or amplitude of the pendulum, the dot (•) indicating the time derivative. The moment of inertia (J_0) consists of two parts

$$J_0 = J_L + J_s , \qquad 5.2$$

with

$$J_L = \int_{m_L} r^2 dm_L \qquad 5.3\text{ a}$$

being the moment of inertia of the empty disk of mass (m_L) and (J_s) being the moment of inertia of the sorbent material (m_s) within the ring slit. Assuming this material to be homogenously distributed within the slit – in case of pellets their characteristic dimensions should be at least one order of magnitude smaller than the width $|R_a - R_i|$ or height (d) of the slit – this moment of inertia can be written as

$$J_s = \int_{m^s} r^2 dm^s = \frac{1}{2}(R_i^2 + R_a^2) m^s \qquad 5.3\text{ b}$$

The retarding moment (D_0) of the torsional wire included in the equation of motion (5.1) can be calculated from the wire's length (L_w), its diameter ($2r_w$), the elasticity modulus (E) and the Poisson number (μ) by the relations [5.2, 5.3]

$$D_0 = G J_w / L_w \qquad 5.4$$

$$G = E/(2(1+\mu)) \qquad 5.5$$

$$J_w = \pi r_w^4 / 2 \qquad 5.6$$

The coefficient of internal damping of the wire (R_0) included in Eq. (5.1) in practice must be determined from calibration experiments with either the empty or the loaded pendulum in vacuum, cp. Eq. (5.12). Experiments have shown that (R_0) may change over the years which can be due to irreversible plastic deformation or corrosion if measurements in chemically aggressive sorptive gases have been performed [5.2].

After the fading away of initial disturbances the motion of the rotational pendulum in vacuum always can be described by a simple damped harmonic oscillation. For initial conditions $\alpha(0) = \alpha_0$, $\dot{\alpha}(0) = 0$, we have for the angular amplitude

$$\alpha(t) = \alpha_0 e^{-\Delta_0 \omega_0 t}\left[\cos(\omega_0 t) + \Delta_0 \sin(\omega_0 t)\right] \qquad 5.7$$

Here α_0 is the initial amplitude of the pendulum, $\omega_0 = 2\pi/T_0$ is the angular frequency with T_0 indicating the period of the oscillations. The parameter Δ_0 is the logarithmic decrement defined by the relation as

$$2\pi\Delta_0 = \ln\frac{\alpha(t)}{\alpha(t+T_0)} \qquad t \geq 0 \qquad 5.8$$

The frequency (ω_0) and decrement (Δ_0) are related to the parameters of the pendulum (J_0, D_0, R_0) by the equations

$$\omega_0^2 = \frac{D_0}{J_0} - \left(\frac{R_0}{2J_0}\right)^2 \qquad 5.9$$

$$\Delta_0\omega_0 = \frac{R_0}{2J_0} \qquad 5.10$$

Conversely, we have from (5.9, 5.10)

$$J_0 = \frac{D_0}{\omega_0^2(1+\Delta_0^2)}, \qquad 5.11$$

$$R_0 = 2\Delta_0\omega_0 J_0 \qquad 5.12$$

These relations can be used to calculate numerically J_0 and R_0 from measured data (ω_0, Δ_0) and D_0 which is known from wire's geometric and thermophysical data via equations (5.4-5.6).

For the motion of the empty pendulum ($m^s = 0$) in vacuum we have analogously to (5.1) the equation of motion

$$J_L \ddot{\alpha} + R_0 \dot{\alpha} + D_0 \alpha = 0 \tag{5.13}$$

leading for the same initial conditions as above to the amplitude function

$$\alpha(t) = \alpha_0 e^{-\Delta_L \omega_L t} \left[\cos(\omega_L t) + \Delta_L \sin(\omega_L t)\right] \tag{5.14}$$

and correspondingly to (5.11), (5.12) to the relations

$$J_L = \frac{D_0}{\omega_L^2 (1 + \Delta_L^2)}, \tag{5.15}$$

$$R_0 = 2\Delta_L \omega_L J_L \tag{5.16}$$

Hence we get by dividing equations (5.12) and (5.16) for the ratio of the moments of inertia (J_L/J_0) the expression

$$\frac{J_L}{J_0} = \frac{\omega_0 \Delta_0}{\omega_L \Delta_L} \tag{5.17}$$

which can be determined from (calibration) measurements of the empty (L) and the loaded (0) pendulum in vacuum.

2.2.2 The Motion of the Pendulum in Sorptive Gas

The motion equation of the rotational pendulum loaded with sorbent of mass (m^s) in a gaseous atmosphere is [5.1]:

$$J \ddot{\alpha} + R_0 \dot{\alpha} + D_0 \alpha = M(t) \tag{5.18}$$

The moment of inertia (J) of the pendulum can be written as

$$J = J_0 + J_{osc} \tag{5.19}$$

with J_0 as defined by (5.2) and

$$J_{osc} \equiv J^a + J^f + \Delta J^f \quad 5.20$$

The additional moment of inertia (J_{osc}) is made up of three parts:

- the moment of inertia (J^a) of the mass adsorbed (m^a),
- the moment of inertia (J^f) of the mass of gas (m^f) in the free space between the sorbent pellets in the ring slit,
- and the moment of inertia (ΔJ^f) of the mass (Δm^f) of a boundary layer of the fluid's flow above and below the oscillating disk, cp. Fig. 5.4.

We now assume the masses m^a and m^f moving with the disk to be homogenously distributed within the ring slit. Then we have for J^a and J^f correspondingly to Eq. (5.3) the expressions

$$J^a = \int_{m^a} r^2 dm^a = \frac{1}{2}(R_a^2 + R_i^2)m^a \quad 5.21$$

$$J^f = \int_{m^f} r^2 dm^f = \frac{1}{2}(R_a^2 + R_i^2)m^f \quad 5.22$$

Also let us introduce the "effective" or "equivalent" mass Δm^f for the moment of inertia ΔJ^f by

$$\Delta m^f \equiv \Delta J^f / \left((R_a^2 + R_i^2)/2\right) \quad 5.23$$

Then J_{osc} according to (5.20) can be written as

$$J_{osc} = \frac{1}{2}(R_a^2 + R_i^2)(m^a + m^f + \Delta m^f) \quad 5.24$$

Combining Eqs. (5.2, 5.3, 5.19, 5.24) we get for the sum of masses accelerated with the disk, m^a and m^f plus the equivalent mass Δm^f corresponding to the fluid's flow boundary layer moment of inertia, in addition to the empty disk m_L and sorbent sample m^s, divided by the sample mass m^s the expression

$$\Omega_{osc} \equiv \frac{m^a + m^f + \Delta m^f}{m^s} = \frac{\frac{J}{J_0} - 1}{1 - \frac{J_L}{J_0}} \qquad 5.25$$

The ratios of moments of inertia in the numerator and the denominator on the r.h.s. of this equation need to be related to measurable quantities of the motion of the pendulum, i. e. the angular frequency (ω) and the logarithmic decrement (Δ) of the corresponding damped harmonic oscillations. To achieve this we again have to consider the equation of motion of the pendulum (EOM) (5.18). It includes on its r.h.s. the torque or damping moment exerted by the fluid flow on the pendulum. Physically this is caused by the internal friction of the fluid, i. e. in Newtonian fluids to which we are here restricted, by the velocity gradient in the fluid's flow perpendicular to the surface of the sorbent loaded disk. Basically, the torque (M(t)) is an unknown quantity. However, since the motion of the disk and the fluid flow initiated by it are strongly coupled systems, in addition to (5.18) there should be a relation between the amplitude of the pendulum (α(t)) and the torque (M(t)) rooted in the equations of motion of the fluid and its boundary conditions, i. e. the shape and physical properties of the disk. Assuming rotationally symmetric flow of the sorptive fluid, M(t) can be represented as a surface integral

$$M(t) = -\eta \int_{O} r^2 \left(\mathbf{n}_r \frac{\partial}{\partial r} + \mathbf{n}_z \frac{\partial}{\partial z} \right) \Omega \mathbf{df} \qquad 5.26$$

with the velocity field of the (instationary) fluid flow represented as

$$\mathbf{v}(r,t,z) = r\Omega(r,t,z)\mathbf{\Phi} \qquad 5.27$$

Here η is the dynamic viscosity of the fluid $\mathbf{n}_r$ and $\mathbf{n}_z$ being unit vectors in r- and z-direction of cylindrical coordinates with the wire of the pendulum as z-axis and $\mathbf{\Phi}$ is the unit vector in the azimuthal direction. The quantities (O) and (df) indicate the surface of the pendulum and a vectorial differential of it, respectively.

Restricting ourselves to slow motions of the pendulum, i. e. small Reynolds numbers

$$Re = \frac{\omega_0 R_a^2}{\nu} \leq 100 \qquad 5.28$$

and "thick" disks for which the inequalities hold

$$d >> \delta, \quad \frac{1}{2}(R_i + R_a) >> \delta, \tag{5.29}$$

with the characteristic length of the system

$$\delta^2 = \frac{\nu}{\omega_0}, \tag{5.30}$$

and the kinematic viscosity of the fluid

$$\nu = \eta / \rho(p, T), \tag{5.31}$$

the Navier-Stokes-equations for the velocity field **v** reduce to

$$\partial_t v_\alpha = \nu \Delta v_\alpha, \quad \alpha = 1, 2, 3 \tag{5.32}$$

They are complemented by the continuity equation **(ρ=const)** (summation convention)

$$\partial_\alpha v_\alpha = 0, \tag{5.33}$$

the boundary conditions

a) on fixed surfaces:

$$\mathbf{v} = 0 \tag{5.34}$$

b) on the surface of the pendulum

$$\mathbf{v} = r\dot{\alpha}(t)\,\mathbf{\Phi} \tag{5.35}$$

and the initial condition

$$\mathbf{v}\,(r, z, t = 0) = 0 \tag{5.36}$$

Now in view of (5.26), (5.27) and (5.35) one can expect a linear nonlocal or integral relation between M(t) and $\dot{\alpha}(t)$ to exist. This indeed is the case [5.1]. Taking Laplace transforms of this relation and also of the EOM (5.18), we get two linear algebraic equations from which both the Laplace transform of the torque (M(t)) and the moment of inertia (J) of the pendulum can be calculated if the amplitude of the pendulum ($\alpha(t)$) has been determined experimentally. As the relaxational motions of the pendulum for small amplitudes ($\alpha \leq 30^o$) can be described by a damped harmonic oscillation, we have

$$\alpha(t) = \alpha_0 e^{-\Delta_E \omega_E t}\left(\cos(\omega_E t) + \Delta_E \sin(\omega_E t)\right), \qquad 5.37$$

the index "E" at the frequency (ω_E) and the decrement (Δ_E) of the oscillation indicating "experimentally determined". By this reasoning we get for the ratio of the moments of inertia J and J_0 the expression [5.1, 5.2]:

$$\frac{J}{J_0} = \frac{1+\Delta_0^2}{1+\Delta_E^2}\left(\frac{\omega_0}{\omega_E}\right)^2 \qquad 5.38$$

Hence we have for (Ω_{osc}) in view of (5.17), (5.25) and (5.38)

$$\Omega_{osc} = \frac{\dfrac{1+\Delta_0^2}{1+\Delta_E^2}\left(\dfrac{\omega_0}{\omega_E}\right)^2 - 1}{1 - \dfrac{\Delta_0 \omega_0}{\Delta_L \omega_L}} \qquad 5.39$$

Thus we recognize that in order to determine the relative increase (Ω_{osc}) of the mass of the pendulum due to gas adsorption as defined by the l.h.s. of (5.25), three types of measurements are necessary leading to the characteristic quantities of

- the motion of the empty disk in vacuum (ω_L, Δ_L),
- the motion of the disk filled with sorbent in vacuum (ω_0, Δ_0),
- the motion of the disk filled with sorbent in (m^s) in sorptive gas (ω_E, Δ_E).

Once (Ω_{osc}) is known from Eq. (5.39), the mass adsorbed (m^a) can be calculated from (5.25). The sorptive gas mass (m^f) included in Ω_{osc} for pellet-like sorbents has to be calculated via the helium-volume (V^s_{He}) of the sorbent material from the relation

$$m^f = \rho^f(p,T)(V^* - V^s_{He}), \qquad 5.40$$

with V^* being the ring slit volume

$$V^* = \pi(R_a^2 - R_i^2)d \tag{5.41}$$

and $\rho = \rho\,(p,T)$ indicating the sorptive gas density.

The effective mass (Δm^f) of the boundary layer of the pendulum occurring during its motion has to be determined by calibration experiments using non-adsorbing pellets – preferably quartz glass spheres – having a similar surface as the sorbent pellets [5.2].

Combining equations (5.25), (5.40) we have for the Gibbs excess mass adsorbed

$$m_{GE}^a = \Omega_{osc} m^s - \rho^f (V^* - V_{He}^s) - \Delta m^f \tag{5.42}$$

with (Ω_{osc}) to be calculated from oscillations measurement by (5.39) and the density (ρ^f) to be determined from (T, p)-Measurements using the EOS of the sorptive gas.

Applying the formalism presented above to measurements performed with non-adsorbing material (m^g), for example glass spheres or cylinders of the size of the original adsorbent pellets, leads similar to (5.42) to the relation

$$0 = \Omega_{osc}^g m^g - \rho^f (V^* - V^g) - \Delta m^f \tag{5.43}$$

which allows to write m_{GE}^a as

$$m_{GE}^a = \Omega_{osc} m^s - \Omega_{osc}^g m^g + \rho^f (V_{He}^s - V^g) \tag{5.44}$$

Here (V^g) is the net volume of the glass material of mass (m^g). It most easily can be measured using buoyancy effects at a microbalance. The experimental quantity (Ω_{osc}) again has to be calculated from Eq. (5.39) referring now to measurements with the pendulum filled with non-adsorbing material (glass) [5.2, 5.7]

Auxiliary remarks:

1. The slow motion/laminar flow condition (5.28) does not hold in the low pressure region ($p \to 0$) as then $\nu \to 0$ and hence $Re \to \infty$. In

practice we experienced that for CH_4 as sorptive gas at T = 300 K for pressures $p \leq 0.5$ MPa oscillations become irreproducible which probably is mainly due to turbulent secondary flows initiated by the motion of the pendulum. To avoid theses, it is recommended to place thin plates above and below the pendulum, cp. Fig. 5.4.

2. The analytic method leading to Eq. (5.38) in principle also allows one to determine the kinematic viscosity ($\nu = \eta/\rho$) of the sorptive fluid [5.1, 5.2]. This can be of interest if gas mixtures are used for which viscosity data often are scarce.

3. To avoid certain difficulties with the rational pendulum, cp. Sect. 4.2, it should be mentioned that on principle the pendulum can be substituted by a floating rotator. By this we understand a cylinder rotating freely, i. e. floating in either vacuum or a gaseous atmosphere within another hollow cylinder and bearing on top a permanent magnet coupled to a magnetic suspension, cp. Chaps. 3, 4, and at its bottom a bowl filled with sorbent material [5.4], Fig. 5.5.

An instrument of this type has been designed a couple of years ago for gas viscosity measurements [5.5, 5.6]. After initializing rotator's motion electromagnetically, a rotational relaxation motion of the rotator represented by a sequence of time intervals t_n, n = 1, 2, 3... needed for n = 1, 2, 3... rotations can be observed.

This motion can be represented by its angular velocity $\dot{\alpha}(t)$ which is

- for vacuum

$$\dot{\alpha}(t) = \dot{\alpha}_0 \, e^{-t/T_0} \tag{5.45}$$

- for a gaseous atmosphere

$$\dot{\alpha}(t) = \dot{\alpha}_0 \, e^{-t/T_E} \tag{5.46}$$

with T_0, T_E being characteristic relaxation times to be determined from the respective sets (t_n) via a Gaussian minimization procedure. The parameters T_0, T_E depend – among other quantities – on the moment of inertia of the rotator and hence of its (cylindrically symmetric distributed) mass. By analogous reasoning as for the pendulum, one can present the mass ratio

$$\Omega_{rot} = \frac{m^a + m^f + \Delta m^f_{rot}}{m^s} \qquad 5.47$$

as

$$\Omega_{rot} = \frac{\left(1+(\eta/\hat{\eta})\widehat{F}_z\right)T_E - T_0}{T_0 - T_L} \qquad 5.48$$

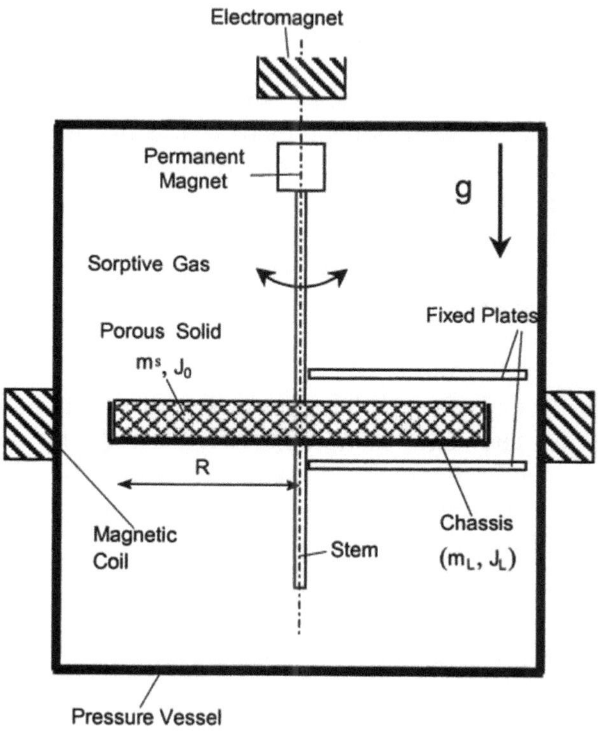

Figure 5.5. Floating rotator.

Here (m^a) is the mass adsorbed on the sorbent mass (m^s), m^f is the gas mass between the sorbent pellets moved along with the sorbent and Δm^f_{rot} is the "equivalent" mass of gas in a boundary layer on the surface of the rotator. In Eq. (5.48) the various relaxation times refer to relaxational motions of the

- empty rotator without sorbent material in Vacuum (T_L),
- loaded rotator with sorbent material (m^s) in vacuum (T_0) and
- loaded rotator in sorptive gas atmosphere (T_E).

Also η is the dynamic viscosity of the sorptive gas and $\hat{F}_z$ is a characteristic parameter of the rotator - sorbent system to be determined from calibration measurements with a non-adsorbing gas, preferably helium or argon, via the equation

$$\hat{F}_z \cong \hat{F}_{zHe} = \frac{T_0}{T_{EHe}} - 1 \geq 0, \qquad 5.49$$

T_{EHe} indicating the rotator's relaxation time in helium in a thermodynamic state with dynamic viscosity $(\hat{\eta} = \hat{\eta}_{He})$.

If (m^f) has been determined from helium measurements and (Δm^f_{rot}) is known from calibration measurements with a dense, i. e. non-adsorbing loading material in the bowl, m^a can be calculated from Ω_{rot} via (5.47).

We finally would like to mention that for a non-adsorbing dense rotator the viscosity of the surrounding gas can be calculated from relaxational time measurements by the formula

$$\eta = \hat{\eta} \frac{T_0 - T_E}{T_0 - \hat{T}_E} \bullet \frac{\hat{T}_E}{T_E} \qquad 5.50$$

Here $(\hat{\eta}, \hat{T}_E)$ indicate the dynamic viscosity and the relaxation time of rotator's motion in an arbitrary reference state [5.4].*)

2.3 Uncertainties or Errors of Measurement

We here present a formula allowing one to calculate numerically the dispersions or mean square deviation (MSD) of Gibbs excess masses of adsorbates (σ^2_{mGE}) which have been measured oscillometrically and calculated by Eq. (5.44). The Gauss law of error propagation leads to the expressions [5.17, 5.18]:

*) For more details and actual information on rotator-viscometers the interested reader should contact Rubotherm Company at Bochum and / or the Institute of Thermo- and Fluiddynamics at the University of Bochum, Bochum, Germany.

$$\sigma_{mGE}^2 = (m^s)^2\sigma_{\Omega osc}^2 + \Omega_{osc}^2\sigma_{ms}^2 + (m^g)^2\sigma_{\Omega oscg}^2 + (\Omega_{osc}^g)^2\sigma_{mg}^2 + (V_{He}^s - V^g)^2\sigma_{\rho f}^2 + \rho^{f2}(\sigma_{VsHe}^2 + \sigma_{Vg}^2) \qquad 5.51$$

The dispersions $(\sigma_{\Omega osc}, \sigma_{\Omega oscg})$ have to be calculated via Eq. (5.39) from experimental uncertainties $(\sigma_\Delta, \sigma_\omega)$ of the logarithmic decrement (Δ) and of the frequency (ω) of the various oscillations. In the scheme below all parameters (x) and related dispersions (σ_x) together with numerical data of their ratio (σ_x/x) according to our laboratory experience are listed [5.7, 5.16].

x	Δ	ω	Ω_{osc}	Ω_{osc}^g	m^s	m^g	ρ^f	V_{He}^s	V^g
σ_x	σ_Δ	σ_ω	$\sigma_{\Omega osc}$	$\sigma_{\Omega oscg}$	σ_{ms}	σ_{mg}	σ_{ρ}^{f}	σ_{VsHe}	σ_{Vg}
(σ_x/x)	$3\cdot10^{-3}$	$5\cdot10^{-5}$	$5\cdot10^{-3}$	$5\cdot10^{-3}$	10^{-4}	10^{-4}	10^{-3}	10^{-3}	10^{-3}

These numerical values of dispersions (σ_x) provided one can expect relative uncertainties of Gibbs excess masses measured with the rational pendulum to be limited by $(\sigma_{mGE} / m_{GE}^a) \leq 2\%$. This is about the accuracy of standardized volumetric / manometric measurements, cp. Chap. 2. To realize accurate oscillometric measurements it is most important to determine the (Δ) and (ω) parameters related to the various oscillations of the pendulum as accurate as possible, cp. Eq. (5.39). For this it is necessary to choose an initial amplitude (α_0) of about 2-3 times the size of (α_1, α_2) but less than 60 ° as for higher values of (α_0) measurable deviations of the oscillation from ideal harmonic behavior occur. Also it is necessary to observe at least 10 periods within an oscillation if not more, to get reliable (Δ, ω)-data. To ensure this, the use of diodes with characteristic detection times less than 0.1 ms is recommended [5.7].

2.4 Examples

The rotational pendulum depicted in Fig. 5.1 has been used to perform gas adsorption equilibria measurements on a variety of sorbent materials and sorptive gases. Some geometric and physical parameters of the instrument are given in Table 5.1, cp. also Fig. 5.4.

We here present only adsorption measurements of He and CO_2 on activated carbon (AC) Norit R1 Extra at 293 K. Physico-chemical parameters of this sorbent material have been given in Chap. 3, Sect. 2.3. Also, some porosity parameters referring to pellets and powder of this AC are given below [5.24].

Figure 5.6 shows a top view of the ring slit on the disk of the pendulum filled with activated carbon Norit R1 Extra powder. Uniform distribution of the powder within the slit is mandatory to get reproducible results of oscillometric measurements.

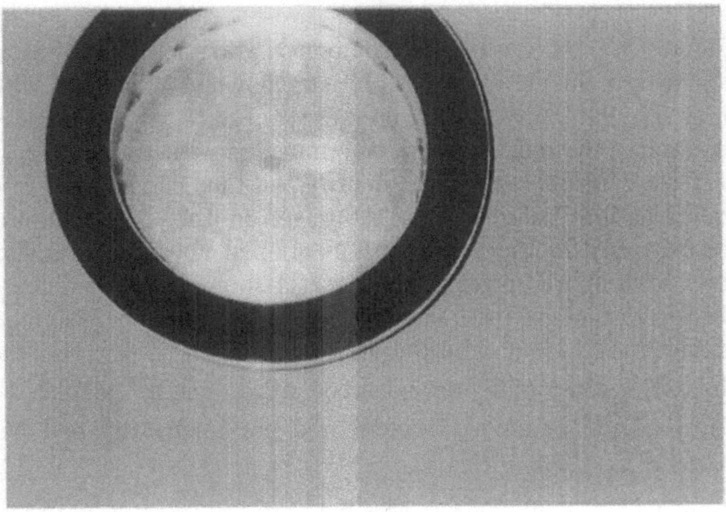

Figure 5.6. Ring slit of pendulum (R_i=3.75cm, R_a=5.5cm) filled with activated carbon powder (Norit R1 Extra).

Table 5.1. Geometrical and physical parameters of the rotational pendulum Figure 5.1.

Disk		Wire		
Radius	R_a = 75 mm	Length	L_w	= 369 mm
Height	d = 11 mm	Diameter	$2r_w$	= 0.125 mm
Material	Al	Material	FeCrNi	188
Distance between disk and fixed plates			b	= 5 mm
Empty Disk Oscillation in Vacuum:				
Moment of inertia	J_o = $4.38 \cdot 10^{-4}$ kg m^2			
Period of oscillation*)	T_o = 62.8 s			
Angular frequency *)	ω = $0.1001 \cdot 10^{-4}$			

*) These parameters do not depend on the remnant gas pressure (p) in the instrument provided $p < 10^{-3}$ mbar = 0.1 Pa. However, they are sensitive to temperature. Values given here refer to 293 K.

Table 5.2. Porosity parameters of AC Norit R1 Extra [5.2, 5.7, 5.24].

BET-surface area	1,407 m^2/g
	cm^3/g
Helium volume	0.3511
Macropore volume	0.47
Micropore volume	0.616
Bulk volume of pellets	2.72
Bulk volume of powder	1.90

In Figure 5.7 the results of combined gravimetric and oscillometric adsorption measurements of Norit R1 Extra powder in contact with helium with purity 99.9990 % ≡ 5.0 for pressures up to 4 MPa at T = 293 K are sketched. Data of the reduced mass determined gravimetrically (Ω_{grav}) show a linear decrease with increasing gas density and thus can be used to calculate the so-called helium volume of the AC given in Tab. 5.2, assuming helium not to be adsorbed at the conditions mentioned above. The reduced mass gained by oscillometric measurements (Ω_{osc}) shows a nearly linear increase due to the increase of gas mass included in the free space between the powder particles and moved along with the oscillations of the disk. From these data, the Gibbs excess mass of helium adsorbed m^a_{GEosc} can be calculated by using the helium volume of the AC determined gravimetrically and leading by definition to $m^a_{GEgrav} = 0$.

As can be seen the m^a_{GEosc}-data deviate only slightly, i. e. within experimental uncertainties which are about three times the size of the graphical symbols used, from the abscissa axis. Hence both measurement methods lead to the same experimental results thus proving consistency of oscillometric measurements, i. e. the key equation (5.39).

In Figure 5.8 data of gas adsorption equilibria of carbon dioxide (CO_2) on activated carbon Norit R1 Extra taken at 293 K for pressures up to 7 MPa (equivalent to gas densities of ca. 140 kg/m^3) are presented. Data have been taken by both oscillometric and gravimetric measurements [5.2]. As the adsorption capacity of the carbon at high pressures (> 5 MPa) is approaching saturation, the reduced masses (Ω_{grav}, Ω_{osc}) of both types of measurement can be interpolated by straight lines, cp. Fig. 5.8. These reflect the fact that for gravimetric measurements at adsorption saturation the buoyancy of the sorbent-sorbate system increases linearly with increasing gas density reducing the weight by ($-\rho^f V^S$) whereas for oscillometric measurements the mass of gas (m^f) moved along with the pendulum in the interstitial space (V^f) between the pellets increases linearly by ($\rho^f V^f$) and the mass of the boundary layer surrounding the pendulum at high pressures – according to calibration measurements – nearly remains constant [5.7, 5.10].

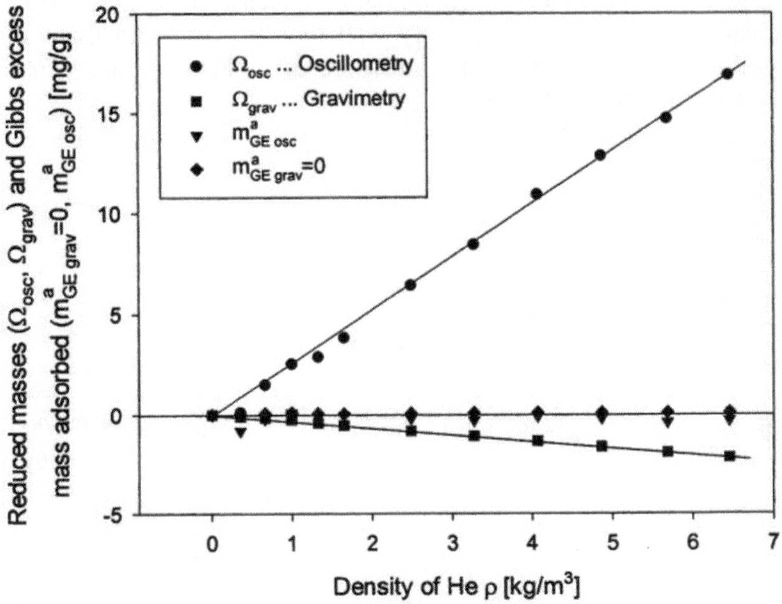

Figure 5.7. Reduced masses (Ω_{osc}, Ω_{grav}) resulting from oscillometric and gravimetric adsorption measurements of He on activated carbon (Norit R1 Extra) at 293 K. Gibbs excess masses adsorbed ($m^a_{GEgrav} \cong 0$, m^a_{GEosc}) are calculated from (Ω_{osc}, Ω_{grav})-data.

In Figure 5.8 also data of the Gibbs excess mass adsorbed (m^a_{GE}) calculated from gravimetric measurements (Ω_{grav}) are depicted. These data show a slight decrease with increasing sorptive gas density at high density values indicating (again) problems in interpreting thermodynamically Gibbs excess masses at high gas pressures, cp. Chapter 1. The absolute masses of CO_2 adsorbed (m^a_{grav}, m^a_{osc}) have been calculated from (Ω_{grav}, Ω_{osc}) data using Eqs. (2.9, 2.10). They show a monotonous increase with increasing sorptive gas density and coincide remarkably well thus proving – in this example – that oscillometric and gravimetric measurements lead (within experimental uncertainties) to identical results.

In Figure 5.8 also data of the Gibbs excess mass adsorbed (m^a_{GE}) calculated from gravimetric measurements (Ω_{grav}) are depicted. These data show a slight decrease with increasing sorptive gas density at high density values indicating (again) problems in interpreting thermodynamically Gibbs excess masses at high gas pressures, cp. Chapter 1. The absolute masses of CO_2 adsorbed (m^a_{grav}, m^a_{osc}) have been calculated from (Ω_{grav}, Ω_{osc}) data using

Eqs. (2.9, 2.10). They show a monotonous increase with increasing sorptive gas density and coincide remarkably well thus proving – in this example – that oscillometric and gravimetric measurements lead (within experimental uncertainties) to identical results.

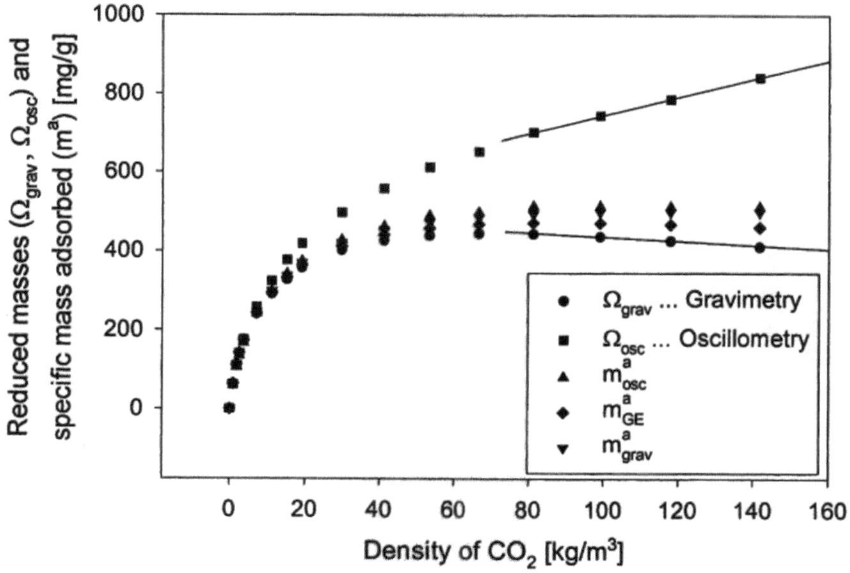

Figure 5.8. Reduced masses resulting from oscillometric and gravimetric adsorption measurements (Ω_{osc}, Ω_{grav}) of CO_2 on activated carbon (Norit R1 Extra) at 293 K. Gibbs excess masses (m^a_{GE}) and absolute masses adsorbed (m^a_{osc} , m^a_{grav} -data).

3. OSZILLOMETRIC - GRAVIMETRIC MEASUREMENTS OF GAS ABSORPTION IN SWELLING MATERIALS

3.1 Introductory Remarks

Polymeric materials can absorb considerable amounts of gas, for example CO_2 especially at elevated pressures (p) and temperatures (T) above the so-called glass transition temperature [5.19]. This often causes changes in size and volume of the polymer, which have to be taken into account in industrial processing situations, for example in gas separation processes using polymeric sorbent materials [5.20]. Sorption phenomena of swelling polymers cannot be measured adequately by either gravimetric or volumetric methods, since the volume of the material needed for buoyancy or dead space corrections respectively, depends not only on pressure and temperature of the sorptive gas but also on the amount of gas absorbed. Today there are several techniques described in literature for measuring polymeric swelling phenomena [5.21-23], usually determining the change in one or two dimensions of a polymer sample. However, these methods require a specific shape or state of the polymeric material, i. e. a thin film or a liquid drop and, despite the geometry, often assume isotropic swelling of the polymer sample.

By combining oscillometric and gravimetric measurements it is possible to simultaneously determine the mass of gas (m^a) absorbed in a polymeric material of mass (m^s) and the volume (V) of the sample material in the absorption state thus providing together with (p) and (T) the basic data for the thermal equation of state of the polymer loaded with gas [5.8], [5.26]. The polymer samples which can be investigated in this way can be either highly viscous liquids or dense solid materials formed as cylindrical rings, pulverous flakes or pellets. Isotropic swelling is not mandatory for the method to be applied.

In this section we first will describe the experimental device, Sect. 3.2, then provide formulas to calculate m^a and V^{as} of a polymer sample from measured data Sect. 3.3, and finally present an example namely sorption of CO_2 in (swelling) polycarbonate (Makrolon 2400), Sect. 3.4.

3.2 Experimental

An instrument for combined oscillometric-gravimetric measurements of sorption equilibria in swelling (polymeric) materials has been designed and built at IFT during 1997-2000, [5.2, 5.7, 5.8]. It mainly consists of the rotational pendulum already described in Sect. 2.1 and a microbalance installation as presented in Chap. 3, Sect. 2.1.1. A schematic diagram of the instrument is given in Figure 5.9 followed by photos of the ring slit of the pendulum filled with polycarbonate pellets, Fig. 5.10 and a front view of the installation in our laboratory (PB-A 0126/1) in 1998, Fig. 5.11. Detailed information of the instrument including specifications of vessels, tubes, valves, and materials are given in the literature [5.2, 5.7] and references cited, cp. [Tom94], [Sch99a].

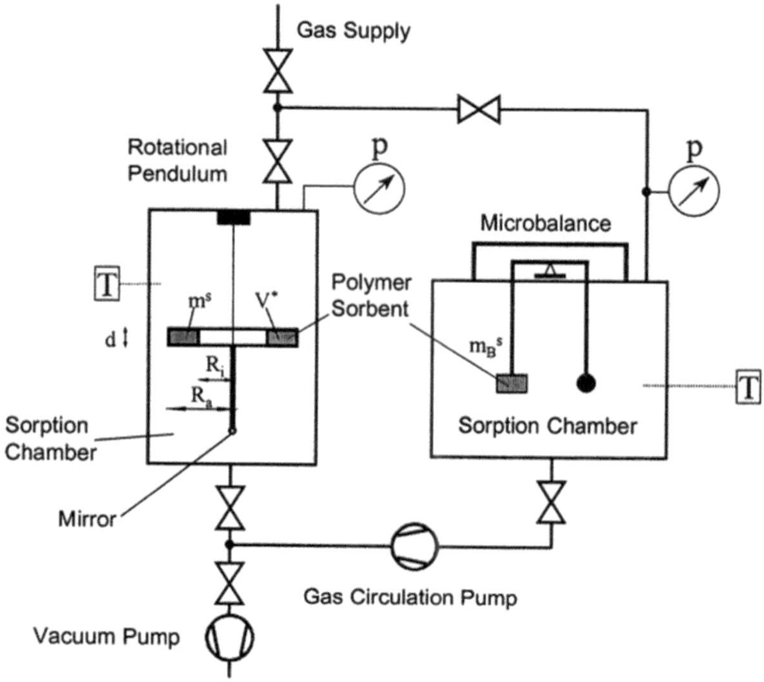

Figure 5.9. Experimental setup for oscillometric-gravimetric measurements.

Figure 5.10. Ring-slit of pendulum (diameter: 15cm) filled with pellets of polycarbonate (Makrolon 2400).

Combined oscillometric and gravimetric measurements provide a basis to determine simultaneously the mass and the volume of a swelling sorbent material like polymers or resins in a sorptive gas atmosphere. However measurements seem in practice to be restricted to determinations of sorption equilibria of these materials as the kinetics of mass uptake often is very slow. To give an example we mention that in case of sorption of CH_4 on pellets of Makrolon 2400 at 35 °C, p = 2 MPa, it took more than 4 days till equilibrium was reached.

Figure 5.11.
Rotational Pendulum for combined oscillometric-gravimetric sorption measurements of gases in swelling (polymeric) materials. © IFT University of Siegen, 1999.

According to our experience, data of oscillometric measurements were highly reproducible, but were more sensitive to changes in temperature than gravimetric measurements. Relative uncertainties in oscillometry are comparable to those in gravimetry. The pendulum can be used in a fairly large range of pressure and temperature comparable to that of commercial thermogravimetric instruments available today. For proper design the condition that the disk of the pendulum must be "thick" to make sure inequalities (5.29) hold, should be observed. Otherwise, the pendulum cannot be used for low gas pressures, i. e. p < 1 MPa for our instrument [5.7]. Choice of sealing material (teflon, gylon, silver) turned out to be most important and for a while has been a source of frustration.

Also it should be mentioned that by adding a gas chromatograph or a mass spectrometer to the instrument, Fig. 5.9, cosorption equilibria of gas mixtures in swelling sorbents can be measured. The theory of measurements of this type is based on the theory of oscillometric-gravimetric sorption measurements of pure gases, Sect. 3.3. In addition concentrations of the sorptive gas originally supplied to the system (y_i^*) and those in sorption equilibrium (y_i, i=l...N) have to be taken into account.

Finally we would like to point out that by combined oscillometric-gravimetric measurements of adsorption equilibria in rigid, i. e. non-swelling materials the volume of the sorbent-sorbate system (V^{as}) in the limit of saturation can be determined. An example for this is given in Fig. 5.8 referring to the adsorption of CO_2 on AC Norit R1 Extra at 293 K. Here V^{as} can be calculated bydifferentiating **Ω_{osc}-data** to the gas density, i. e. by the slope of the straight line interpolating these data [5.7].

3.3 Theory

In this section we will provide the reader with the basic equations allowing one to calculate from combined oscillometric and gravimetric measurements both the mass of gas (m^a) adsorbed or absorbed in a swelling sorbent material of mass (m^s) and its volume (V^{as}) in the sorbate state at given pressure (p) and temperature (T) of the sorptive gas and sorbent material. We start by mentioning the result of oscillometric measurements, cp. Sect. 2.2, namely the relation (5.25)

$$\Omega_{osc} = \frac{1}{m^s}(m^a + m^f + \Delta m^f) \qquad 5.52$$

Here the (dimensionless) quantity Ω_{osc} is calculated from 3 oscillometric experiments via equation (5.39). On the r.h.s. of(5.52) m^a indicates the mass of gas sorbed within the mass of sorbent (m^s), m^f is the mass of gas included in the interstitial volume (V^f) of gas between the sorbent pellets which also is moved along with the pendulum, cp. Fig. 5.4, i. e. we have

$$m^f = \rho^f V^f \qquad 5.53$$

with

$$V^f = V^* - V^{as} \tag{5.54}$$

Here (ρ^f) is the density of the sorptive gas, V^* is the geometric volume of a ring slit including the sorbent material in the swelling state and (V^{as}) is a (still unknown) volume of the sorbent / sorbate mass ($m^s + m^a$) in the same state. The mass (Δm^f) belongs to a surface boundary layer of the sorptive gas outside the volume V^* and also moved along with the pendulum. This quantity can be determined from either measurements with non-sorbing (dummy) pellets made of glass, or from combined oscillometric-gravimetric measurements with a non-swelling, i. e. rigid sorbent material [5.2, 5.7]. Data of this quantity have been determined at our pendulum for different gases in a certain range of pressure and temperature and easily can be correlated and / or interpolated [5.25]. However, these quantities naturally depend on the geometry of the pendulum used and hence are not physical parameters related to the sorbent / sorptive system. Hence we do not present them here.

For swelling processes of polymeric spherical or cylindrical pellets the ratio (b) of the pellet volume (V^{as}) to the bulk volume (V^*) often remains constant.*) As this ratio easily can be determined by He-pycnometer measurements in an unswollen reference state, i. e. at ambient conditions, we have

$$b = \frac{V^{as}}{V^*} = \frac{V_0^s}{V_0^*} \tag{5.55}$$

Here (V_0^s) is the volume of the sorbent pellets of mass (m^s) in the reference state and (V_0^*) their bulk volume in the same state which by the way in principle can be chosen arbitrarily. To give an example we mention that for simple cubic packing of N^3 spherical pellets of radius r in a cubic box of side length (2rN) we have

$$b = \frac{4\pi}{3} r^3 N^3 / (2rN)^3 = \frac{\pi}{6} = 0.524 \tag{5.56}$$

Note that in this case – as in many others – b does not depend on the radius (r) of the pellets.

*) This does not hold in case that pellets due to sorption of gas undergo plastic deformations or even melt, then forming a homogenous phase with no interstitial volume.

Introducing the ratio of volumes (b) into equation (5.54), the volume (V^f) of the interstitial gas phase at the pendulum can be rewritten as

$$V^f = \left(\frac{1}{b} - 1\right) V^{as} \tag{5.57}$$

From gravimetric measurements – using preferably a magnetic suspension balance – one can determine the quantity

$$\Omega_{grav} = m_B^s + m_B^a - \rho^f V_B^{as} \tag{5.58}$$

Here m_B^s is the mass of sorbent material included in the basket at the magnetic suspension, cp. Figs. 3.4, 4.12, 5.9, m_B^a is the mass of gas sorbed in (m_B^s) and V_B^{as} is the volume of the sorbent / sorbate mass($m_B^s + m_B^a$) in the swollen state. To facilitate rotation we introduce the ratio (γ) of sorbent masses at the balance (m_B^s) and the pendulum (m^s) defined by

$$m_B^s = \gamma m^s \tag{5.59}$$

Due to the extensivity property of (m^a) and (V^{as}) we have also

$$m_B^a = \gamma m^a \tag{5.60}$$

$$V_B^{as} = \gamma V^{as} \tag{5.61}$$

In view of (5.59-5.61) equation (5.58) can be rewritten as

$$\Omega_{grav} = \gamma(m^a + m^s - \rho^f V^{as}) \tag{5.62}$$

Similarly we have from (5.52) and (5.57)

$$\Omega_{osc} m^s = m^a + \left(\frac{1}{b} - 1\right) \rho^f V^{as} + \Delta m^f \tag{5.63}$$

Equations (5.62, 5.63) are two linear equations for m^a and V^{as} having the solutions

$$m^a = \frac{1}{\gamma}(1-b)\Omega_{grav} + \left[(1+\Omega_{osc})b - 1\right]m^s - b\Delta m^f \qquad 5.64$$

$$V^{as} = \frac{b}{\rho^f}\left((1+\Omega_{osc})m^s - \frac{1}{\gamma}\Omega_{grav} - \Delta m^f\right) \qquad 5.65$$

These relations allow one to calculate (m^a, V^{as}) from the experimental quantities (Ω_{osc}, Ω_{grav}), known parameters (γ, b), sorbent mass (m^s), sorptive gas density (ρ^f), and sorbent and instrument related gas boundary layer mass (Δm^f). Actually, this later quantity can be determined from measurements with non-sorbing pellets ($m^a = 0$) from equation (5.64) or measurements using non-swelling sorbents ($V^{as} = V_0^s$) from equation (5.65).

Calculations of uncertainties of (m^a, V^{as})-data are straightforward starting from eqs. (5.64, 5.65). Numerical examples have shown that accurate measurements at the pendulum, cp. eq. (5.39), are most important to get small dispersions of (m^a) and (V^{as}), the relative values of these quantities (σ_{ma} / m^a), (σ_{Vas} / V^{as}) being normally between 1 % and 5 %, [5.16], [5.17].

3.4 Example

The solubility of carbon dioxide in a special polycarbonate (PC) (Goodfellow, UK) has been investigated experimentally by oscillometric-gravimetric measurements at T = 293 K for pressures up to 6 MPa, [5.7, 5.8].

In Table 5.3 some thermophysical properties of the PC are listed together with the respective data of polymethylmethacrylate (PMMA) (Makrolon 2400) which will be discussed later in Sect. 4.4, [5.19, 5.28, 5.29]. The table is followed by a sketch of the chemical structure of the basic elements of the PC and PMMA.

In Figure 5.12 results of measurements are presented. The oscillometric data (•) show a monotonic increase with the gas pressure (p), while the gravimetric data (■) are strongly curved, clearly indicating the influence of buoyancy. At pressures above 5 MPa the microbalance reading becomes even negative, i. e. the effect of buoyancy becomes larger than the increase in mass due to absorption of CO_2.

Tab. 5.3. Thermophysical data of polycarbonate (PC) (Goodfellow, UK) and polymethyl-methacrylate (PMMA) (Makrolon 2400), [5.7].

	PC	PMMA
Bulk volume $\left(V_0^s / m^3 / kg\right)$	0.83	0.84
Molar mass $\left(\overline{M} / g / mol\right)$	35,000	120,000
Diameter of pellets (d/mm)	3	0.6
Glass transition temperature (T_G/°C)	149	100
Melting temperature (T_m/°C)	267	160

PC

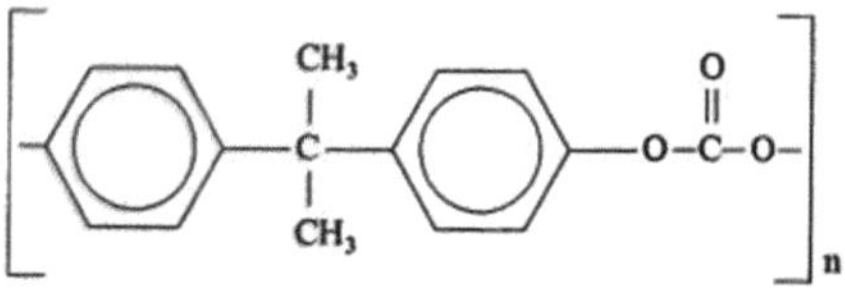

PMMA

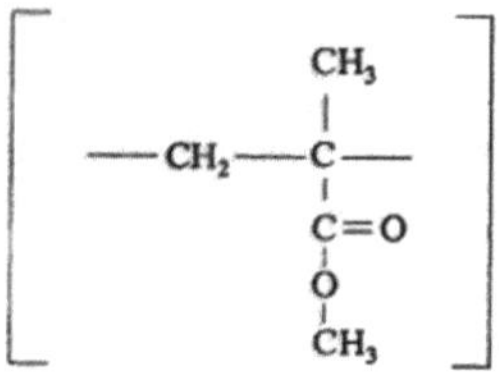

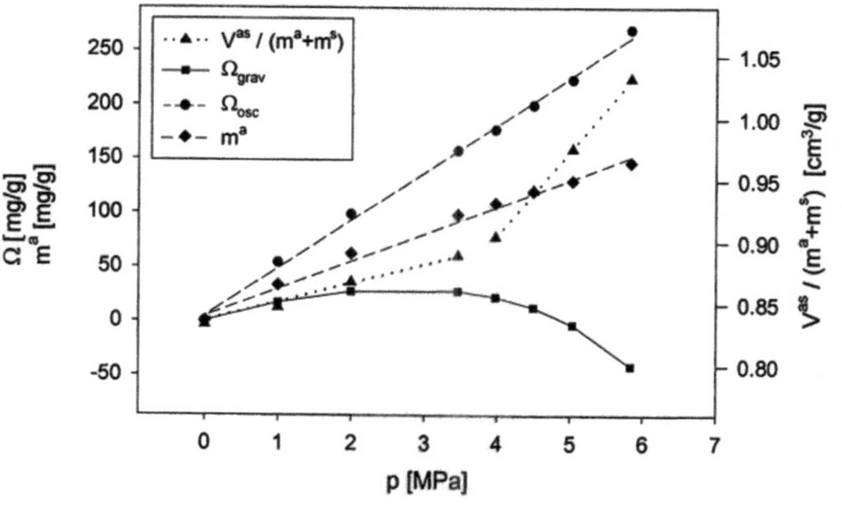

Figure 5.12. Swelling and sorption isotherm of polycarbonate / CO_2 at 293 K.

The swelling isotherm (▲), i. e. the (specific) data of the volume of the PC-pellets as a function of the gas pressure (and temperature) exhibits two almost linear branches intersecting at about 4 MPa. At lower pressures ($p < 4$ MPa) the specific volume (Vas/(ms + ma)) increases only slightly with the CO_2-load. At higher pressures ($p > 4$ MPa) an increase of the steepness of the isotherm can be observed, this possibly indicating a phase change transition (glass transition) due to sorption of CO_2. The sorption isotherm (♦) determined from the (Ω_{osc}, Ω_{grav})-data indicates a simple Langmuir type behavior. Equilibration times ranged from 4 hours at low gas pressures ($p < 1$ MPa) up to 24 hours at high pressures ($p = 6$ MPa) approaching saturation of the subcritical CO_2 gas. Relative uncertainties of all data are about 3 %. More accurate measurements seem to be possible but will be laborious and very sensitive to disturbances.

4. OSCILLOMETRIC – MANOMETRIC MEASUREMENTS OF GAS ABSORPTION IN SWELLING MATERIALS

4.1 Introductory Remarks

In this section we will discuss in brief another method to measure gas solubilities in swelling materials like polymers or resins. It consists of a combination of oscillometric measurements using a rotational pendulum and a gas expansion experiment, i. e. volumetric or manometric measurements of the sorptive gas. Hence this procedure can be called the oscillometric-manometric method [5.27]. The main advantage of this method compared to oscillometric-gravimetric measurements is that it does not need a microbalance which often includes sophisticated mechanical and electronic elements. Instead only a storage vessel for the sorptive gas is needed, the pressure vessel surrounding the oscillating disk of the pendulum serving also as sorption vessel, cp. Fig. 5.13 and also Fig. 2.1. Hence measurements of this type may be easily automated. However, they only will lead to less accurate results compared to those of oscillometric-gravimetric measurements, as manometric measurements are less accurate than gravimetric [5.16, 5.27]. Still, results can be very useful for industrial purposes, namely providing basic data for processes like pre-swelling, dying, or impregnating polymers with supercritical gases like carbon dioxide [5.31, 5.32].

In what follows we first will briefly describe the experimental device, Sect. 4.2, then the basic equations are given from which the mass (m^a) of gas sorbed in a sorbent of mass (m^s) and the volume (V^{as}) of the sorbent/sorbate system can be calculated, Sect. 4.3. In Sect. 4.4 we present as an example data (p, T, m^a, V^{as}) for CO_2 sorbed in polymethylmethacrylate (PMMA) (Makrolon 2400) at 308 K for pressures up to 6 MPa, [5.27].

4.2 Experimental

Figure 5.13 shows a schematic diagram of an installation designed for oscillometric-manometric measurements of gas solubilities in swelling materials [5.27]. The instrument consists of a rotational pendulum complemented by a gas storage vessel and – for measurement of cosorption equilibria with gas mixtures – a gas chromatograph. In addition a system for gas supply and evacuation of the apparatus as well as thermometers, manometers, a gas circulation pump and thermostats are needed. These elements are connected by proper tubing and valves as indicated in the figure. Stainless steel of high quality including electropolishing of all inner surfaces to reduce and stabilize wall adsorption is highly recommended. More details of the instrument and especially of the pendulum are given in [5.2, 5.7, 5.27].

Figure 5.14 shows a photo of a training instrument designed, built, and operated at IFT, University of Siegen, during 1999-2001. The pendulum (left) is covered by a plexiglass vessel allowing observation of the rotational oscillations of the pendulum. It is coupled not only to a gas storage vessel but also to a microbalance (center) and an impedance analyzer (right), cp. Chap. 6, to perform various kinds of combined measurements. This simple instrument allowed also to measure gas adsorption in nearly rigid non-swelling porous materials like zeolite or activated carbon. Results of test measurements were identical within experimental uncertainties with those gained by traditional manometric or gravimetric measurements.

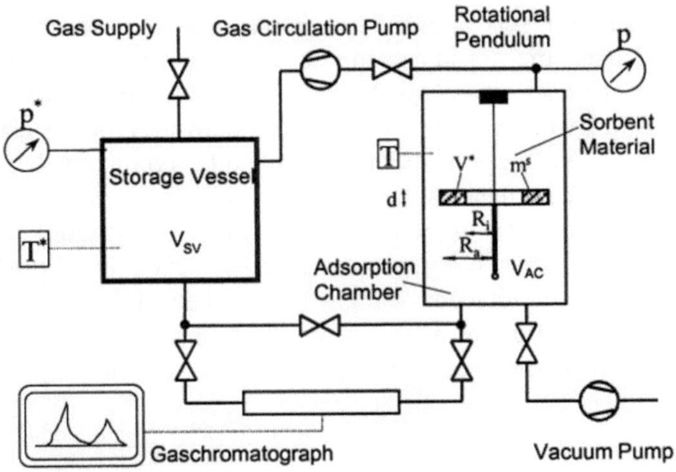

Figure 5.13. Experimental setup for oscillometric-manometric measurements.

Figure 5.14. Training instrument for oscillometric, volumetric, gravimetric, and dielectric measurements of gas adsorption equilibria in rigid and swelling sorbent materials. The pendulum (left) is covered by a plexiglass vessel allowing direct optical observations of its rotational oscillations.
© IFT University of Siegen, 1999.

As the pressure measurements in oscillometric-manometric experiments only are needed to calculate the density of the sorptive gas ($\rho^f = \rho^f$ (p, T)) in a sorption equilibrium state, it seems to be worthwhile to consider direct measurements of (ρ^f), for example by using the buoyancy of a sinker at a magnetic suspension balance. Experiments of this type can be called "oscillometric-densimetric measurements". They can be recommended to measure solubilities of gas mixtures in swelling (polymeric) sorbing materials. A scheme for such measurements is sketched in Figure 5.15. Equations to calculate the total mass of gas (m^a) sorbed in the sorbent mass (m^s) and the volume (V^{as}) of the mass (m^a + m^s) are the same as those given for oscillometric-manometric measurements in the next section (4.3).

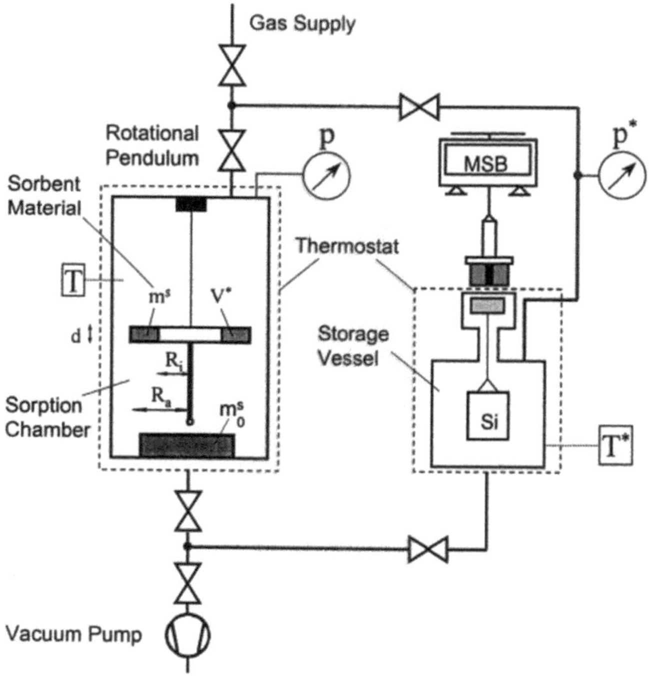

Figure 5.15. Experimental setup for oscillometric-densimetric measurements. MSB...magnetic suspension balance, Si...sinker for gas density measurements by buoyancy effect.

We also would like to draw reader's attention to the possibility to combine oscillometric, gravimetric, and manometric measurements in a single experiment. The schematic diagram of such an instrument is given in Figure 5.16. It basically consists of a rotational pendulum, a microbalance (two beams as indicated in Fig. 5.16 or single beam as sketched in Fig. 5.15) and a storage vessel for the sorptive gas. This instrument would allow cosorption measurements of binary sorptive gas mixtures with non-isomeric components in a swelling sorbent material without the need to analyze the gas phase in sorption equilibrium, i. e. without needing a gas chromatograph or a mass spectrometer. As we have been prepared to start measurements of this type early in 2002 at IFT, University of Siegen, we are sorry to say that we have not been able to actually realize them due to the lack of man power and time.

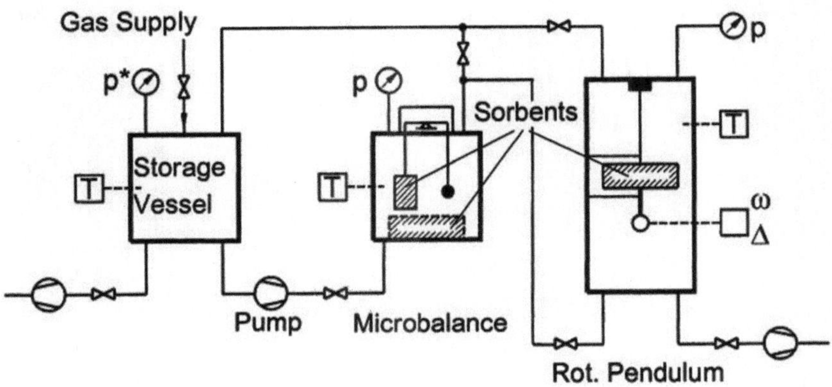

Figure 5.16. Scheme of an instrument for manometric-gravimetric-oscillometric measurements of binary coadsorption equilibria in swelling materials (polymers, resins etc.) without sorptive gas analysis by GC or MS.

Finally we want to advise the reader interested in oscillometric measurements of masses to consider design and experiments with a training instrument operating either at ambient air conditions or being included in a glass or plexiglass cover allowing really to observe rotational oscillations of the pendulum.*) Photographs of such an instrument used in our laboratories during 1996-98 are given in Figures 5.17, 5.18 below.

The training instrument allows one to study effects of changes in sizes and masses of the disk and the wire of the pendulum. Results of these experiments often can be described by few dimensionless numbers characterizing the pendulum and provide a good starting point for proper design of a pendulum for scientific or industrial measurements [5.2, 5.7].

*) As a first step, studying the famous clock "Atmos", manufactured by Jaeger-LeCoultre, Geneva, Switzerland, is highly recommended.

Figure 5.17. Rotational Pendulum.
Training instrument for measurement of gas adsorption equilibria near ambient temperature and pressure conditions.

Figure 5.18. Rotational Pendulum. Training instrument (rear) and optical signal detection system (foreground). © IFT University of Siegen, 1999.

4.3 Theory

The mass of gas (m^a) adsorbed or absorbed in a swelling sorbent material of mass (m^s) and volume (V^{as}) in the sorbate state at given pressure (p) and temperature (T) of the sorptive gas can be calculated from combined oscillometric and volumetric/manometric measurements. To derive the respective equations we consider again the result of oscillometric measurements, i. e. equation (5.25)

$$\Omega_{osc} = \frac{1}{m^s}(m^a + m^f + \Delta m^f) \qquad 5.70$$

The dimensionless quantity (Ω_{osc}) can be calculated from frequencies and decrements of various rotational oscillations performed by the pendulum via eq. (5.39). All quantities on the r.h.s. of eq. (5.70) have the same meaning as explained in Sect. 3.3. Also eqs. (5.53-5.57) have to be taken into account in what follows. The mass balance of the sorptive gas (m^{f*}) originally included in the storage vessel of volume (V_{SV}) is

$$m^{f*} = m^f_B + m^a \qquad 5.71$$

with the bulk sorptive gas mass

$$m^f_B = \rho^f \left(V^*_B - V^{as}\right). \qquad 5.72$$

Here the volume (V^*_B) is defined according to Fig. 5.13 by

$$V^*_B = V_{SV} + V_{AC} \qquad 5.73$$

However, it should be noted that in this quantity also the volumes and dead spaces of all the tubes, valves, and circulation pump(s) within the experimental installation which are filled with gas during the sorption process should be included, cp. Fig. 5.13. In fact this quantity in practice must be determined experimentally by – for example – nitrogen gas expansion measurements.

The quantity (m^a) in eqs. (5.70, 5.71) indicates the mass of sorptive gas sorbed in the sorbent mass (m^s). From eqs. (5.71, 5.72) we get

$$\Omega_{vol} \equiv m^a - \rho^f V^{as} = m^{f*} - \rho^f V^*_B. \qquad 5.74$$

Similarly we have by combining eqs. (5.70, 5.57) the relation

$$m^a + \left(\frac{1}{b} - 1\right)\rho^f V^{as} = \Omega_{osc} m^s - \Delta m^f \qquad 5.75$$

As equations (5.74), (5.75) provide two linear relations for (m^a, V^{as}) we get

$$m^a = b(\Omega_{osc} m^s - \Delta m^f) - (1-b)(\rho^f(p,T)V_B^* - m^{f*}) \qquad 5.76$$

$$V^{as} = \frac{b}{\rho^f}(\Omega_{osc} m^s + \rho^f(p,T)V_B^* - m^{f*} - \Delta m^f) \qquad 5.77$$

From these equations the quantities (m^a, V^{as}) can be calculated from measured values of (Ω_{osc}, p, T), leading via the thermal equation of state of the sorptive gas to its density $(\rho^f = \rho^f(p, T))$. All the other quantities included on the r.h.s. of (5.76, 5.77), namely $(b, m^s, \Delta m^f, m^{f*}, V_B^*)$ can be determined from auxiliary and calibration type experiments at the pendulum and the gas storage system respectively. Note that eqs. (5.76), (5.77) also holds for oscillometric-densimetric measurements mentioned in Sect. 4.2 in brief.

Experimental uncertainties of (m^a, V^{as})-data can be calculated straightforwardly from eqs. (5.76, 5.77) using Gauss' law of error propagation. In our experiments relative values of uncertainties (σ_{ma}/m^a), (σ_{Vas}/V^{as}) always turned out to be about 5 %, thus indicating that oscillometric-volumetric measurements tend to be somewhat less accurate than oscillometric-gravimetric ones, cp. Sect. 3.3. As already indicated, this situation can be improved by adding a microbalance to the installation Fig. 5.13 allowing one to measure the density of the sorptive gas by the buoyancy of a sinker, cp. Fig. 5.15, [5.27].

4.4 Example

The solubility of carbon dioxide (Messer-Griesheim, 5.0) in polymethylmethacrylate (PMMA) Makrolon 2400 (Bayer AG), cp. Tab. 5.3 and [5.29], and the swelling of the polymer have been measured at T=308 K using the oscillometric-manometric method [5.27]. Results are presented in Figure 5.19. The reduced masses $(\Omega_{osc} \equiv m^a - \rho^f V^{as})$ resulting from manometric measurements, cp. Eq. (5.74), are indicated by (■). As clearly can be seen, these data increase at low pressures (p < 3.5 MPa) to have a maximum at p = 3.5 MPa, then to decrease and even assume negative values for pressures ($p > 5$ MPa). The later effect is due to the swelling of the polymer/gas system increasing in this way "the dead space" for the sorptive gas phase. The reduced masses gained by oscillometric measurements (Ω_{osc}) (▲) increase monotonously with increasing gas pressure (p). So do the data for the specific volume ($V^{as}/(m^a + m^s)$) of the polymer/gas system (●) which can be correlated linearly by

$$\frac{V^{as}(p,T,m^a+m^s)-V^s(p=0,T,m^s)}{V^s(p=0,T,m^s)}=k(T)p\frac{V^f_{mol}}{V^f_{Nmol}} \qquad 5.78$$

Here $V^s(p = 0, T, m^s)$ is the volume of the PMMA at $p \to 0$, T, k(T) is the temperature dependent Henry coefficient of the swelling isotherm, V^f_{mol} is the molar volume of the sorptive gas (CO_2) at (p, T) and $V^f_{Nmol} = 22{,}4$ l/mol is the standard molar volume of an ideal gas at $p = 1$ atm, $T = 273{,}15$ K.

The data for the gas mass sorbed (●) also increase with increasing pressure showing deviation from purely linear behavior. Fitting of these data is possible by using the isotherm

$$\frac{m^a(p,T,m^s)}{m^s}=k(T)p+m_\infty(T)\frac{b(T)p}{1+b(T)p} \qquad 5.79$$

which is a superposition of a linear term reflecting bulk sorption of gas in the polymer, and a Langmuirian term describing adsorption of a limited amount of gas on the surface of "micro cavities" formed by the chain like molecules of the polymer [5.33]. Equilibration times of the sorption process lasted from 6 hours at low gas pressures ($p \simeq 1$ MPa) up to 2 days at high gas pressures ($p \simeq 6$ MPa). Relative uncertainties of all data are about 5 %. A comparison of the swelling volume data with some data from the literature is given in Fig. 5.20. Agreement within experimental uncertainties – as far as those can be guessed for literature data – seems to be reasonable.

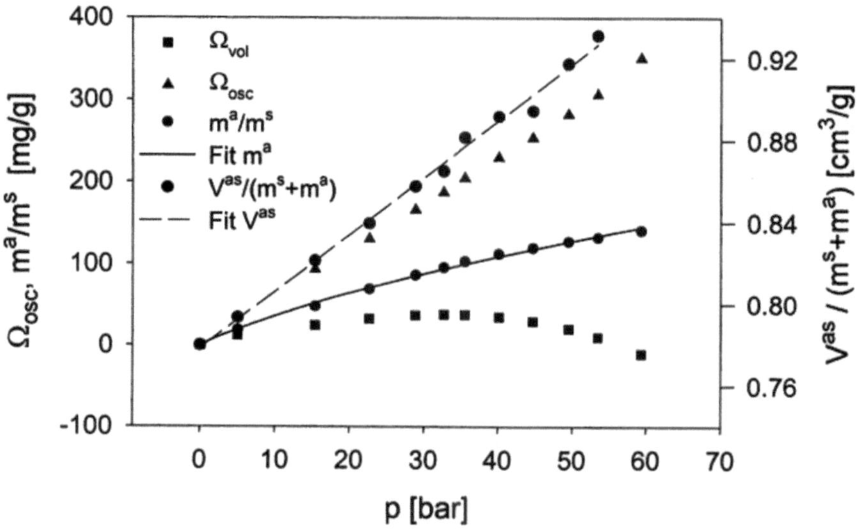

Figure 5.19. Swelling of volume (V^{as}) and sorption isotherm (m_a) of CO_2 on polymethylmethacrylate Makrolon 2400 (PMMA, Bayer) at 308 K.

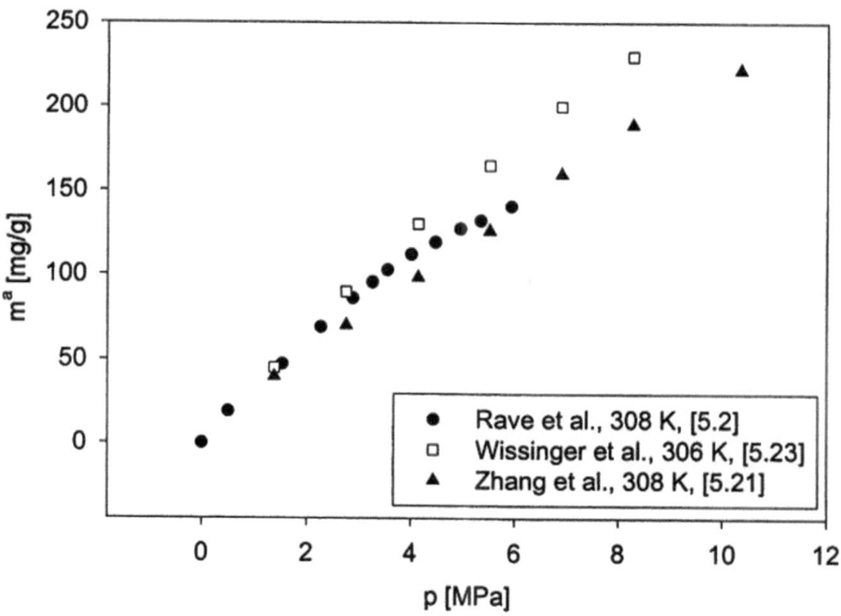

Figure 5.20. Comparison of swelling isotherms for the system Makrolon 2400 (PMMA)/CO_2 at about 308 K.

5. PROS AND CONS OF OSCILLOMETRY

In this section some of the advantages and disadvantages of sorption measurements by using slow rotational oscillations of a pendulum coming from our experience are described in brief.

5.1 Advantages

Oscillometry is based on the inertia of masses observed and neither on their extension as in manometry nor on its weight as in gravimetry. Hence, from a physical point of view oscillometry delivers basically new information on the masses observed and can be operated even under zero gravity conditions.

1. Versatility
   For swelling sorbents like polymers and resins, combined oscillometric-gravimetric or oscillometric-manometric measurements allow one to determine simultaneously the total mass ($m^s + m^a$) and the volume (V^{as}) of the sorbent / sorbate system. In this way, for example, a shift in the glass transition temperature of a polymer due to gas absorption can be observed.

2. Measuring techniques
   For rigid sorbent materials, combined oscillometric-gravimetric or oscillometric-manometric measurements in a saturation state of an adsorption system also allow one to determine both the total mass ($m^s + m^a$) and the Volume (V^{as}) of the system without using the so-called helium volume hypothesis, cp. Fig. 5.8 and Chap. 1.

   By choosing proper materials for the pendulum, for example titanium for the disk, gold alloy for the torsional wire etc., oscillometric measurements can be performed in a fairly broad range of temperature and pressure and even in corrosive gas atmospheres where other techniques may fail or simply become inapplicable.

5.2 Disadvantages

1. Availability
   Torsional pendulums designed for adsorption measurements of gases on solids are not available commercially today. Also principles for optimal design aiming at high sensitivity and reproducibility of data have not yet been formulated though hints for this are given in [5.2, 5.7].

2. Calibration
   As the mechanical properties of the torsional wire (retarding moment, internal damping, length) depend on temperature, calibration of the pendulum is necessary for all temperatures at which measurements should be performed.

3. Pressure limitations
   The pendulum should not be used for adsorption measurements at low pressures ($p \rightarrow 0$), as then not only the sorptive gas density $\rho \rightarrow 0$, but also the kinematic viscosity $\nu \rightarrow 0$ and hence the Reynolds-number may increase considerably ($Re \rightarrow \infty$). This indicates that for $p \rightarrow 0$ turbulent secondary flows may occur leading to non reproducible oscillations of the pendulum. The low pressure limit of the pendulum strongly depends on its geometry (thick disk criterion (5.29)). It can be shifted to lower values by increasing the height of the ring slit on the disk, i. e. using slim hollow cylinders filled with pellets as sorbent material.

4. Duration of measurements
   Measurements of gas solubilities in continuous materials like soft matter or highly viscous liquid polymers filled in the ring slit of the pendulum are normally very slow and may last for days and even weeks. Hence, it is often not easy to decide whether thermodynamic equilibrium really has been attained. Accompanying gravimetric and / or dielectric impedance measurements can be very helpful to decide this question, cp. Chapts. 3, 6.

5. Kinetics
   The kinetics of a sorption process of gas in a sorbent material is hardly observable by monitoring rotational oscillations. At least our experiments aiming at this problem have been unsuccessful [5.7].

6. LIST OF SYMBOLS

A collection of symbols used in Chapter 5 including SI-units is given.

$b = V_0^s / V_0^*$	1	ratio of bulk volume of pellets of mass (m^s) to the volume of the same mass (m^s) at same temperature and pressure in a dense bulk state
D_0	Nm	retarding moment of torsional wire
d	m	height of ring slit on pendulum
E	N/m^2	elasticity module of wire material
g	m/s^2	gravity of earth
J	kg m^2	moment of inertia of loaded pendulum in sorptive gas
J^a	kg m^2	moment of inertia of mass adsorbed (m^a) on sorbent (m^s)
J^f	kg m^2	moment of inertia of sorptive gas between sorbent pellets within the ring slit of the pendulum
ΔJ^f	kg m^2	moment of inertia of gaseous boundary layer surrounding the pendulum
J_L	kg m^2	moment of inertia of the empty rotational pendulum
J_s	kg m^2	moment of inertia of sorbent material (m^s)
J_0	kg m^2	moment of inertia of pendulum filled with sorbent mass (m^s) in vacuum
L_w	m	Length of a torsional wire

M(t)	Nm	torque or mechanical moment of frictional forces exerted by the sorptive gas on the pendulum
m^a	g	absolute mass of gas adsorbed on m^s
m^a_{GE}	g	Gibbs excess mass adsorbed
m^f	g	mass of sorptive fluid
Δm^f	g	mass included in the gaseous boundary layer surrounding the rotational pendulum
m^g	g	mass of non-adsorbing (glassy) material
m^s	g	mass of sorbent material
m^{f*}	g	mass of sorptive gas included in a storage vessel of volume (V_{SV}) prior to a gas sorption experiment, Fig. 5.13
m^f_B	g	mass of adsorptive gas in adsorption equilibrium of oscillometric-manometric experiments
m_L	g	mass of an empty rotational pendulum
$\mathbf{n}_r$		unit vector in radial direction
$\mathbf{n}_z$		unit vector in z-direction
R_0	kgm^2/s	coefficient of internal damping of a torsional wire
R_i	m	internal radius of a ring slit on a disk
R_a	m	external radius of a ring slit on a disk
Re	1	Reynolds-number of sorptive gas flow around the pendulum
r	m	radius coordinate

r_w	m	radius of a torsional wire
T_E	s	relaxation time of the experimentally observed relaxation of a rotator loaded with sorbent material and floating freely in a sorptive gas
T_L	s	relaxation time of the rotational motion of an empty rotator in vacuum
T_0	s	Eq. (5.8): period of rotational oscillations, Eq. (5.45): relaxation time of the rotational motion of a rotator floating freely in vacuum
T_{EHe}	s	relaxation time of a rotator loaded with sorbent material and floating freely in helium, gas
$\hat{T}_E$		value of T_E referring to a certain reference state of the sorptive gas
t	s	time coordinate
v	m/s	velocity field of fluid flow around the pendulum
V^f	m^3	volume of sorptive gas phase
V^g	m^3	volume of non-sorbing (glassy) material of mass (m^g)
V^{as}	m^3	volume of the adsorbent/adsorbate system of mass ($m^a + m^s$)
V^*	m^3	volume of the ring slit on the disk of the pendulum
V_{AC}	m^3	volume of an adsorption chamber
V_{SV}	m^3	volume of a gas storage vessel

V^s_{He}	m^3	volume of a sorbent material determined by (low pressure, high temperature) helium measurements
V^s_0	m^3	volume of sorbent pellets of mass (m^s) in a reference state
V^*_0	m^3	bulk volume of sorbent pellets of mass (m^s) in (same) reference state
$V^*_B = V_{SV} + V_{AC}$	m^3	volume of gas storage vessel (V_{SV}) and adsorption vessel(V_{AC}) in oscillometric-manometric measurements, Fig. 5.13.
$\alpha_0 = \alpha\ (t = 0)$	1	initial amplitude of pendulum
α_i, i = 1, 2	1	azimuthal amplitudes of the oscillation of a rotational pendulum
$\alpha(t)$	1	amplitude or angular displacement of a pendulum
$\gamma = \frac{m^s_B}{m^s}$	1	ratio of sorbent masses at balance (m^s_B) and at the pendulum (m^s)
Δ_0	1	logarithmic decrement of damped oscillations of the pendulum loaded with sorbent in vacuum
Δ_E	1	logarithmic decrement of experimentally observed damped harmonic oscillations of the pendulum in the gas
Δ_L	1	logarithmic decrement of damped oscillations of the empty pendulum in vacuum
ΔJ^f	kgm^2	moment of inertia of the mass included in a boundary layer around the pendulum and moved along with the pendulum

Δm^f	g	mass of a gaseous boundary layer surrounding the pendulum during its motion in the sorptive gas atmosphere
$\delta = (\nu/\omega_0)^{1/2}$	m	characteristic length of pendulum / sorptive gas system
$\mathbf{\Phi}$	1	unit vector in the azimuthal direction of the disk of the pendulum
η	Nsm^{-2}	dynamic viscosity of a sorptive gas
$\hat{\eta}$	Nsm^{-2}	dynamic viscosity of a sorptive gas in a certain reference state, cp. (5.50)
μ	1	Poisson number of wire material
$\nu = \eta / \rho$	m^2 / s	kinematic viscosity of a sorptive gas
ρ^f	kg / m^3	density of sorptive gas
σ_x^2	$[x]^2$	mean square deviation, dispersion or square of uncertainty of quantity (x)
Ω_{grav}	g	reduced mass of a sorbent / sorbate system determined by gravimetric measurements, cp. (5.58)
Ω_{osc}	1	mass adsorbed (m^a) plus mass of sorptive gas (m^f + Δm^f) moved along with the pendulum during its oscillations per mass of sorbent (m^s), cp. (5.25)
Ω_{rot}		relative increase of sorbent mass (m^s) due to gas adsorption (m^a) during its motion on a freely floating rotator in a gaseous atmosphere (m^f) with boundary layer (Δm^f_{rot}), cp. eq. (5.47)

Ω^g_{osc}	1	mass of sorptive gas ($m^f + \Delta m^f$) moved along with the pendulum during its motion of mass of non-adsorbing material (m^g) per unit of mass (m^g)
ω_E	s^{-1}	angular frequency of experimentally observed oscillations of the pendulum
ω_L	s^{-1}	angular frequency of oscillations of the empty pendulum in vacuum
ω_0	s^{-1}	angular frequency of oscillations of the loaded pendulum in vacuum

REFERENCES

[5.1] **Keller J. U.**
Theory of Measurement of Gas-Adsorption Equilibria by Rotational Oscillations, Adsorption, 1 (1995), p. 283-290.

[5.2] **Rave H.**
Messung von Sorptionsgleichgewichten von Gasen an Feststoffen mit Hilfe langsamer Schwingungen eines Rotationspendels,
Fortschritt-Berichte VDI, Reihe 3, Verfahrenstechnik, Nr. 660,
VDI-Verlag, Düsseldorf, 2000.

[5.3] **Krall A. H.**
Oscillating-Body Viscometry and its Applications to Phase-Separating Liquid Mixtures, Dissertation, University of Maryland, College Park, Maryland, 1992.

[5.4] **Keller J. U.**
Theory of Measurement of Gas-Adsorption Equilibria by a Floating Rotator,
Unpublished manuscript (E 9937, 9-6-1998), available from the author upon request, Siegen/Berlin, 2001.

[5.5] **Docter A., Lösch H. W., Wagner W.**
Entwicklung und Aufbau einer Anlage zur simultanen Messung der Viskosität und der Dichte fluider Stoffe,
Fortsch. - Ber. VDI Reihe 3 Nr. 494, Düsseldorf, VDI Verlag, 1997.

[5.6] **Docter A., Lösch H. W., Wagner W.**
A New Apparatus for Combined Measurements of the Viscosity and Density of Fluids for Temperatures from 233 K to 523 K at Pressures up to 30 MPa,
Int. Journal of Thermophysics, 20 (1999), p. 485-505.

[5.7] **Rave H.**
Messung von Sorptionsgleichgewichten von Gasen an Feststoffen mit Hilfe langsamer Schwingungen eines Rotationspendels,
PHD-Thesis, Institute of Fluid- and Thermodynamics, University of Siegen, Siegen, 1999.

[5.8] **Keller J. U., Rave H., Staudt R.**
Measurement of Gas Adsorption in a Swelling Polymeric Material by a Combined Gravimetric-Dynamic Method,
Macromol. Chem. Phys., 200 (1999), 2269-2275.

[5.9] **Young H. D.**
Physics
Extended Version with Modern Physics,.
Addison-Wesley Publ. Comp., 8th Ed., New York etc., 1992.

[5.10] **Ward M. D., Buttry D. A.**
In Situ Interfacial Mass Detection with Piezoelectric Transducers,
Science, 249(1990), p. 1000.

[5.11] **Kestin J., Paul R., Shankland J. R.**
A High-Temperature, High Pressure Oscillating-Disk Viscometer for Concentrated Ionic Solutions, Bunsen-Ber. Phys. Chem., 84 (1980), 1255.

[5.12] **Michels A., Gibson O.**
The measurement of the viscosity of gases at high pressures – The viscosity of nitrogen to 1,000 atms,
Proc. Royal Soc. London, Series A 134 (1931), p. 288.

[5.13] **Krall A. H., Sengers. V., Kestin J.**
Viscosity of liquid toluene at temperatures from 25 °C to 150 °C and pressure up to 30 MPa, J. of Chemical & Engineering Data, 37 (1992), p. 349.

[5.14] **Torklep K., Øye H. A.**
J. Chem. Eng. Data, 27 (1982), 387.

[5.15] **Torklep K., Øye H. A.**
J. Physics E: Sci. Instrum., 12 (1979), 875.

[5.16] **Djam A.**
Fehlerrechnung für die experimentelle Bestimmung von Adsorptionsgleichgewichten mit verschiedenen Messmethoden,
Studienarbeit Inst. Fluid- und Thermodynamik, Universität Siegen, 1998.

[5.17] **Fiedler M.**
Fehlbetrachtung der Messwerterfassung eines Rotationspendels,
Studienarbeit, IFT, Universität Siegen, Siegen, 1999.

[5.18] **International Organization for Standardization, ISO**
Guide to the expression of uncertainty in measurement,
2nd Printing, 1995, CH-1211 Geneva

[5.19] **ULLMANN'S Encyclopedia of Industrial Chemistry, 6th Edition**
Parker D. et al.
Polymers, High-Temperature, Wiley-VCH, New York etc., 2001.

[5.20] **Toshima N.**
Polymers for Gas Separation, VCH Verlagsgesellschaft, Weinheim, 1992.

[5.21] **Zhang Y., Gangwani K. K., Lemert R. M.**
Sorption and Swelling of Block Capolymers in the Presence of Supercritical Fluid Carbon Dioxide, J. Supercritical Fluids, 11 (1997), p. 115.

[5.22] **Lockemann C. A., Riede Th., Magin P.**
An Experimental Method to Determine the Sorption and Swelling Behaviour of Solids at High Pressures, article in: High Pressure Chemical Engineering, Elsevier Science B. V., Amsterdam, 1996.

[5.23] **Wissinger R. G., Paulaitis M. E.**
Swelling and Sorption in Polymer-CO_2 Mixtures at Elevated Pressures,
J. Polym. Sci. B, Polym. Physics, 25 (1987), 2497-2510.

[5.24] **Norit**
Produktinformationen zur Aktivkohle Norit R1 Extra,
Norit N. V., Postbus/ P.O.Box 105, 3800 AC Amersfoort, Netherlands, 1986.

[5.25] **Schwarzer R.**
Messung des Sorptions- und Schwellverhaltens von Polycarbonat und Polymnethylmethacrylat in CO_2 mit einer kombinierten oscillometrisch-statischen Messmethode. Studienarbeit, IFT, Universität Siegen, Siegen, 1999.

[5.26] **Rave H., Staudt R., Keller J. U.**
Measurement of Gas-Adsorption Equilibria via Rotational Oscillations, J. Thermal Analysis and Calorimetry, 55 (1999), p. 601-608.

[5.27] **Rave H., Kölsch O., Schwarzer R., Staudt R., Keller J. U.**
Oscillometric-volumetric measurements of gas solubitilities in swelling glassy polymers,
in preparation (2003). Article can be obtained from the authors upon request.

[5.28] **Elias H.-G.**
An Introduction to Polymere Science, VCH Verlagsgesellschaft, Weinheim, 1997.

[5.29] **NN**
Makrolon: Das Polycarbonat mit der Kompetenz des Erfinders, Bayer AG, Produktinformation, Leverkusen 1999.

[5.30] **Echte A.**
Handbuch der technischen Polymerchemie, VCH Verlagsgesellschaft, Weinheim, 1993.

[5.31] **Kikic I., Lora M., Cortesi A.**
Sorption of CO_2 in Biocompatible Polymers: Experimental Data and Qualitative Interpretation.
Article in: Smits P. J., Peters C. J., de Loos Th. W. (Hrsg.), Proc. 8th Intern. Conf. on Properties and Phase Equilibria for Product and Process Design, Noordwijkerhout, Netherlands 1998.

[5.32] **v. Schnitzler J., Eggers R.**
Mass Transfer Phenomena in Polymers during Treatment in Supercritical CO_2-Atmosphere.
Article in Proc. of the 5th Meeting on Supercritical Fluids, Nice, 1998.

[5.33] **Fleming G. K., Koros W. J.**
Dilation of Polymers by Sorption of Carbon Dioxide at Elevated Pressures. 1. Silicone Rubbers and Unconditioned Polycarbonate.
Macromolecules, 19 (1986), p. 2285-2291.

This page intentionally left blank

Chapter 6

IMPEDANCE SPECTROSCOPY

Abstract The physical principles and basic experimental techniques of impedance spectroscopy, i. e. static or frequency dependent dielectric permittivity measurements of sorbent/sorbate systems are given. These measurements can be used to characterize the state of a sorbent material in industrial adsorption processes. Combined with either manometric or gravimetric measurements of adsorption equilibria leading to calibration curves, permittivity measurements also allow fairly simple and quick measurements of gas adsorption equilibria. Kinetic processes and catalytic reactions inside a sorbent/sorbate system also can be observed. Pros and cons of dielectric measurements are discussed. List of Symbols. References.

1. INTRODUCTION

If a dielectric material or a weakly electrical conducting material like most activated carbons and zeolites is exposed to either a static or an alternating electric field, the molecular state of the material is changed [6.1]. The nuclei of the atoms and molecules within the material are shifted somewhat in the direction of the electric field whereas the electrons are displaced a little bit in the opposite direction. Hence dipole moments are introduced in the material either by orientation of polar molecules like (H_2O, H_2S, CO) etc. in the direction of the electric field or by separating electric charges within non-polar molecules like (N_2, O_2, CH_4) etc. These atomic or molecular dipole moments are fixed within the electric field, their sum within the unit volume being called the dielectric polarization of the material [6.1, 6.2]. The dielectric polarization can be measured by capacitance or impedance measurements of capacitors filled with the material, cp. Sect. 2.1, Fig. 1. Applying oscillating electric fields of variable frequency (ν / Hz) one gets curves $C = C(\nu)$ which are characteristic not only for the sorbent material in vacuum but also for the sorbent / sorbate system which is realized after exposing the sorbent to a sorptive gas atmosphere [6.9-6.11]. Though a molecular interpretation of the resulting curves normally is not easy, they can be used in industrial adsorption processes to check the actual state of a sorbent material and to detect for example preadsorbed dirt molecules, unusual CO-production or hot spots within an adsorption reactor.

Combining the dielectric measurements with either manometric, gravimetric or oscillometric measurements of gas adsorption equilibria states, one gets calibration curves allowing one the determination of Gibbs excess adsorbed masses by purely electric measurements which normally can be performed fairly quickly and on site in industrial situations.

It also should be possible to detect single components of a multicomponent adsorbate or molecules at different adsorption places within an energetically heterogeneous sorbent material by impedance measurements using appropriate (high) frequency electric fields leading to resonance phenomena of the respective molecules. Unfortunately, these frequencies often are – for example for water adsorbed on zeolite – in the GHz- and even THz-region requiring sophisticated impedance analysing technology. Therefore, presently they do not seem to be suited for industrial measurements. Also, the pore spectrum of a sorbent material, i. e. micropores ($d < 2$ nm) or mesopores ($2 \text{ nm} < d < 50$ nm) cannot easily be detected by dielectric permittivity measurements as these would require high frequency electric fields at $\upsilon \simeq 10^{17}$ Hz, corresponding to heat radiation or near infrared radiation fields which normally are strongly absorbed by the material [6.1].

Dielectric measurements of gas adsorption systems can be performed fairly quickly, typically within a few seconds [6.3]. Hence the kinetics of adsorption processes being "slow" on this time scale can be observed. Indeed these processes are sometimes "invisible" to purely manometric or even gravimetric measurements. As examples we mention internal diffusion, reorientation or catalytically induced chemical reaction processes of admolecules within a sorbent material. The mass of the adsorbed phase normally is constant during processes of this type, whereas the dipole moment of the admolecules and hence their polarization changes, cp. Sect. 3.2.

We here restrict to consider only gas adsorption systems exposed to weak electric fields, these being considered as sensors for the system without changing its macroscopic properties. However, it should be mentioned that in principle the adsorption properties of a sorbent material are changed by the electric field. This so-called electro-adsorptive effect is important in microsystems as used, for example, in advanced gas sensing devices [6.4].

Today impedance measurements are used in many fields of science and engineering, among them biosciences and medicine to analyze and characterize complex material structures [6.5-6.7].*)

The material presented in this chapter is organized as follows: In Sect. 2 the basic experiments necessary for dielectric permittivity measurements is given, followed by an outline of the theory of dielectric polarization, which considers the uncertainties of measurements and gives several examples related to gas adsorption equilibria on microporous solids. In Sect. 3 combined dielectric-manometric and dielectric-gravimetric measurements of adsorption equilibria are considered briefly. In Sect. 4 the pros and cons of dielectric measurements are discussed. A list of Symbols used is given in Sect. 5 followed by the references cited.

2. DIELECTRIC MEASUREMENTS OF GAS ADSORPTION SYSTEMS

2.1 Experimental

An instrument to measure the dielectric permittivity of a gas adsorption system basically consists of an electric capacitor (plates, cylinders, spheres) placed within an adsorption vessel. The vessel should be placed within a thermostat (water, oil etc.) and provided with tubes for gas supply and evacuation. Also manometers and thermometers are needed to measure the gas pressure (p) and temperature (T) inside the chamber. The capacitor is filled with sorbent material (powder, pellets, continuous matter etc.) which can be considered to be homogenous as long as its characteristic length – for example the diameter of cylindrical pellets – is small compared to a characteristic length of the capacitor.

The capacitor is connected to an impedance analyzer (Hewlett Packard HP 4192A) which allows the application of an alternating voltage (5 Hz - 13 MHz) to measure either the static or the complex impedance and / or capacitance of the sorbent / sorbate system. A schematics of such an instrument is sketched in Figure 6.1 for a plate capacitor. A photo of a cylindrical capacitor which has been used for many years at IFT is presented in Figure 6.2.

*) Information on new developments in the field can be found in the "Dielectric Newsletter" edited by G. Schaumburg, Novocontrol GmbH, 56414 Hundsangen, Germany, www.novocontrol.com.

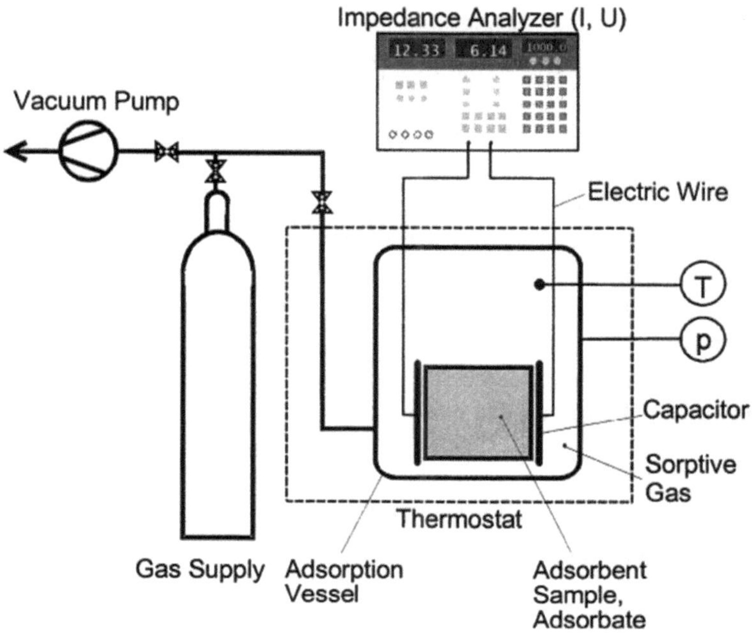

Figure 6.1. Installation for measurements of the static or frequency dependent dielectric permittivity of gas adsorption systems. © IFT University of Siegen, 1988.

Assuming a small constant electric voltage (U) to be applied to the plates of the capacitor producing a weak electric field ($E < 1$ V/m), the static capacitance (C_s) of the system can be determined by the relation

$$Q = C_s U \tag{6.1}$$

with

$$Q = \int_0^{\infty} I(t)dt \tag{6.2}$$

being the total charge on either of the plates of the capacitor which is the time integral of the electric current (I (t)) in the wires connecting the capacitor to an electric power source inside the impedance analyzer.

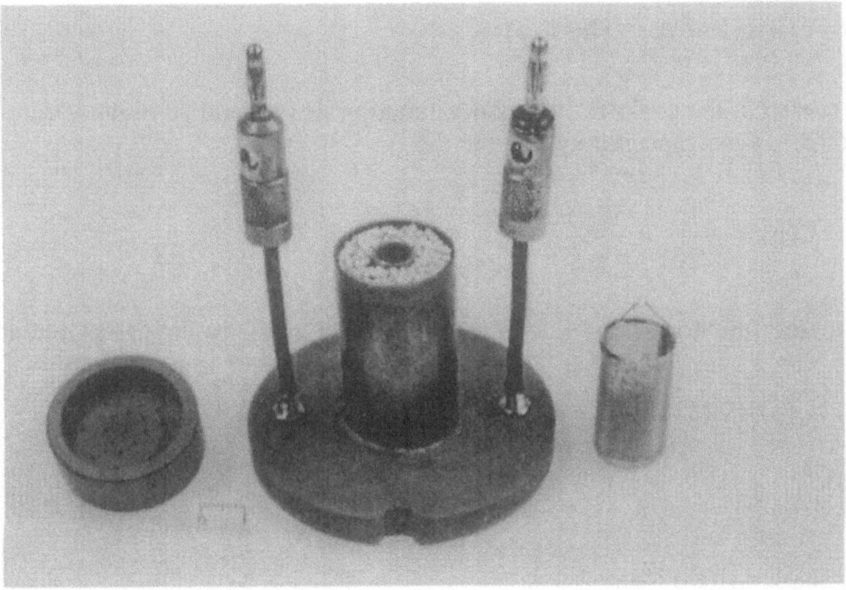

Figure 6.2. Cylinder capacitor filled with zeolite pellets for dielectric permittivity measurements of gas adsorption equilibria.
Data of cylinders: height: h = 48.5 mm, diameter: d_i = 10 mm, d_a = 28 mm, material: Cu, thickness: 1 mm. © IFT University of Siegen, 2000, [6.8].

Similar to Eq. (6.1) we have for periodically oscillating electric voltages ($U(t) = U_0 e^{i\omega t}$, $\omega = 2\pi\nu$) and loads ($Q(t) = Q_0 e^{i(\omega t-\varphi)}$, $\omega = 2\pi\nu$) and the relation [6.1, 6.12]

$$\hat{Q}(p) = C(p)\hat{U}(p) . \quad 6.3$$

Here

$$\hat{Q}(p) = \int_0^\infty Q(t)e^{-pt}dt, \quad \text{Re}\,p > 0 \quad 6.4$$

$$\hat{U}(p) = \int_0^\infty U(t)e^{-pt}dt, \quad \text{Re}\,p > 0 \quad 6.5$$

are the Laplace transforms of the charge ($Q(t)$) and voltage ($U(t)$) at the capacitor, and $p = p_1 + ip_2$ is a complex frequency with $p_1 = \text{Re}\,(p) > 0$, $p_2 = \text{Im}\,(p) = \omega = 2\pi\nu$, [6.12]. The frequency dependent complex capacitance defined by Eq. (6.3)

$$C(p) = \mathrm{Re}\, C(p) + i\, \mathrm{Im}\, C(p) \qquad 6.6$$

is related to the complex impedance function (Z (p)) and admittance functions (Y (p)) of the capacitor system by

$$C(p) = \frac{1}{pZ(p)} = \frac{Y(p)}{p}. \qquad 6.7$$

Here the function (Z (p) = 1 / Y (p)) is defined by the generalized Ohm's law

$$\hat{U}(p) = Z(p)\, \hat{I}(p) \qquad 6.8$$

with

$$\hat{I}(p) = \int_0^\infty I(t) e^{-pt} dt, \qquad 6.9$$

being the Laplace transform of the electric current (I (t)).

Choosing for the complex frequency $p = p_1 + ip_2$, the values $p_1 = 0$, $p_2 = 2\pi\nu$, the functions

$$C_1(\nu) = \mathrm{Re}\, C(p = 2\, \pi i \nu),\; C_2\, (\nu) = \mathrm{Im}\, C(p = 2\, \pi i \nu) \qquad 6.10$$

resulting from the measured complex capacitance C = C(p), (6.6) can be used to characterize the capacitor itself and any kind of adsorption system included in it. The real part ($C_1(\nu)$) of the complex capacitance (C(p)) describes reversible interactions between the atoms and molecules of the system and the electric field applied leading to periodic polarization of the molecules. The imaginary part ($C_2(\nu)$) of the complex capacitance (C(p)) describes irreversible interactions between the electric field and the atoms and molecules, namely Ohmian heating of the material due to its often small but nontheless existing electric conductivity [6.1, 6.13].

As an example, data for the capacitor shown in Fig. 6.2 taken in vacuum ($p_{AIR} < 10^{-2}$ Pa) and for gases (N_2, p = 12.3 MPa; CO, p = 13.5 MPa) at T = 298 K are shown for frequencies up to 13 MHz in Figure 6.3. Note that the real part of the capacitance ($C_1(\nu)$) of the system is given in units 1 pF = 10^{-12} Farad.

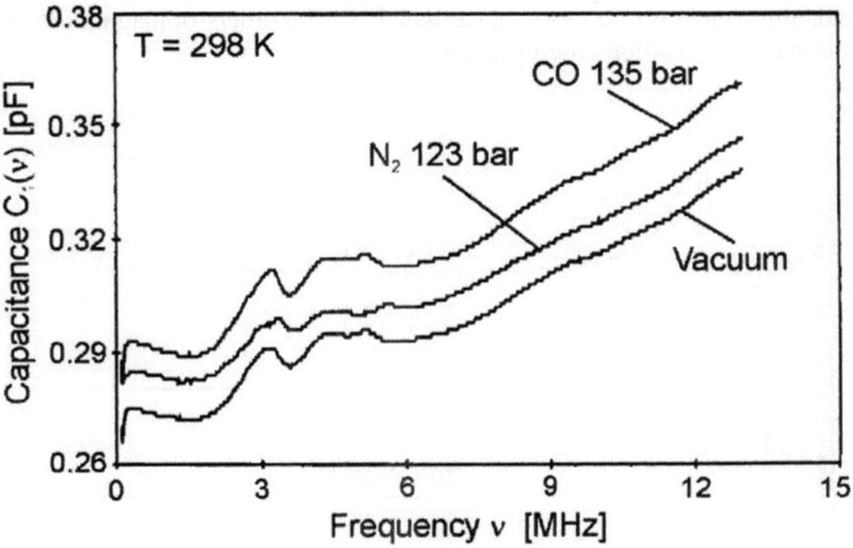

Figure 6.3. Capacitance spectrum $C_I(\nu)$ of the empty capacitor (lower curve) and of non-polar (N_2) and polar (CO) sorptive gases at elevated pressures and T = 298 K, [6.8, 6.13].

The capacitance of the system with and without gases increases nearly monotonously with increasing frequency of the electric field. As the capacitance is, roughly speaking, a measure of polarization, i. e. the electric dipole moment within the unit volume of the capacitor, one would expect a certain increase of the capacitance of the capacitor in vacuum due to the presence of gas. This indeed is the case, as the increase in capacitance for nitrogen having only induced dipole moments is lower than the increase in capacitance for carbon monoxide filling, the molecules of this gas having a small permanent dipole moment ($\mu_{CO} = 0.1$ Debye).

In Figure 6.4 dielectric spectra of three different sorbent materials in vacuum are shown for frequencies $\nu \leq 13$ MHz. The upper curve refers to activated carbon (Norit R1) at 298 K, showing at 500 kHz kind of resonance, i. e. a small peak which probably is due to active atomic groups located at the edges of the graphite layers of the carbon [6.13]. The middle curve refers to a hybrid sorbent material (Envisorb b^+, Engelhard) consisting of a mixture of activated carbon and zeolite which has been activated thermally. It shows a small resonance peak at 200 kHz and a capacitance which is nearly one order of magnitude smaller than that of the activated carbon. The lower curve presents capacitance measurements of dealuminated Y-zeolite (DAY, Degussa). Atoms in this material are so strongly connected to each other that their periodic oscillations are not much influenced by the external electric field, thus leading to a capacitance which is two orders of magnitude smaller than that of the carbon filled capacitor.

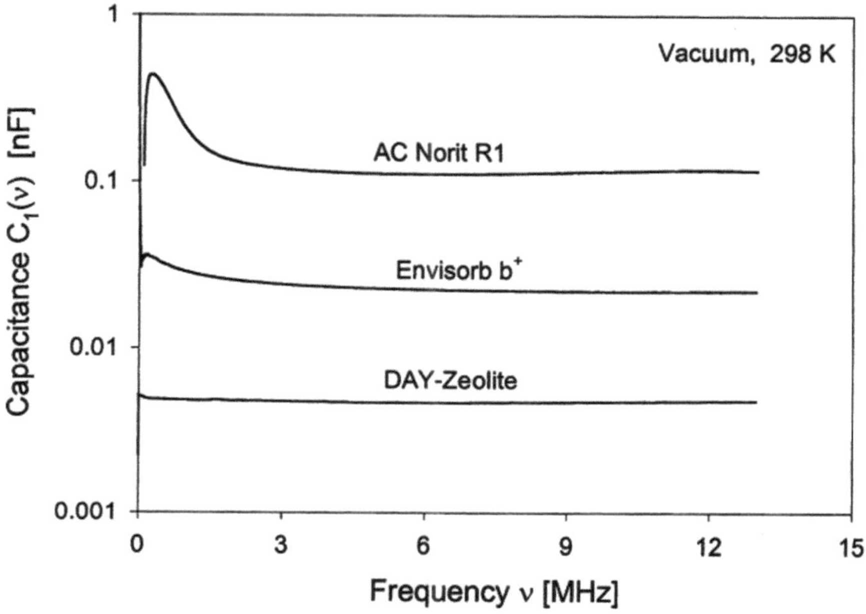

Figure 6.4. Capacitance spectra C_l (ν) of activated carbon Norit R1 (upper curve), of a hybrid sorbent material (Envisorb b^+, Engelhard) consisting of a mixture of activated carbon and zeolite (middle curve), and of dealuminated Y-zeolite (DAY, Degussa) in vacuum at T = 298 K, [6.13].

In Figure 6.5 the impedance spectra ($C_l = C_l$ (ν, p, T = const)) of molecular sieve MS 13X (UOP) exposed to hydrogen sulfide (H_2S) as sorptive gas at T = 298 K for different pressures (0.01 MPa < p < 0.4 MPa) are shown [6.10]. Measurements have been performed in the frequency range (1 kHz < ν < 100 kHz) using an impedance analyzer of type HP 4192 A. The capacitance of the sorbent / sorbate system decreases with increasing frequency (ν) of the field – so-called normal dispersion [6.6]. However, the capacitance increases with increasing gas pressure, i. e. amount of gas adsorbed. This is due to the increase of the number of admolecules within the system each of which having an electric dipole moment (μ_{H2S} = 0.9 Debye). Also it should be noted that the changes in the capacitance with changing gas pressure or amount of H_2S adsorbed are much larger for low frequencies (ν ≅ 1 kHz) than for higher ones (ν ≅ 100 kHz). This gives an indication of which frequency should be used to observe breakthrough curves or to check the state of a porous sorbent in industrial adsorption processes [6.14, 6.15].

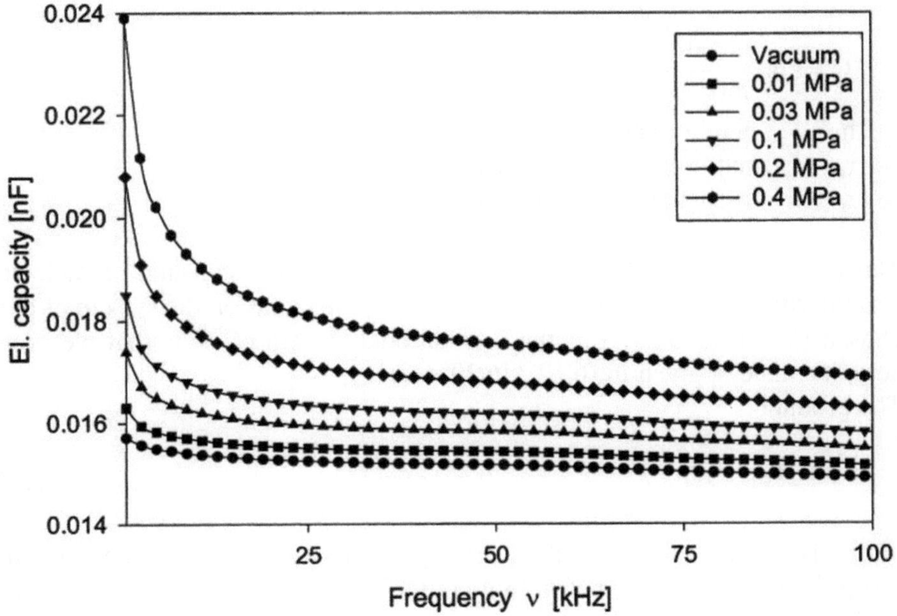

Figure 6.5. Impedance spectra of molecular sieve MS13X (UOP) including hydrogen sulfide (H_2S) adsorbed at various gas pressures (0.01 MPa < p < 0.4 MPa) and T = 298 K for frequencies (1 kHz < ν < 100 kHz), [6.10].

Measurements of the capacitance (C) of an electric capacitor allow one to calculate the relative dielectric permittivity or dielectric constant (ε_r) of the material included in the capacitor. For plate capacitors we have, cp. Figs 6.1, 6.8, [6.1, 6.2]:

$$C = \varepsilon_0 \varepsilon_r \frac{A}{d} \tag{6.11}$$

A ... area of plates
d ... distance of plates

6.12

$\varepsilon_0 =$ $8.854187818 \cdot 10^{-12}$ As / (Vm)

For cylindrical capacitors we have, cp. Fig. 6.2:

$$C = \varepsilon_0 \varepsilon_r \frac{2\pi h}{\ln(r_a / r_i)} \qquad 6.13$$

h ... height of cylinder
r_a ... radius of outer cylinder
r_i ... radius of inner cylinder

The parameter ε_0 in (6.11, 6.13) is the electric field constant or dielectric permittivity of the vacuum. Its numerical value depends on the system of units chosen [6.3, 6.21-6.24]. For the international system of units that we prefer here, its value is given in (6.12), [6.20, 6.21].
The product

$$\varepsilon = \varepsilon_0 \, \varepsilon_r \qquad 6.14$$

often is called absolute dielectric permittivity, whereas the quantity $\varepsilon_r = (\varepsilon / \varepsilon_0)$ is the relative permittivity of the material included in the capacitor. It should be emphasized that ε_r is in a thermodynamic sense an intensive quantity of state. For a rigid, inert sorbent / sorbate system it is a function of the relative amount of mass adsorbed (m^a / m^s) or the sorptive gas pressure (p) and the temperature (T) of the system, hence

$$\varepsilon_r = \varepsilon_r(m^a / m^s, T) \cong \varepsilon_r(p, T) \qquad 6.15$$

These equations represent different forms of the dielectric equation of state (DEOS) for the sorbent / sorbate system. It can be used to characterize the system and is of importance for electrostriction phenomena and / or the electro-adsorptive effect – especially for continuous non-rigid materials, [6.16, 6.17, 6.4]. The dielectric permittivity (ε_r) is a phenomenological measure of the interaction between the electric field and the material within the capacitor. Numerical values of ε_r for static electric fields range as follows [6.5-6.7, 6.20, 6.21]:

	ε_r
Vacuum	1
Gases	$\simeq 1$
Liquids	2-100
Solids	2-10,000

As an example in Figure 6.6 we present data of the static ($\nu = 0$) dielectric permittivity of molecular sieve MSKE 154 including nitrogen (N_2 (5.0)) as an adsorbate taken at T = 294.7 K for gas pressures up to 10 MPa, [6.18]. These data can easily be correlated by a dielectric (adsorption) isotherm of Langmuirian type

$$\varepsilon_r - \varepsilon_r^s = \frac{p}{A + Bp} \tag{6.16}$$

or likewise in the linearized form

$$\frac{p}{\varepsilon_r - \varepsilon_r^s} = A + Bp \tag{6.17}$$

as in Fig. 6.6. The parameters are: A = 0.1 GPa, B = 12.8 for pressures given in [p] = GPa = 10^9 Pa. The parameter $\varepsilon_r^s = \varepsilon_r^s(p = 0, T)$ refers to the permittivity of the (empty) molecular sieve in vacuum.

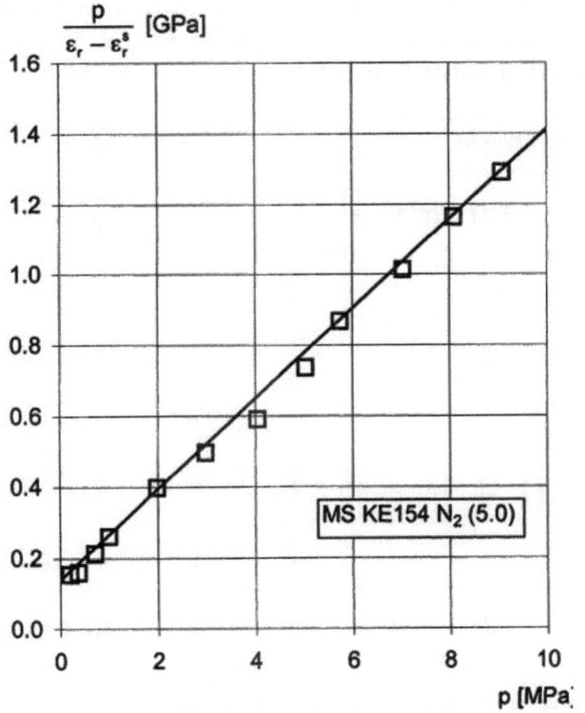

Figure 6.6. Dielectric isotherm of nitrogen (N_2) adsorbed in molecular sieve KE 154 at T = 294.7 K for pressures up to 10 MPa, [6.18].

Calculations of the dielectric polarizability (α^a) of the adsorbed nitrogen (N_2) from the data presented in Fig. 6.6 indicate that admolecules are in an "intermediate state" between gas and liquid, cp. Sects. 2.2, 3.2.

Dielectric permittivity measurements also can be used to detect selectively strongly polar components like water (electric dipole moment μ_{H2O} = 1.85 Debye, [6.20, 6.21]) being adsorbed from a mixture carrier gas – polar adsorptive gas on zeolite or Silicagel. In Figure 6.7 data are presented of the static dielectric permittivity (ε_r) of molecular sieve MS 544 H (Bayer) in contact with a) pure argon (Ar) and b) an argon-water vapor mixture including $4 \cdot 10^{-3}$ %mass of water [6.19]. Measurements were performed at T = 293.6 K in the pressure range 0.1 MPa < 12 MPa. Data for pure argon (□) clearly show that as expected there is nearly no change of the dielectric permittivity (ε_r) of the molecular sieve due to argon adsorption of less than 1 % mass for pressures up to 10 MPa. However, in case of humid argon (Δ) the permittivity increases considerably with increasing gas pressure, the difference between respective data at the same pressure (p) being related to the amount of water being adsorbed on the molecular sieve, cp. Sects. 2.2, 3.2. Hence dielectric measurements seem to be very well suited for checking the state of a sorbent material, for example its dryness, i. e. whether water has been preadsorbed or not [6.3, 6.10].

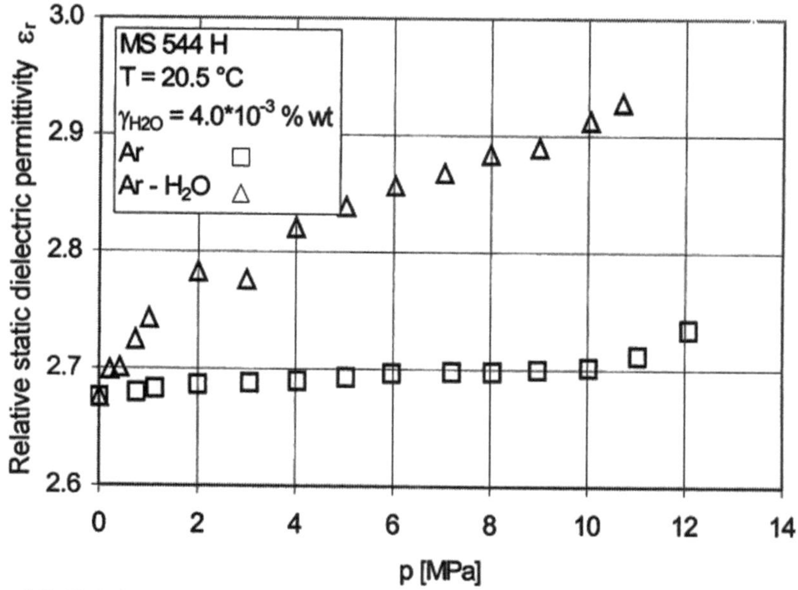

Figure 6.7. Relative static dielectric permittivity of molecular sieve MS 544 M (Bayer) exposed to a) pure argon (□) and b) argon/water vapor mixtures (Δ) at T = 293.6 K for water concentrations of $4 \cdot 10^{-3}$ % weight, equivalent to 90 vpm and gas pressures up to 12 MPa, [6.19].

2.2 Theory

The main goal of this section is to elucidate the relation between dielectric measurable quantities like the capacitance (C) of a capacitor filled with a sorbent/sorbate system or its dielectric permittivity (ε) and the amount of adsorbate (n^a, m^a) adsorbed on the sorbent. For this we will present in Sect. 2.2.1 some of the basic concepts of the theory of dielectric materials in static or oscillating electric fields and then to describe in Sect. 2.2.2 the relation between the static dielectric permittivity (ε_{rs}) and the number of particles or molecules of a dielectric material (n^a). Finally we will provide the reader in Sect. 2.2.3 with simple models for the complex dielectric permittivity reflected in linear electric networks, allowing the calculation of the static dielectric permittivity (ε_{rs}) from the complex permittivity ($\varepsilon_r(p)$) or equivalent impedance measurements [6.1, 6.3, 6.12].

2.2.1 Basic Concepts

We consider again a flat plate capacitor connected with a current meter and a voltmeter to measure (static or time dependent) electric currents (I/Ampere) and potential differences (U/Volt), Fig. 6.8. The capacitor is assumed to be filled with homogenously distributed sorbent material and a sorptive gas, cp. Fig. 6.1, leading to a sorbate phase within the sorbent. Due to the electromotive force (U) applied to the capacitor an electric field is created within the material system which is described by its field strength (E) given by

$$E = \frac{U}{d} \qquad 6.18$$

with (d) being the distance between the plates of the capacitor. In physical terms (E) is the force K exerted by the electric field on the unit charge to be measured in vacuum in a small slit parallel to the vector field (E). Hence, for an arbitrary electric charge (q) we have [6.1, 6.24]:

$$K = qE,$$

Another consequence of the electric field (E) is that local electric charges within the material are somewhat shifted either in or against the direction of the field depending on whether these charges are positive (atomic nuclei) or negative (electrons).

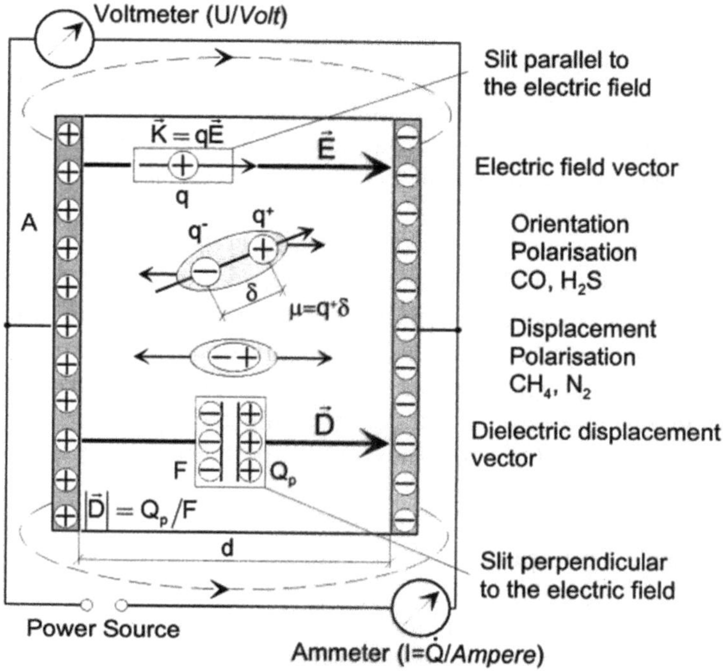

Figure 6.8. Plate capacitor filled with quasi-homogeneously distributed dielectric material including non-polar and polar molecules. Any electric field (E, D) applied will separate electric charges within a non-polar molecule (≅0.1 Å) thus leading to displacement or "induced polarization" of the molecule. Also polar molecules will be oriented (somewhat) in the direction of the (E, D)-field leading to "orientation polarization" of the molecule [6.1, 6.2, 6.6, 6.7].

This phenomenon is described by the dielectric displacement vector field (D).*) It can be measured in slits normal to the E- or D-field and is given by

$$D = \frac{Q}{A} \qquad 6.19$$

Definitions (6.18, 6.19) hold for both static and alternating electric fields. According to Eqs. (6.1), (6.11) and in view of (6.18, 6.19) the fields (E) and (D) are related by the equation

*) We here are restricted to isotropic or quasi-isotropic materials in which E- and D-fields always are parallel. For the thermodynamic properties of anisotropic materials in electric fields the reader is referred to [6.16, 6.17].

$$D = \varepsilon_0 \varepsilon_{rs} E \tag{6.20}$$

with ε_{rs} being the static relative dielectric permittivity of the material in the capacitor. This relation can be converted to

$$\varepsilon_0 \varepsilon_{rs} = \frac{d}{A}\frac{Q}{U} \tag{6.21}$$

allowing to calculate (ε_{rs}) from directly measured quantities (U) and (Q), cp. (6.2).

As it is experimentally not easy to provide electric voltages to the capacitor which are exactly constant in time, it is useful to apply periodically alternating electromotive forces and to extrapolate experimental data to zero frequencies, i. e. constant potential differences. Combining Laplace transformation of Eqs. (6.18-6.20) with Equation (6.2) and the generalized Ohm Law (6.8), we get

$$\varepsilon_0 \varepsilon_r (p) = \frac{d}{A}\frac{1}{pZ(p)} \tag{6.22}$$

Here $p = p_1 + ip_2$, $p_1 \geqq 0$, $p_2 = 2\pi\nu$, is the complex Laplace variable with frequency (ν), $\varepsilon_r = \varepsilon_r(p)$ is the complex relative permittivity of the material system. Also Z(p) is its complex impedance function defined by Eq. (6.8) as

$$Z(p) = \frac{\hat{U}(p)}{\hat{I}(p)} \tag{6.23}$$

$\hat{U}$ and $\hat{I}$ indicating Laplace transformations of the measured electric potential (U(t)) and current (I(t)). Sometimes it is convenient also to use the reciprocal of Z(p), namely the admittance function $Y(p) = 1/Z(p) = \hat{I}(p)/\hat{U}(p)$. Then we have instead of (6.22) for $\varepsilon_r(p)$ the equation

$$\varepsilon_0 \varepsilon_r (p) = \frac{d}{A}\frac{Y(p)}{p} \tag{6.24}$$

Relations analogous to (6.22, 6.24) apply if instead of a plate capacitor a cylindrical capacitor is used, cp. (6.13, 6.7):

$$\varepsilon_0\varepsilon_r(p) = \frac{\ln(r_a r_i)}{2\pi h}\frac{1}{pZ(p)} = \frac{\ln(r_a r_i)}{2\pi h}\frac{Y(p)}{p} \qquad 6.25$$

The complex relative permittivity ($\varepsilon_r(p)$) of a sorbent/sorbate system determined by either (6.22) or (6.25) can be split into its real and imaginary part

$$\begin{aligned}\varepsilon_0 &= \mathrm{Re}\,\varepsilon_r + i\,\mathrm{Im}\,\varepsilon_r \\ &= \varepsilon'(p) + i\varepsilon''(p)\end{aligned} \qquad 6.26$$

The curve $\varepsilon'' = \varepsilon''(\varepsilon')$, or in parametric representation $\varepsilon' = \varepsilon'(\omega), \varepsilon'' = \varepsilon''(\omega)$ with $p = i\omega$, provides a characteristic temperature and pressure dependent curve of the sorbent/sorbate system [6.10, 6.14, 6.15]. Examples will be given in Sect. 2.4.

2.2.2 Polarization of Dielectrics

Exposing a dielectric material to an electric field (E, D) always leads to polarization, i. e. a separation of electric charges forming a macroscopic dipole moment (P_e) characteristic to the system and the applied electric field (E, D) applied, cp. Fig. 6.8:

$$P_e = p_e N_T \qquad 6.27$$

Here (N_T) is the number of “particles”, i. e. molecules, atoms, or ions included in the system and p_e the averaged (molecular) dipole moment per particle. For weak electric fields ($E \leq 1V/cm$) p_e is proportional to the electric field strength (E) giving

$$p_e = \alpha E \qquad 6.28$$

with (α) being the so-called polarizability of a single particle or molecule. According to Debye and Langevin [6.22], this quantity always can be represented as sum of the induced polarizability (α_{ind}) caused by displacement of electrons and ions in the electric field and orientational polarizability (α_{ori}) caused by a reorientation of the permanent dipole moment (μ) of thermally moving polar molecules in the electric field, giving

$$\alpha = \alpha_{ind} + \alpha_{ori} \tag{6.29}$$

where

$$\alpha_{ori} = s\frac{\mu^2}{kT} \tag{6.30}$$

Here s is a numerical "shape factor" being s = 1/3 for simple cubic crystals and freely rotating molecules [6.22, 6.24]. For non-polar admolecules ($\mu = 0$) numerical values of $\alpha = \alpha_{ind}$ can be determined by combined dielectric and, for example, gravimetric measurements (cp. examples given in Sect. 3.2). For polar admolecules the same is true at least at low frequencies ($\nu < 1$ MHz) of the electric field where often $\alpha_{ind} << \alpha_{ori}$ and hence can be neglected compared to α_{ori}. For both types of admolecules numerical data of (α) are between their values for the gas and the liquid phase of the adsorptive and normally depend on the amount adsorbed, i. e. degree of saturation. Hence they can give an indication of the nature of the site where the admolecule is adsorbed and also on the structure of the sorbate phase [6.3].

Combining Eqs. (6.27, 6.28) we can write the macroscopic dipole moment of the material system in the capacitor as

$$P_e = \alpha N_T E = \alpha n_T V_c E = P V_c = \Omega_{DE} E \tag{6.31}$$

Here the number of particles per unit volume

$$n_T = \frac{N_T}{V_C} \tag{6.32}$$

and the dielectric polarization, i. e. the dipole moment per unit volume of the material

$$P = p_e n_T = \alpha n_T E \tag{6.33}$$

have been introduced, $V_C = A \cdot d$ indicating the volume of the plate capacitor, cp. Fig. 6.8. The quantity

$$\Omega_{DE} = \alpha N_T = \alpha n_T V_c \tag{6.34}$$

serves as an abbreviation and can be called macroscopic polarizability of the material system [6.25].

According to Clausius and Mossotti [6.1-6.3, 6.22, 6.24] the polarization (P) can be represented for static electric fields (E = const) as

$$P = 3\varepsilon_0 \frac{\varepsilon_{rs} - 1}{\varepsilon_{rs} + 2} E \qquad 6.35$$

Here ε_{rs} is the static relative dielectric permittivity of the system. For gases, where $\varepsilon_{rs} \rightarrow 1$, Eq. (6.35) reduces to

$$P = \varepsilon_0 (\varepsilon_{rs} - 1) E \qquad 6.36$$

Combining Eqs. (6.33), (6.35) we have for the polarizability of the particles in unit volume of the material

$$\alpha n_T = 3\varepsilon_0 \frac{\varepsilon_{rs} - 1}{\varepsilon_{rs} + 2} \qquad 6.37$$

From this relation we get in view of (6.32)

$$\alpha_m n = 3\varepsilon_0 \frac{\varepsilon_{rs} - 1}{\varepsilon_{rs} + 2} V_C \qquad 6.38$$

with the molar polarizability of the material

$$\alpha_m = \alpha N_A = \left(\alpha_{ind} + s \frac{\mu^2}{kT} \right) N_A \qquad 6.39$$

and the mole number

$$n = N_T/N_A, \qquad 6.40$$

N_A being Avogadro's number.

Likewise we get from (6.37)

$$\alpha_s m = 3\varepsilon_0 \frac{\varepsilon_{rs} - 1}{\varepsilon_{rs} + 2} V_C \qquad 6.41$$

with the specific polarizability per unit of mass

$$\alpha_s = \alpha n_T \frac{1}{\rho} = \alpha_m / M \tag{6.42}$$

and

$$m = \rho V_C \tag{6.43}$$

being the total mass of the material in the capacitor.

Equations (6.38) and (6.41) combine molecular properties (α_m, α_s) and the mole number (n) or the mass (m) of the dielectric material with its static relative permittivity (ε_{rs}) and the volume (V_C) of the capacitor. Given (ε_{rs}, V_C) and n or m from gravimetric or manometric measurements, the polarizabilities (α_m or α_s) can be calculated. Given (ε_{rs}, V_C) and models for α_m or α_s, n or m can be calculated.

The macroscopic dipole moment (P_e) of the material in the capacitor is by definition an extensive quantity. Hence it easily can be generalized to multi-component or multiphase systems, assuming all components or phases being homogenously distributed within the capacitor. Considering a sorption system consisting of three quasi-homogenous phases: sorbent (s), sorbate (a), and sorptive gas (f), we get from (6.38) and (6.41) respectively

$$\sum_{i=a,f,s} \alpha_m^i n^i = 3\varepsilon_0 \frac{\varepsilon_{rs} - 1}{\varepsilon_{rs} + 2} V_C \tag{6.44}$$

$$\sum_{i=a,f,s} \alpha_s^i m^i = 3\varepsilon_0 \frac{\varepsilon_{rs} - 1}{\varepsilon_{rs} + 2} V_C \tag{6.45}$$

Here n^i, m^i, i = a, f, s are the mole numbers and masses of the respective phases: adsorbate (a), fluid adsorptive (f), adsorbent (s).

For the sorbent material in vacuum, i.e. $m^f = m^a = 0$ we have from (6.45)

$$\alpha_s^s m^s = 3\varepsilon_0 \frac{\varepsilon_{rs} - 1}{\varepsilon_{rs} + 2} V_C \tag{6.46}$$

Neglecting contributions to the l.h.s of (6.45) from the sorptive gas phase, i. e. restricting to low gas pressures and assuming $(\alpha_s^f m^f << \alpha_s^a m^a)$, we get from (6.45) and (6.46)

$$\alpha_s^a m_{GE}^a = 3\varepsilon_0 \left(\frac{\varepsilon_{rs} - 1}{\varepsilon_{rs} + 2} - \frac{\varepsilon_{rs}^s - 1}{\varepsilon_{rs}^s + 2} \right) V_C \qquad 6.47$$

From this equation the Gibbs excess mass adsorbed (m_{GE}^a) can be determined from measured data of ($\varepsilon_{rs}, \varepsilon_{rs}^s$) and a reasonably accurate model for α_s^a or gravimetric calibration measurements, cp. Sect. 3.

2.2.3 Models for the Complex Permittivity of Dielectric Sorbent-Sorbate Systems

Let us consider again a sorption system consisting on a sorbent-sorbate phase and a sorptive gas located between the plates or cylinders of a capacitor, Fig. 6.8. This system is an electric network which for small applied voltages (U(t)) can be interpreted as a Linear Passive System (LPS). That is: a stimulus (U(t)) applied to the system creates a response, the electric current I(t), which is linearly related to U(t). However it may exhibit a phase shift and also lead to energy dissipation, i. e. Ohmian heat which, as a consequence of the Second Law of Thermodynamics at finite ambient temperature, never can completely be reverted again to electric energy. Linear Passive Systems can be found quite frequently in Physics. A mathematical theory of such systems has been developed by H. König and J. Meixner in the 1960's, [6.27] and later on extended and applied to various stochastic processes, i. e. statistical physics by J. U. Keller, [6.28].

As the theory of LPS provides a fairly general and physically sound basis to develop models for the dielectric behavior of sorbent-sorbate systems, we here will present some of its main results. These then are used to develop an electric network model which includes the well-known Debye model of dielectric materials as a special case [6.24]. Another approach to model dielectric properties of solid sorption systems has been discussed by Coelho in [6.29], which however will not be considered here.

Any voltage (U(t)) applied externally to the material-filled capacitor, Fig. 6.8, will create a current (I(t)), which generally speaking is a functional of the voltage (U(t)). Hence we have a relation

$$I(t) = L\{U(s), -\infty < s \le t\} \qquad 6.50$$

The functional (L) is assumed to have very general properties, as follows:

1. Domain Property

To every "stimulus" $U(t) \in C^{(2)}$, there exists a "response" of the system $I(t) \in C^{(0)}$. Here $C^{(0)}$ is the class of all continuous functions, $C^{(2)}$ that of all functions having (at least) two continuous derivatives.

2. Linearity

If $I_1 = L(U_1)$ and $I_2 = L(U_2)$ are the response functions of two arbitrary stimuli (U_1, U_2) for any set of constants (a_1, a_2) the following relation holds

$$\begin{aligned} L(a_1 U_1 + a_2 U_2) &= a_1 LU_1 + a_2 LU_2 \\ &= a_1 I_1 \quad + a_2 I_2 \end{aligned} \qquad 6.51$$

3. Time Shift Invariance

If I = L(U) according to (6.50) and τ is an arbitrary parameter we have also

$$I(t-\tau) = L\{U(s-\tau), -\infty < s \le t\} \quad \ldots \text{all } \tau, t \qquad 6.52$$

4. Passivity Property

For any pair of (real valued) stimulus function (U(t)) and related response function (I(t)) the inequality holds*)

$$\int_{-\infty}^{t} I(s)U(s)ds \ge 0 \quad \ldots \text{all } t \qquad 6.53$$

This inequality basically is a consequence of the Second Law of Thermodynamics [6.27c, 6.30].

Note: It is not necessary to add to postulates (1-4) an extra requirement for causality, i. e. if $U(s) = 0$ for $s < t_0$, then also $I(t) = L(U(s), -\infty < s < t \le t_0) = 0$ for all $t \le t_0$. Such a property can be concluded from postulates (1-4), [6.30].

*) For complex valued functions (U(t), I(t)) inequality (6.53) has to be replaced by

$$\mathrm{Re} \int_{-\infty}^{t} I(s)U^*(s)ds \ge 0,$$

the operator (Re) indicating the real part of the integral and U(s)* being the conjugate complex function of U(s).

From the above postulates one can conclude that for the Laplace transformations $\hat{U}(p),\hat{I}(p)$ of the voltage U(t) and current I(t), cp. Eqs. (6.5, 6.9) the linear relation

$$\hat{I}(p) = Y(p)\hat{U}(p) \qquad 6.54$$

holds, cp. also Eq. (6.8). Here the function Y=Y(p) is the admittance function of the LPS. It is the inverse of the impedance function already defined in (6.8), i. e. we have: Y(p) = 1/Z(p). Both functions are real positive functions which are analytic in (Re(p)>0), have always non-negative real parts (ReY(p) ≥ 0, ReZ(p) ≥ 0, all p with Re(p) > 0) and according to Herglotz (1930), [6.27a] can be represented by only two constants and a spectral function as

$$Y(p) = ap + \frac{b}{p} + 2p\int_0^\infty \frac{1+\rho^2}{p^2+\rho^2} d\Phi(\rho) \qquad 6.55$$

$$Z(p) = Ap + \frac{B}{p} + 2p\int_0^\infty \frac{1+\rho^2}{p^2+\rho^2} d\Psi(\rho) \qquad 6.56$$

In (6.55) a, b indicate two non-negative constants

$$a = \lim_{p\to\infty} \frac{Y(p)}{p} \geq 0, \qquad 6.57$$

$$b = \lim_{p\to 0} pY(p) \geq 0 \qquad 6.58$$

The integral in (6.55) is a Stiltjes-integral with the spectral function $\Phi = \Phi(\rho)$ consisting of three different parts

$$\Phi = \Phi_1 + \Phi_2 + \Phi_3 \qquad 6.59$$

The function $\Phi_1 = \Phi_1(\rho)$ is continuous and almost everywhere with finite first derivative. It can be calculated from a model function Y(p) as

$$\Phi_1'(\rho) = \frac{1}{\pi(1+\rho^2)} \lim_{p\to i\rho} \mathrm{Re}\, Y(p) \qquad 6.60$$

Physically it is related to dissipative effects within the LPS (cp. examples given below).

The function $\Phi_2 = \Phi_2(\rho)$ is a pure step function having for rational admittance functions Y(p) a finite number of positive steps. To give an example let us assume Φ_2 has at $\rho = \rho_i, i = 1, 2, 3 ...$ steps of height

$$\Phi_2(\rho_i + 0) - \Phi_2(\rho_i - 0) = \Delta\Phi_i > 0 \qquad 6.61$$

i. e. we have

$$\Phi_2(\rho) = \sum_i E(\rho - \rho_i)\Delta\Phi_i \qquad 6.62$$

with the unit step function

$$E(\rho) = \begin{cases} 0 ... \rho < 0 \\ 1 ... \rho > 0 \end{cases} \qquad 6.63$$

Then the contribution of these steps to the admittance function Y(p) is, according to (6.55),

$$Y(p) = ... + 2p \sum_i \frac{1 + \rho_i^2}{p^2 + \rho_i^2} \Delta\Phi_i + ... \qquad 6.64$$

The third part of the spectral function $\Phi_3 = \Phi_3(\rho)$ is the "pathological portion" of the spectral function $\Phi(\rho)$. It is continuous but almost nowhere differentiable as the trajectories of a classical Brownian particle or of a random walker in the limit of vanishing step width $(\Delta x \to 0)$ and time interval between single steps $(\Delta\tau \to 0)$ with $\lim\left((\Delta x)^2 / \Delta\tau\right) = D$ being finite, or as curves of fractal dimension (D > 1) like Koch's curves [6.31, 6.32]. Since presently no example of a LPS is known where this part of the spectral function Φ is different from Zero, we do not take it into account in model functions for $\varepsilon(p)$, i. e. we assume in what follows $\Phi_3(\rho) = 0$. However, it very well could be that the *kinetics* of adsorbate phases of fractal dimension, i. e. mass transfer and diffusion processes in these phases, is reflected in this "fractal" spectral function (Φ_3) of the system.

Statements analogous to (6.57-6.64) hold for the parameters (A, B, $\Psi(\rho)$) of the impedance function (6.56).

In view of this assumption and Eqs. (6.59, 6.64) the (complex) admittance function (6.55) of an arbitrary sorption system always can be written as

$$Y(p) = ap + \frac{b}{p} + 2p\sum_i \frac{1+\rho_i^2}{p^2+\rho_i^2}\Delta\Phi_i + 2p\int_0^\infty \frac{1+\rho_i^2}{p^2+\rho_i^2}\Phi_1'(\rho)d\rho \qquad 6.65$$

Here the integral is of the simple Riemann-type with the spectral function being monotonously increasing and differentiable, i. e. $\Phi_1'(\rho) \geq 0$, [6.27a].

The dielectric permittivity ($\varepsilon_0\ \varepsilon_r$) of the system is related to Y(p) according to (6.24, 6.25) by

$$\varepsilon_0\varepsilon_r(p) = G\frac{Y(p)}{p} \qquad 6.66$$

Here G is the geometrical factor of the capacitor which is defined by

$$G = \begin{cases} d/A & \text{plate capacitors} \\ \ln(r_a/r_i)/(2\pi h) & \text{cylinder capacitors} \end{cases} \qquad 6.67$$

Combining (6.65) and (6.66) and choosing $p = i\omega$ we get for the frequency dependent permittivity $\varepsilon_r(\omega)$ the representation

$$\varepsilon_0\varepsilon_r(\omega) = G\left\{a - \frac{b}{\omega^2} + 2\sum_i \frac{1+\rho_i^2}{\rho_i^2-\omega^2}\Delta\Phi_i + 2\int_0^\infty \frac{1+\rho^2}{\rho^2-\omega^2}\Phi_1'(\rho)d\rho\right\} \qquad 6.68$$

Since in physically relevant models the parameter (b) always vanishes, i. e. b = 0, cp. (6.58), we get from (6.68) for

a) the static dielectric permittivity ($\omega = 0$)

$$\varepsilon_0\varepsilon_{rs} = G\left\{a + 2\sum_i\left(\frac{1}{\rho_i^2}+1\right)\Delta\Phi_i + 2\int_0^\infty\left(\frac{1}{\rho^2}+1\right)\Phi_1'(\rho)d\rho\right\} \qquad 6.69$$

with $\Phi_1'(\rho) \geq 0 \ldots \text{all } 0 < \rho < \infty$, and for b) the optical permittivity ($\omega \to \infty$)

$$\varepsilon_0\varepsilon_{r\infty} = \varepsilon_0 n_{opt}^2 = Ga \qquad 6.70$$

Here n_{opt} indicates the optical refractive index of the system.

The representation (6.69) of ε_{rs} indicates that this quantity includes contributions from all the resonance frequencies $(\omega_i = \rho_i)$ of the frequency dependent permittivity (6.68). Moreover it also is determined by some dissipative properties of the material reflected in the spectral function $(\Phi_1(\rho))$ and also in the optical permittivity $(\varepsilon_{r\infty} = n_{opt}^2)$, [6.6, 6.29].

Considering the representation (6.66) not only along the imaginary axis, i. e. for $p = i\omega$, as we have done here but in the right half of the complex p-plane, i. e. for $p = p_1 + ip_2$ with $p_1 \geq 0$, would allow the separation of the real and the imaginary parts of ε_r and permit the introduction of the Cole-Cole curves characterizing the dielectric system to be modelled [6.6, 6.7, 6.11, 6.29]. However, owing to space limitations we shall refrain from going into further details.

Instead we want to emphasize that simple electric network models of LPS may include three different elemental systems: capacitors, resistances, and inductances [6.12]. The basic physical relations, admittance functions, elements of the representation theorem (6.55) and corresponding static and optical permittivity are collected in Table 6.1 below. These elements can be combined by series or parallel connections in may different ways. For the admittance functions of the electric network generated in this way, the simple rules hold that

a) the admittance function (Y(p)) of a parallel connection of elements is the sum of the admittance functions $(Y_i(p))$ of the elements, i. e.

$$Y = \sum_i Y_i \qquad 6.71$$

b) the inverse of the admittance function (Y(p)) of a series connection of elements is the sum of the inverses of the admittance functions of the elements $(Y_i(p))$, i. e.

$$\frac{1}{Y(p)} = \sum_i \frac{1}{Y_i(p)}. \qquad 6.72$$

With these rules and Table 6.1 in mind, it is very simple to write down the admittance functions of simple 2-pole electrical network considered as models for the permittivity of sorption systems [6.12].

Table 6.1: Elements of electric network models for the (complex) admittance function Y=Y(p) and the frequency dependent permittivity $\varepsilon_r = \varepsilon_r(\omega)$ of a sorption system.

	Capacitor	Resistor	Inductance
Symbol	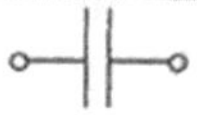		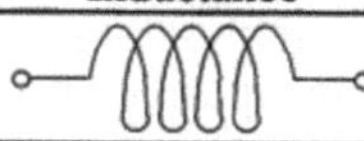
Quantity	C/(As/V)	R/Ω=(V/A)	L/(Vs/A)
Physical Equation	$I = C\dot{U}$	$I = U/R$	$\dot{I} = U/L$
Admittance Y(p) =	pC	$\frac{1}{R}$	$\frac{1}{pL}$
Presentation Theorem (6.55):			
a =	C	0	0
b =	0	0	1/L
$\Phi_1(\rho) =$	0	$\frac{1}{\pi R}$ arct ρ	0
$\Phi_2(\rho) =$	0	0	0
Static Permittivity [6.24, 6.25]:			
$\varepsilon_0\varepsilon_{rs} =$	GC	(- i ∞)	(- ∞)
$\varepsilon_0\varepsilon_{r\infty} =$	GC	0	0

To elucidate the foregoing general statements let us consider the network sketched in Fig. 6.9 as an example. It basically consists of an ideal capacitor (C_∞) and in parallel another capacitor (C) with a series connection to an ideal resistor (R). The capacitance (C_∞) is proportional to the permittivity at very high (optical) frequencies of the applied electric field ($\varepsilon_{r\infty} = n_{opt}^2$) which is the square of the optical refraction index (n_{opt}) of the material. The capacitance (C) is proportional to the difference of the static permittivity and the optical permittivity ($\varepsilon_{rs} - \varepsilon_{r\infty}$). The resistor (R) is related to the dissipative properties of the material: at frequencies $\omega_0 = 1/RC$ the imaginary part of the permittivity (6.80), i. e. the dielectric losses in the medium, are a maximum compared to their respective values at any other frequency (ω) of an alternating electric field ($U(t) \approx e^{i\omega t}$) applied to the system.

According to Table 6.1 and the rules (6.71, 6.72) for admittance functions of networks given above, the admittance function of the so-called Debye-network, Fig. 6.9 is

$$Y(p) = pC_\infty + \frac{pC}{1 + CRp}. \qquad 6.73$$

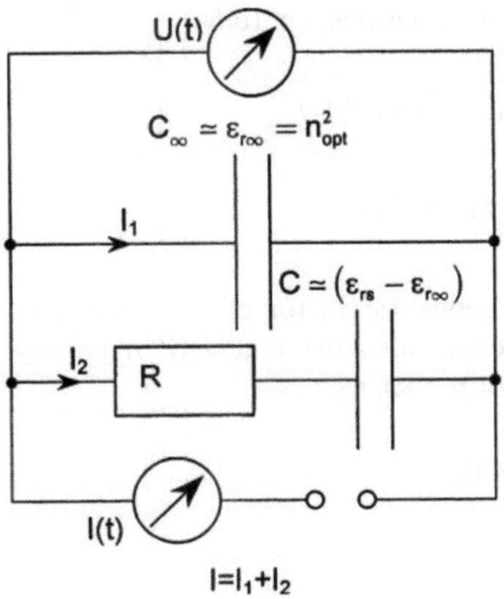

Figure 6.9. Electric network equivalent to the Debye model for the permittivity of dielectric materials [6.3, 6.6, 6.7]. The network consists of a capacitor ($C_\infty \approx \varepsilon_{r\infty} = n_{opt}^2$) related to the "optical permittivity" $\varepsilon_{r\infty}$ and the optical refraction index (n_{opt}) of the system. In parallel there are another capacitor (C) and a resistor (R) whose values determine the static permittivity (ε_{rs}) of the network ($C \approx \varepsilon_{rs} - \varepsilon_{r\infty}$) and the characteristic frequency for a maximum of the dielectric losses, i. e. dissipation ($\omega_0 = 1/RC$) of electric energy in the network.

The parameters a, b, Φ of this function included in the representation theorem (6.55) are, according to (6.57-6.60):

$$a = C_\infty,\ b = 0$$

$$\Phi_1'(\rho) = \frac{1}{\pi R(1+\rho^2)} \bullet \frac{(\tau\rho)^2}{1+(\tau\rho)^2},\ \tau = CR \qquad 6.74$$

$$\Phi_2 = 0, \Phi_3 = 0$$

The permittivity corresponding to (6.73) is, in view of (6.66)

$$\varepsilon_r(p) = \varepsilon_{r\infty} + \frac{\varepsilon_{rs} - \varepsilon_{r\infty}}{1+p\tau},\ \tau = CR. \qquad 6.75$$

Here we have introduced the static permittivity (ε_{rs}) and the optical permittivity ($\varepsilon_{r\infty}$) by the relations, cp. (6.66)

$$\varepsilon_0\varepsilon_{rs} = \lim_{p\to 0}\varepsilon_0\varepsilon_r(p) = G(C_\infty + C). \qquad 6.76$$

$$\varepsilon_0\varepsilon_{r\infty} = \lim_{p\to\infty}\varepsilon_0\varepsilon_r(p) = GC_\infty, \qquad 6.77$$

G representing the geometric factor of the system, cp. (6.67). For sake of completeness we denote also the real and imaginary part of the complex permittivity (6.75) for $p = i\omega$, as

$$\varepsilon_r(p) = \varepsilon'(p) + i\varepsilon''(p). \qquad 6.78$$

with

$$\varepsilon'(\omega) = \varepsilon_{r\infty} + \frac{\varepsilon_{rs} - \varepsilon_{r\infty}}{1+(\tau\omega)^2}. \qquad 6.79$$

$$\varepsilon''(\omega) = \varepsilon_{r\infty} + \frac{\varepsilon_{rs} - \varepsilon_{r\infty}}{1+(\tau\omega)^2}\tau\omega. \qquad 6.80$$

Graphical representations of these functions are given in Fig. 6.10 followed by a Cole-Cole plot, Fig. 6.11 showing the curve $\varepsilon'' = \varepsilon''(\varepsilon')$ which for the Debye model is a half-circle in the $\varepsilon', \varepsilon''$ -plane. These functions (6.79, 6.80) often are used for a first approximate description of the dielectric properties of a material system [6.6, 6.7]. An example of (6.79) for a sorption system already has been given in Fig. 6.5, cp. also data presented in Sects. 2.3, 3.2.

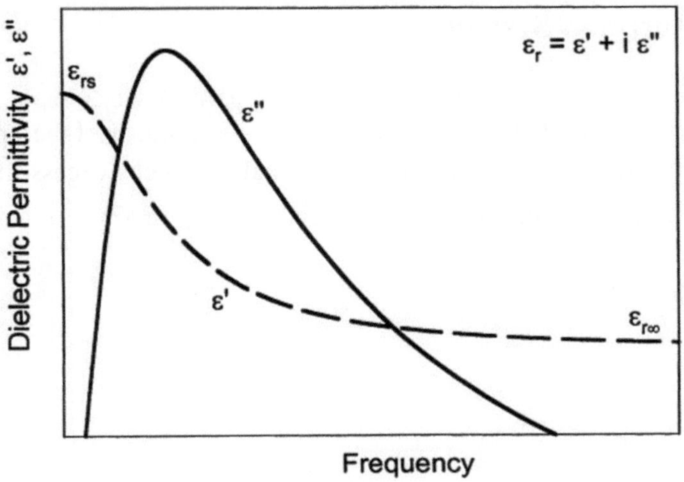

Figure 6.10. Real part $\varepsilon'(\omega)$ and imaginary part $\varepsilon''(\omega)$ of the complex dielectric permittivity $\varepsilon_r = \varepsilon' + i\varepsilon''$ of the Debye model represented by Eqs. (6.79, 6.80) referring to the electric network Fig. 6.9.

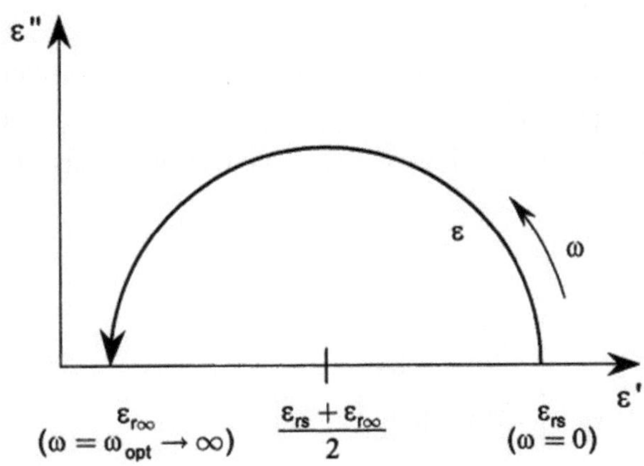

Figure 6.11. Cole-Cole-plot $\varepsilon'' = \varepsilon''(\varepsilon')$ corresponding to the Debye model permittivity Eqs. (6.79, 6.80). The curve is a half circle in the upper $\varepsilon', \varepsilon''$ -plane with diameter $(\varepsilon_{rs} - \varepsilon_{r\infty})$.

2.3 Uncertainties of Dielectric Measurements of Adsorption Systems

Measurements of the (static) dielectric permittivity (ε_{rs}) of a sorbent/sorbate system allow the determination of the product of the specific polarizability of the molecules adsorbed (α_s^a) and the Gibbs excess mass of the adsorbate (m_{GE}^a). Indeed we have from Sect. 2.2, Eq. (6.47)

$$\alpha_s^a m_{GE}^a = 3\varepsilon_0 E_{rs} V_c \tag{6.82}$$

with

$$E_{rs} \equiv \frac{\varepsilon_{rs} - 1}{\varepsilon_{rs} + 2} - \frac{\varepsilon_{rs}^s - 1}{\varepsilon_{rs}^s + 2}. \tag{6.83}$$

Hence we can calculate the following from (6.82)

a) the Gibbs excess mass adsorbed (m_{GE}^a) if the specific polarizability (α_s^a) of the adsorbate is known or vice versa
b) the specific polarizability (α_s^a) if the Gibbs excess mass (m_{GE}^a) has been determined otherwise, cp. Sect. 3.

The uncertainties of both types of measurements represented by the dispersions or mean square deviations ($\sigma_{mGE}, \sigma_{\alpha as}$) of ($m_{GE}^a, \alpha_s^a$) and can be calculated by applying the Gauss law of error propagation to Eqs. (6.82, 6.83) as follows [2.18]:

$$\sigma_{mGE}^2 = \left(\frac{3}{\alpha_s^a} E_{rs} V_C\right)^2 \sigma_{\varepsilon 0}^2 + \left(\frac{3\varepsilon_0}{(\alpha_s^a)^2} E_{rs} V_C\right)^2 \sigma_{\alpha as}^2 + \left(\frac{3\varepsilon_0}{\alpha_s^a} V_C\right)^2 \sigma_E^2 + \left(\frac{3\varepsilon_0}{\alpha_s^a} E_{rs}\right) \sigma_{VC}^2 \tag{6.84a}$$

$$\sigma_{\alpha as}^2 = \left(\frac{3}{m_{GE}^a} E_{rs} V_C\right)^2 \sigma_{\varepsilon 0}^2 + \left(\frac{3\varepsilon_0}{(m_{GE}^a)^2} E_{rs} V_C\right)^2 \sigma_{mGE}^2 + \left(\frac{3\varepsilon_0}{m_{GE}^a} V_C\right)^2 \sigma_E^2 + \left(\frac{3\varepsilon_0}{m_{GE}^a} E_{rs}\right) \sigma_{VC}^2 \tag{6.84b}$$

In these expressions normally the first term on the r.h.s. can be neglected as (ε_0) is a very well known constant and $\sigma_{\varepsilon 0} \cong 10^{-10}$, cp. (6.12), according to (6.83) the dispersion (σ_E) of E is is given by

$$\sigma_E^2 = \frac{9}{(\varepsilon_{rs}+2)^4}\sigma_{\varepsilon rs}^2 + \frac{9}{(\varepsilon_{rs}^s+2)^4}\sigma_{\varepsilon rss}^2 . \qquad 6.85$$

Here the dispersions ($\sigma_{\varepsilon rs}, \sigma_{\varepsilon rss}$) of ($\varepsilon_{rs}, \varepsilon_{rss}$) can be determined from the equations

$$\varepsilon_0 \varepsilon_{rs} = G\frac{Q}{U} . \qquad 6.86a$$

$$\varepsilon_0 \varepsilon_{rs}^2 = G\frac{Q^s}{U^s} . \qquad 6.86b$$

which follow from (6.1, 6.2) and (6.66, 6.67), G indicating a geometrical factor depending of the type of capacitor – plate or cylinder – chosen for measurements. Also (Q, U) and (Q^s, U^s) denote the electric charge and the static voltage applied to the capacitor filled with sorbent material in a sorptive gas atmosphere (6.86a) and vacuum (6.86b) respectively. Applying again the Gauss law we get from (6.86a, b)

$$\sigma_{\varepsilon rs}^2 = \left(\frac{Q}{\varepsilon_0 U}\right)^2 \sigma_G^2 + \left(\frac{G}{\varepsilon_0 U}\right)^2 \sigma_Q^2 + \left(\frac{GQ}{\varepsilon_0 U^2}\right)^2 \sigma_U^2 . \qquad 6.87a$$

$$\sigma_{\varepsilon rss}^2 = \left(\frac{Q^s}{\varepsilon_0 U^s}\right)^2 \sigma_G^2 + \left(\frac{G}{\varepsilon_0 U^s}\right)^2 \sigma_{Qs}^2 + \left(\frac{GQ^s}{\varepsilon_0 U^{s2}}\right)^2 \sigma_{Us}^2 . \qquad 6.87b$$

These expressions must be inserted into (6.85) to lead via (6.84a,b) to the dispersions ($\sigma_{mGE}, \sigma_{\alpha as}$) of ($m_{GE}^a, \alpha_s^a$) respectively. For sake of clarity we list all quantities and parameters to be measured, and their dispersions to be determined or at least reasonably estimated, in a scheme:

ε_0	α_s^a	E	Vc	ε_{rs}	ε_{rs}^s	m_{GE}^a	G	Q	U	Q^s	U^s
$\sigma_{\varepsilon 0}$	$\sigma_{\alpha as}$	σ_E	σ_{VC}	$\sigma_{\varepsilon rs}$	$\sigma_{\varepsilon rss}$	σ_{mGE}	σ_G	σ_Q	σ_U	σ_{Qs}	σ_{Us}

Numerical examples have shown that in order to get accurate data of Gibbs excess masses adsorbed from dielectric measurements it is important to have good calibration curves $\alpha_s^a = \alpha_s^a(\varepsilon_{rs}, \varepsilon_{rs}^s)$ and accurate $(\varepsilon_{rs}, \varepsilon_{rs}^s)$ data at hand, i. e. $(\sigma_{\varepsilon rs}, / \varepsilon_{rs}) < 10^{-4}, (\sigma_{\varepsilon rss}, / \varepsilon_{rs}^s) < 10^{-4}$, [6.26, 6.8, 6.3]. With these dispersions given, relative uncertainties of adsorbed masses of $(\sigma_{mGE} / m_{GE}^a) < 10^{-2}$ normally can be expected. A similar statement holds for measurements of the polarizability of adsorbed molecules (α_s^a) if good calibration curves $m_{GE}^a = m_{GE}^a(\varepsilon_{rs}, \varepsilon_{rs}^s)$ are at hand, cp. (6.82, 6.84b).

2.4 Examples

As already mentioned impedance spectroscopy provides a fairly new method to characterize porous sorbent materials and sorbate phases in them, allowing the observation of both adsorption equilibria and (slow) adsorption processes as well. Also uptake curves and breakthrough curves in industrial adsorption reactors can be measured [6.10, 6.13, 6.14]. In this method the dielectric properties of the material to be characterized in weak alternating electric fields are observed for various frequencies. The results provide qualitative and quantitative information on the dielectric behavior of the material and also its changes due to gas adsorption. Moreover some information on the structure of the adsorbed phase, i. e. its predominant gas or liquid character can be gained as well as some insight into the interaction between the sorbent and the sorbate phase which cannot be detected by purely volumetric or gravimetric measurements, cp. especially Figure 6.29.

In this section we present a few examples of dielectric spectra taken of adsorption systems in both equilibria states and during ad- and desorption processes. A few introductory remarks may be helpful to physically interpret the experimental data.

1. All sorbent materials used were of commercial quality, i. e. consisted of nearly cylindrical pellets with typical lengths and diameters of about 0.5 cm and 0.3 cm respectively. These pellets were randomly poured into the (plate or cylindrical) capacitor, cp. Fig. 6.2, to simulate configurations realized in industrial adsorption reactors.

2. For static electric fields and for alternating fields at low frequencies $(\nu < 1\text{kHz})$, impedance spectra $(C(\nu))$ and permittivities $(\varepsilon(\nu))$ were not reproducible, i. e. were dependant on the individual filling in the capacitor. Only for higher frequencies were reproducible results obtained which were independent not only from the special filling process but also from the size

or type of capacitor and wires between the capacitor and impedance analyzer chosen. To achieve this, it is recommended that four-pole-circuits always be used between the capacitor and impedance analyzer.

3. The impedance spectra shown here are restricted to the frequency range $1\ kHz < \nu < 13\ MHz$ according to the impedance analyzer used (HP 4192A). Hence in the so-called Cole-Cole-plots, Figs. 6.15-6.17, only the low frequencies parts of the curves $\varepsilon'' = \varepsilon'(\varepsilon')$ are shown corresponding to the right hand part of the semi-circle of the ideal Cole-Cole-curve in Fig. 6.11.

The experimental data to be presented in this section can be grouped as follows:

1. Impedance/capacitance measurements of gas adsorption systems in thermodynamic equilibrium:
   Carbon dioxide (CO_2, 4.5) [*)] on dealuminated zeolite of Y-type (DAY, Degussa) (Fig. 6.12);
   Carbon monoxide (CO, 3.7) on molecular sieve (MS13X, UOP) (Fig. 6.13);
   Sorptive gas mixture methane-carbon monoxide (CH_4:CO=90%:10%mol) on activated carbon Norit R1 Extra (Norit, Netherlands) (Fig. 6.14).

2. Cole-Cole curves of complex dielectric permittivities of gas-adsorption systems in equilibrium states:
   Nitrogen (N_2, 5.0) on molecular sieve (MS13X, UOP) (Fig. 6.15);
   Carbon monoxide (CO) on molecular sieve (MS13X, UOP) (Fig. 6.16);
   Hydrogen sulfide (H_2S, 1.8) on molecular sieve (MS13X, UOP) (Fig. 6.17).

3. Impedance/capacitance measurements of sorption equilibria of gases in polymeric materials;
   Carbon dioxide (CO_2) on polycarbonate (Bayer, Leverkusen) (Figs. 6.18,6.19);
   Methane (CH_4, 5.5) on polycarbonate (Bayer, Leverkusen) (Figs. 6.20, 6.21).

4. Impedance/capacitance measurements of gas adsorption and desorption processes:
   Carbon monoxide (CO, 3.7) on activated carbon Norit R1; (Fig. 6.22);

[*)] The numbers indicate the purity or concentration of the gas in the delivered state by the provider (Messer-Griesheim, Air Liquide, and others). Example: 3.7 = 99.97 %, 5.0 = 99.9990 % etc.

Regeneration of activated carbon Norit R1 in helium (He, 5.0)atmosphere. (Fig. 6.23).

Ad 1. Impedance/capacitance measurements of adsorption equilibria.
In Figure 6.12 impedance spectra of dealuminated Y-zeolite (Degussa, Hanau, Germany) in vacuum and in adsorption equilibria with (subcritical) carbon dioxide (CO_2, 4.5) at T = 298 K for pressures up to 2 MPa are shown [6.3]. Data of the real part of the capacitance of the material filled capacitor (Fig. 6.2) calculated from impedance measurements (U(t), I(t)) via Eqs. 6.3, 6.7, are depicted as a function of the frequency of the alternating electric field in the range 1 kHz $< \nu <$ 13 MHz for vacuum ($p < 1$ Pa) and 4 different gas pressures. The spectra display a decrease of the capacitance with increasing frequency, the normal dispersion in the range 1 kHz $< \nu <$ 4 MHz. For higher frequencies there is a small increase of the capacitance with increasing frequency, cp. the ordinate scaling (pF), so-called anomalous dispersion.

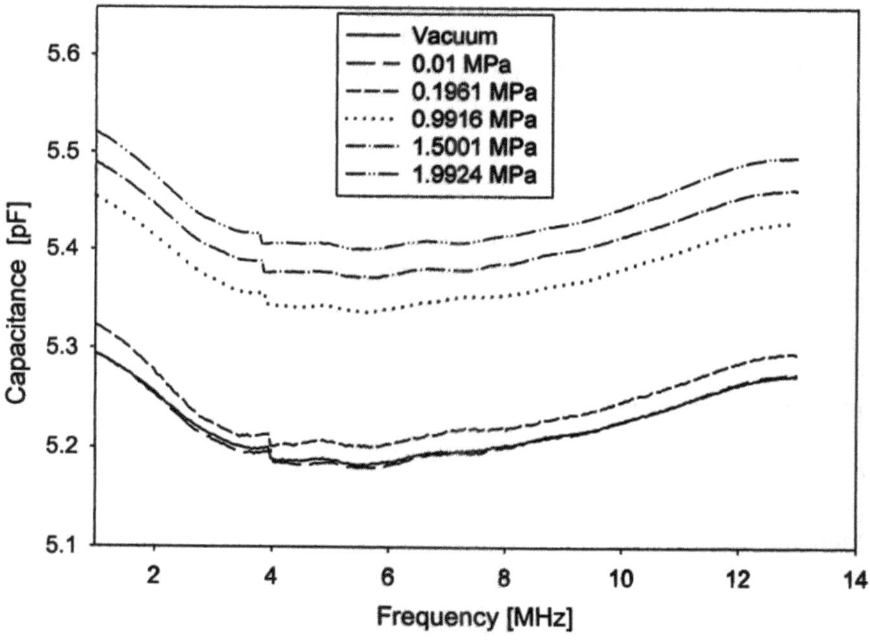

Figure 6.12. Capacitance spectra of zeolite DAY (Degussa) in vacuum and for adsorption equilibria of CO_2 at 298 K for pressures up to 2 MPa and frequencies 1 kHz $< \nu <$ 13 MHz, [6.3].

The later phenomenon is still not well understood from a physical point of view but should be related to special properties of the lattice structure of the sorbent, as all the curves taken have nearly the same geometrical structure, the CO_2-**adsorption** only shifting thems to somewhat higher capacitance values [6.6, 6.7]. The vertical distance of these adsorption spectra curves from the vacuum curve can be considered as a (nearly linear) measures of the amount of CO_2 adsorbed, cp. examples given in Sect. 3.2.

Figure 6.13 refers to impedance measurements taken for the system carbon monoxide (CO, 3.7) on molecular sieve MS13X (UOP, Erkrath, Germany) at T = 295 K for gas pressures up to 12.11 MPa, [6.3, 6.14] and frequencies 1 kHz < ν < 500 kHz. The real part of the capacitance decreases with increasing frequency (normal dispersion), cp. also Fig. 6.11. The spectrum taken in a vacuum deviates remarkably from all the other curves for gas adsorption equilibria at the indicated pressures. This is a hint that the energetically inhomogenous surface of the molecular sieve is in a way homogenized by the CO molecules adsorbed first, i. e. at low surface coverage. This may be due to a kind of damping or freezing of internal degrees of freedom of the lattice of the molecular sieve by the admolecules arriving first [6.3]. At higher sorptive gas pressures, i. e. adsorption loads, admolecules always find nearly the same type of adsorption sites. Hence the capacitance spectra have the same internal structure but otherwise are only shifted to higher capacitance values, this reflecting simply the increasing amount of CO-molecules adsorbed [6.22, 6.33].

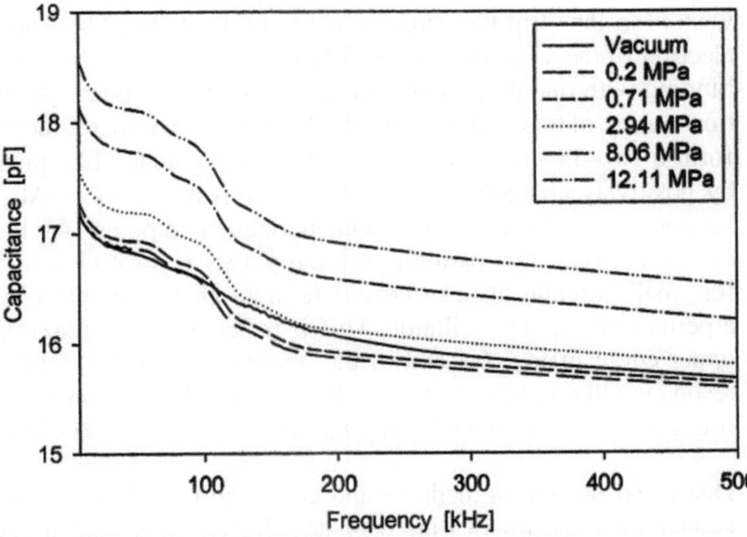

Figure 6.13. Capacitance spectra of molecular sieve MS13X (UOP) in vacuum and for adsorption equilibria of CO at 295 K for pressures up to 12.1 MPa and frequencies 1 kHz < ν < 500 kHz, [6.14].

In Fig. 6.14 capacitance spectra of activated carbon (AC) Norit R1 (Norit, Amersfoort, The Netherlands) in vacuum and for adsorption equilibria for a gas mixture methane – carbon monoxide (CH_4: CO = 90 % : 10 %mol) are shown. Again the real part of the capacitance measured at the cylinder capacitor, Fig. 6.2, filled with the AC and exposed to the sorptive gas at 298 K and for pressures up to 8 MPa is displayed for frequencies 1 kHz < ν < 1 MHz, [6.13].

The vacuum spectrum referring to the nearly empty and thermally activated carbon shows a remarkable maximum at about 200 kHz. It can be interpreted as follows: Microporous carbons exhibit a rather complicated and quasi-chaotic molecular structure [6.33-6.36] which basically consists of aromatic sheets or strips formed by hexagonal rings of C-atoms as in graphite. The sheets often pile up like a heap of disordered carpets, several heaps forming irregular structures with interstitial spaces of slit-like or pore-like character finally called "micropores". The topography, connectivity and material density of this quasi-chaotic and highly disorganized structure depends on the nature of the raw material and the production process, i. e. mechanical, thermal, and chemical treatment during its activation [6.38].

An aromatic string consists of hexagonal rings formed by C-atoms as in a graphite plane. At the edges of such a string active atomic groups including other elements like (O, H, N, P...) often can be found. Every C-atom in the strip has three neighboring C-atoms. The fourth valence electron of the C-atom is a quasi free electron which can build up a π-binding with one of the three other neighboring C-atoms. As the excitation energy of this metallic fourth electron is very small, these electrons form a two-dimensional electron-gas in the graphite-like plane formed by positively charged C-atoms. Hence activated carbon (AC) also can be considered as a medium with permanent dipoles. As the binding energy between two parallel graphite planes is very low, it is possible for small graphite planes to oscillate in an alternating electric field like a permanent dipole oscillator. This is reflected in the maxima of a) the upper curve in Fig. 6.4 and b) the upper curve in Fig. 6.14 showing a resonance-like behavior at ca. 200 kHz, this low frequency indicating oscillations of structures having masses of about 10-100 C-atoms.

Due to adsorption of methane and carbon monoxide the maxima in the capacitance spectra of Fig. 6.14 are shifted somewhat to even lower frequencies indicating an increase of the mass of the oscillating elements due to adsorption. Also the whole structure seems dielectrically to be stabilized by the admolecules as the capacitance decreases with increasing gas pressure, i. e. amount of gas adsorbed [6.13, 6.26].

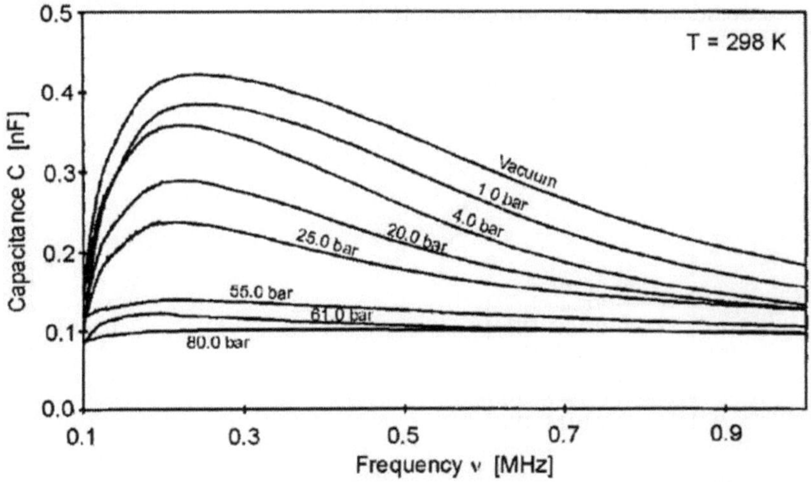

Figure 6.14. Capacitance spectra of activated carbon Norit R1 for vacuum and adsorption equilibria of a gas mixture (CH_4 : CO = 90 % : 10 %mol) at T = 298 K for pressures up to 8 MPa, (10 bar = 1 MPa), and frequencies 1 kHz < ν < 1 MHz taken at the cylinder capacitor, Fig. 6.2.

Ad 2. Cole-Cole-curves of adsorption equilibria.

The (relative) dielectric permittivity (ε_r) of a material system can be calculated from impedance capacitance measurements via Eqs. 6.11, 6.13. As ε_r normally is a complex quantity it is appropriate to consider its real part (ε') and imaginary part (ε'') separately. An elegant way to do this is to consider the functions $\varepsilon' = \varepsilon'(\omega = 2\pi\nu)$, $\varepsilon'' = \varepsilon''(\omega = 2\pi\nu)$, which provide a parametric representation of the curve $\varepsilon'' = \varepsilon''(\varepsilon')$. This was suggested first by K. S. Cole and R. H. Cole. Hence plots of these curves today are called Cole-Cole-plots, an example of which already has been given in Fig. 6.11.

In Fig. 6.15 the Cole-Cole-plots of the complex permittivity of molecular sieve MS13X in a vacuum and for adsorption equilibria of nitrogen (N_2, 5.0) at T = 298 K for pressures up to 1 MPa referring to frequencies 1 kHz < ν < 200 kHz are shown. These curves can be interpreted as a superposition of two semicircles reflecting two basically different oscillating atomic groups in the molecular sieve, cp. Fig. 6.11. This structure is not substantially changed by the adsorption of N_2 as all curves taken at increasing gas pressures, and hence adsorption loads, are nearly identical with the vacuum curve. This was to be expected as the N_2-molecule does not have a dipole moment and hence should not change the dielectric properties of the MS considerably if adsorbed in it [6.3].

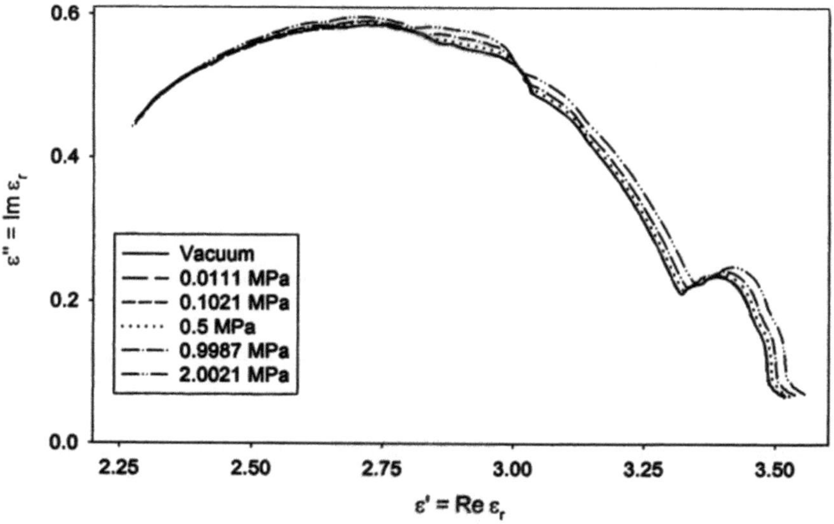

Figure 6.15. Cole-Cole-plot of molecular sieve MS13X (UOP, Erkrath) in vacuum and for adsorption equilibria with nitrogen (N_2, 5.0) taken at 298 K for pressures up to 1 MPa.

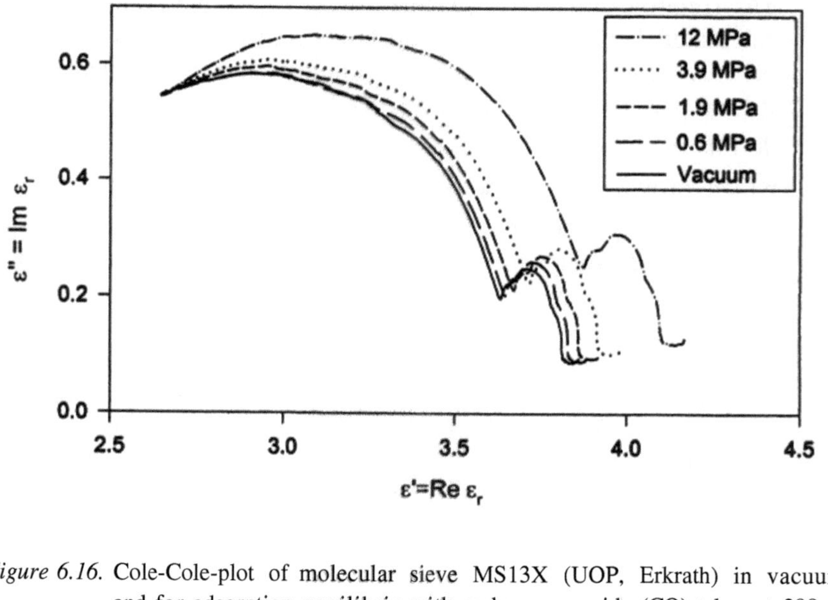

Figure 6.16. Cole-Cole-plot of molecular sieve MS13X (UOP, Erkrath) in vacuum and for adsorption equilibria with carbon monoxide (CO) taken at 298 K for pressures up to 12 MPa., [6.3].

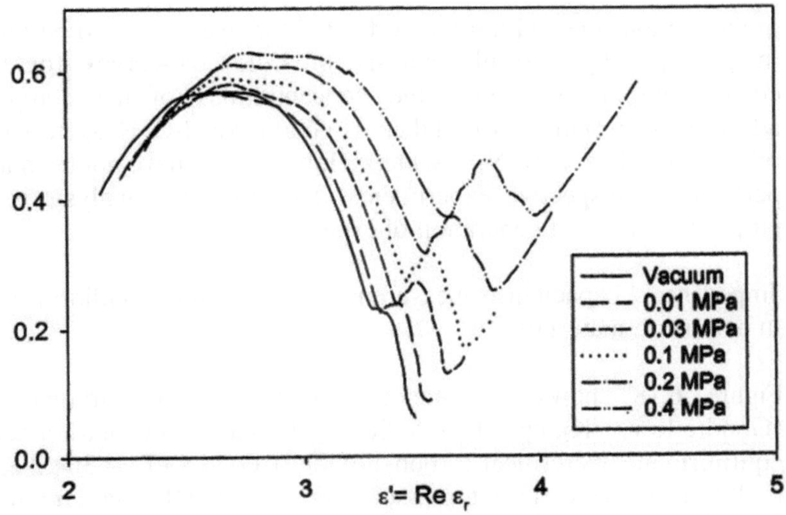

Figure 6.17. Cole-Cole-plot of molecular sieve MS13X (UOP, Erkrath) in a vacuum and for adsorption equilibria with hydrogen sulfide (H_2S, 1.8) taken at 298 K for pressures up to 0.4 MPa, [6.3, 6.10].

In Figures 6.16, 6.17 Cole-Cole-plots of adsorption equilibria of carbon monoxide (CO) and hydrogen sulfide (H_2S) at 298 K for various pressures and frequencies $1\ kHz < \nu < 1\ MHz$ are shown [6.3, 6.10, 6.13]. As molecules in both gases are polar with dipole moments ($\mu_{CO} = 0.1$ D, $\mu_{H2S} = 0.9$ D) the permittivity of the sorbent is changed considerably upon adsorption of these molecules. This is reflected in a shift of the Cole-Cole-curves away from the (lower) vacuum curve. However, for high frequencies approaching the optical range, all the curves merge again (CO) and partly intersect (H_2S).

The (lower) curves presenting the vacuum permittivity of MS13X in Figs. 6.16, 6.17 differ from each other considerably. This is due to a somewhat different activation procedure of the material samples used in the experiments. However, the basic structure of both curves is identical indicating an overlapping of two somewhat confined semicircles related to two different oscillation / relaxation mechanisms in the empty sorbent. Upon adsorption the low frequency, i. e. right hand portion of the Cole-Cole-curves, is growing indicating intensifying of the respective oscillation mechanism. Even occurrence of a third mechanism at very low frequencies possibly being related to clusters of adsorbed molecules is indicated by the sharply rising nearly straight lines right hand portions of the curves. Diagrams of this type can serve in a twofold way: to characterize a sorbent material and / or its state of activation in a vacuum and also to deliver characteristic “fingerprints” of the admolecules within the sorbent. Moreover it should be possible in principle by careful analysis of Cole-Cole-curves for binary coadsorbates to determine the concentrations of the components adsorbed, especially if one of them is polar (like H_2O, H_2S, etc.) and the other is not (like CH_4, N_2, O_2, etc.). However, we have not been able to perform the respective experiments. Therefore this analysis is to be pursued by the next generation of experimenters.

Ad 3. Impedance / capacitance measurements of sorption equilibria of gases in polymeric materials

Figure 6.18 shows capacitance spectra of a polycarbonate (PC) (Goodfellow, UK), cp. Chap. 5, Sect. 3.4, in a vacuum and for sorption equilibria in supercritical carbon dioxide (CO_2, 4.5)(T_C = 304.2 K) at T = 308 K for 3 different pressures up to 4.5 MPa and frequencies 0.1 MHz < ν < 13 MHz. The real part of the complex capacitance (ReC) calculated from impedance measurements, by Eqs. (6.6-6.8), is sketched on the ordinate axis as function of the frequency (ν) of the electric field. As can be seen from the data, for low frequencies (0.1 MHz < ν < 6 MHz) the capacitance decreases with increasing frequencies (normal dispersion). However, for higher frequencies (6 MHz < ν <13 MHz) it increases again, to a certain extent. The vacuum spectrum (lower curve) and all the other sorption spectra (upper curves) are of the same character, this indicating that the molecular structure of the material is not basically changed by the sorption process.

The lower frequency portion of the spectra in Fig. 6.18 is given in Fig 6.19. Here the relative dielectric permittivity (ε_r) of the polymer-sorption system calculated from the capacitance (ReC), by Eq. (6.13), is sketched for frequencies 5 kHz < ν <100 kHz. Again all spectra are nearly of the same simple character, i. e. constant, but shifted to higher ε-values with increasing gas pressure, i. e. amount of CO_2 sorbed in the polymer. This again indicates that the polymer contrary to activated carbons or zeolites, in both unloaded and CO_2-loaded states behaves similarly, that is, like a rigid body with a nearly constant dielectric permittivity.

The impedance measurements of this system have been combined with oscillometric-gravimetric measurements leading to both the volume of the swelling polymer and the mass of CO_2 sorbed at a given gas pressure, cp. Chap. 5. These combined measurements show that there are nearly linear relations between the dielectric permittivity and the gas pressure as well as the volume of the polymer in the sorption state and the gas pressure, cp. data and Figures given in Sect. 3.2.

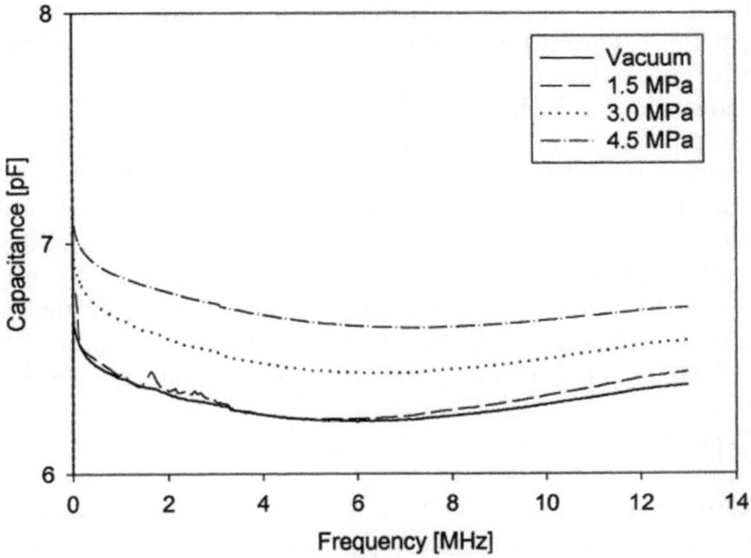

Figure 6.18. Capacitance spectra (ReC(ν)) of polycarbonate (Goodfellow, UK), cp. Chap. 5, Sect. 3.4, in vacuum and for sorption equilibria of supercritical CO_2 at T= 308 K for pressures p = 0 MPa, 1.5 MPa, 3.0 MPa, 4.5 MPa and frequencies 0.1 MHz < ν < 13 MHz, [6.3, 6.10].

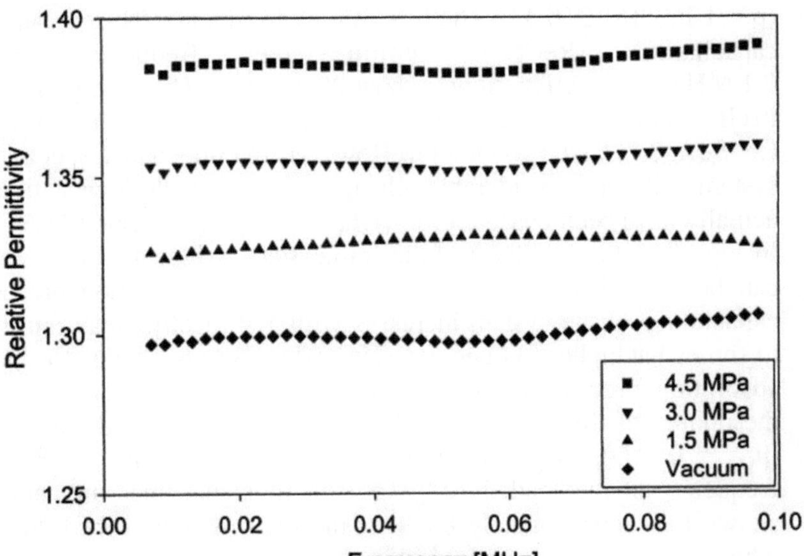

Figure 6.19. Dielectric permittivity spectra (Reε_r(ν)) of polycarbonate (Goodfellow, UK), in vacuum and for sorption equilibria of supercritical CO_2 at T = 308 K for pressures p = 0 MPa, 1.5 MPa, 3.0 MPa, 4.5 MPa and (low) frequencies 5 kHz < ν < 100 kHz, [6.3, 6.10].

The results of similar measurements of the same polymeric sorbent material extended now to methane (CH_4, 5.5) are depicted in Figures 6.20, 6.21. However, in this system the spectra again taken for frequencies 0.1 MHz < ν < 13 MHz show some anomalies for different gas pressures and both high and low frequencies. In Fig. 6.20 an intersection of the spectra for gas pressures 0.5 MPa and 1 MPa at around 12 MHz can be observed. This is quite unusual and probably is due to experimental mistakes. However, the intersection of the spectra for 0.5 MPa and 1 MPa at lower frequencies (50 kHz < ν < 60 kHz) in Figure 6.21 was highly reproducible and may be related to some structural changes within the polymer occurring in the pressure interval (0.5 MPa < p < 1 MPa) at experimental temperature T = 308 K, [6.3].

Ad 4. Capacitance measurements of gas adsorption and desorption processes

Adsorption processes which are slow on a time scale (defined as the time it takes to measure an impedance spectrum within a reasonable frequency interval) can be monitored by impedance analysis. In Figure 6.22 such spectra are shown for the adsorption of carbon monoxide (CO, 3.7) on activated carbon (AC) Norit R1 taken at T = 298 K for a gas pressure of p = 7.5 MPa, the initial state referring to vacuum (p < 1 Pa), [6.13, 6.33]. The time it took to monitor the real part of the capacitance (ReC(ν)) within the frequency interval 0.1 MHz < ν <1 MHz, Δt was about 20 minutes. The adsorption process itself, monitored simultaneously with a magnetic suspension balance, Cp. Sect. 3.2, lasted about 24 h. However, the impedance spectra of the system showed considerable changes for a much longer time and actually have been observed for 41 h, or 2460 minutes, or 123 Δt.
In Fig. 6.22 the spectrum of the initial state is indicated by "Vac.". It can be seen that within the first hour of the adsorption process the capacitance of the system increases with a shift of its maximum value to somewhat higher frequencies. This effect is possibly due to a primary adsorption process of the CO-molecules at locations in macro- and mesopores of the AC where, due to their permanent dipole moment (μ_{CO} = 0.1 D), they add to both the capacitance and the resonance frequency of the loaded AC. Now microbalance measurements have shown that approximately 80 % of the total mass adsorbed during the whole process is adsorbed within the first hour. In view of this, the changes of the capacitance spectrum which were observed afterwards for nearly 40 h are doubtless due to an internal diffusion or dissoziation process of the CO molecules taking place without uptake of more molecules from the gas phase.

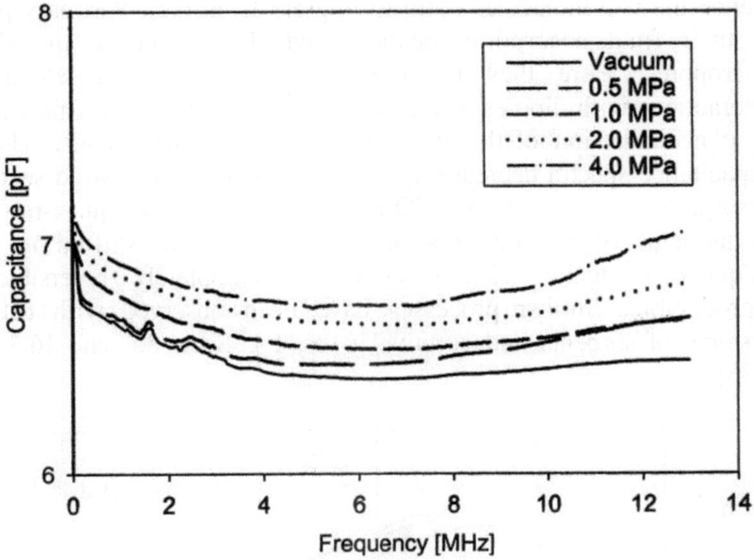

Figure 6.20. Capacitance spectra (ReC(ν)) of polycarbonate (Goodfellow, UK), cp. Chap. 5, Sect. 3.4, in vacuum and for sorption equilibria of CH_4 at T = 308 K for pressures p = 0 MPa, 0.5 MPa, 1 MPa, 2 MPa, 4 MPa and frequencies 0.1 MHz < ν < 13 MHz, [6.3].

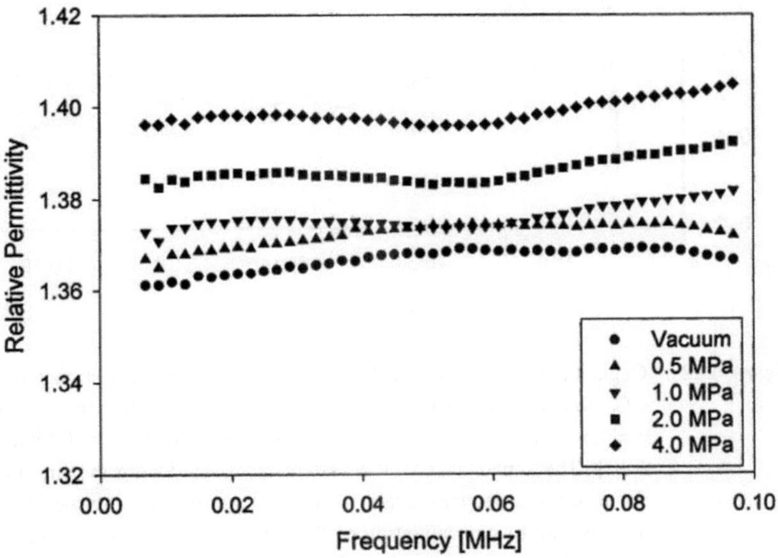

Figure 6.21. Dielectric permittivity spectra (Reε_r(ν)) of polycarbonate (Goodfellow, UK), in vacuum and for sorption equilibria of (CH_4, 4.5) at T = 308 K for pressures p = 0 MPa, 0.5 MPa, 1 MPa, 2 MPa, 4 MPa and (low) frequencies 5 kHz < ν < 100 kHz, [6.3].

Hence the CO-molecules seem to migrate from their "primary places" to their final adsorption locations, which may be in the slit-like micropores. Here they often will experience strong electrostatic interactions with dipoles of the active groups of the AC. This will lead to a neutralization of the local dipoles and electric fields. Thus the capacitance spectra decrease again with this diffusion based secondary adsorption process of the CO-molecules. Also the quasi-resonance frequencies of the maxima of these curves are shifted to lower frequencies again. This process was completely reversible and reproducible. Similar processes have been observed with different systems of sorbents and sorptive gases, cp. Sect. 3.2 and [6.3, 6.13, 6.33].

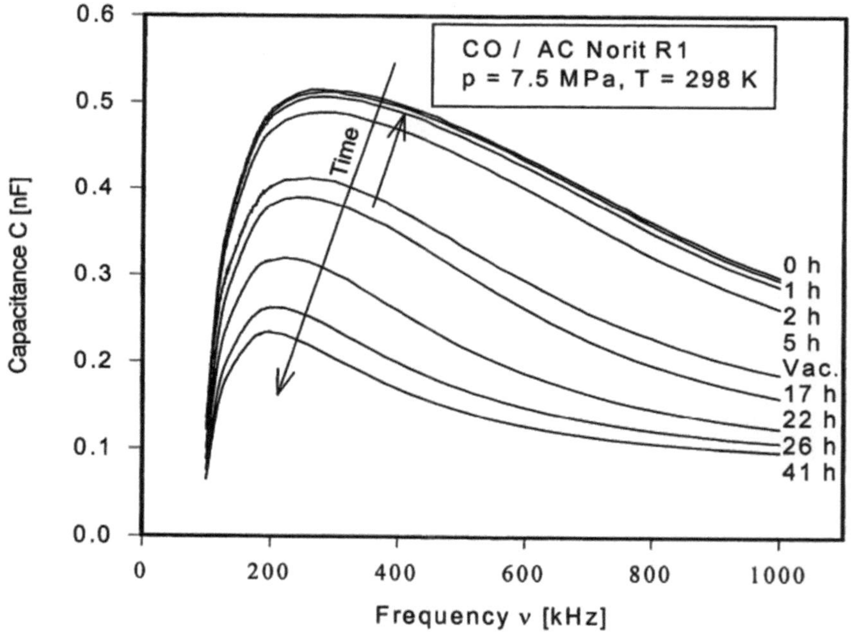

Figure 6.22. Capacitance spectra taken during 41 hours of an adsorption process of CO-molecules on activated carbon (AC) Norit R1 from vacuum to p = 7.5 MPa at T = 298 K for frequencies 0.1 MHz < ν < 1 MHz, [6.3, 6.13]. The capacitance of the system increases during the first hour of the process to decrease again for time of monitoring, i. e. nearly 40 hs. This is due to neutralization of the dipoles of the active groups of the AC by the dipoles of the incoming CO-molecules.

In Figure 6.23 capacitance spectra taken during a desorption process of atmospheric gases from activated carbon (AC) Norit R1 are shown. The real part of the complex and frequency dependent capacitance (ReC(ν)) of the system calculated from impedance measurements via Eqs. (6.6-6.8) is sketched as a function of the frequency (ν) in the interval 0.1 MHz < ν < 1 MHz. The process actually studied was a regeneration process of the AC which had been exposed to (humid) ambient air for several days. These measurements were done in order to study the influence of preadsorbed gases on the sorption capacitance of the carbon [6.3, 6.13].

At the beginning, the spectrum is nearly a flat line and does not show any significant structure. Seemingly all the dipoles of the active atomic groups of the AC have been neutralized by the adsorption of polar molecules from the ambient air like (H_2O, CO, H_2S, SO_2, NO_x etc.). Then the reactivation process was started. It consisted of three steps:

a) heating of the sample from ambient temperature (298 K) to 353 K in a helium atmosphere, i. e. reducing the gas pressure to vacuum ($p < 1$ Pa) but keeping the AC sample at 353 K;

b) after 8 hours release of the helium atmosphere, i. e. reducing the gas pressure to vacuum ($p < 1$Pa) but keeping the AC sample at 353 K;

c) cooling of the sample from 298 K in vacuum.

The spectra taken at the ends of each of these steps are shown in Fig. 6.23. As can be seen, during the reactivation process the AC is cleaned by the desorption of polar (and non-polar) impurity molecules, thus leading to the formation of new polar groups within the AC. These groups seem to be sensitive to a frequency of about 240 kHz as the spectra in this region show a maximum which is growing during the activation process. The height of this maximum actually can be used to characterize the state of activation of the carbon [6.3, 6.13].

Figure 6.23. Capacitance spectra of the reactivation process of activated carbon (AC) Norit R1 which had been exposed to ambient air for several days.
cp. next page
Regenerating gas: He, 5.0; Temperature: 353 K;
Duration: 8 hs.
Evacuation of He-gas ($p < 1$ Pa) and cooling down to ambient temperature (298 K). During the activation process polar molecules (H_2O, CO, H_2S, NO_x etc.) are desorbed from the AC leaving "unsaturated" polar groups in the AC which form the capacitance spectrum with a maximum at ca. 240 kHz, [6.3, 6.13].

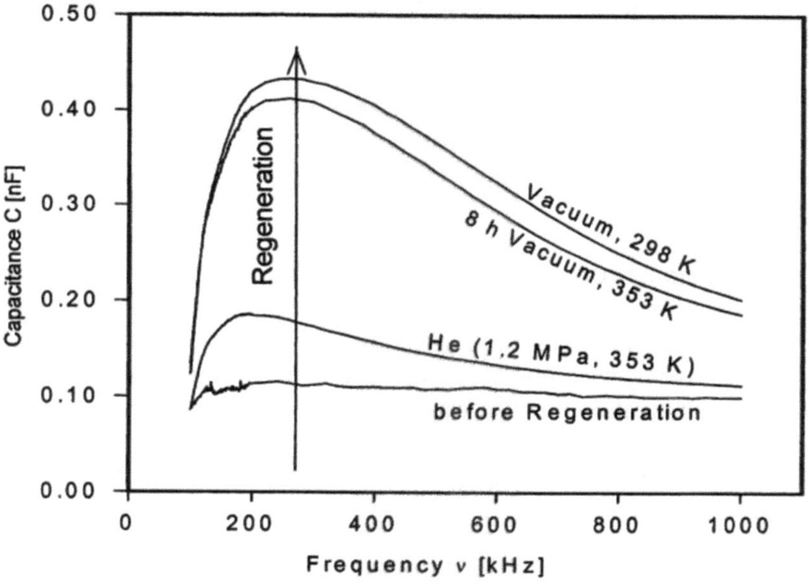

3. DIELECTRIC-MANOMETRIC AND DIELECTRIC-GRAVIMETRIC MEASUREMENTS OF PURE GAS ADSORPTION EQUILIBRIA

3.1 Experimental

Impedance measurements of adsorption systems easily can be combined with both volumetric/manometric and gravimetric measurements, and in case of swelling polymeric sorbent materials also with oscillometric instruments, cp. Chaps. 2, 3, 5. An experimental installation for combined dielectric-manometric measurements of gas adsorption equilibria is sketched in Figure 6.24. The instrument basically consists of a gas storage vessel (volume V_{SV}) and an adsorption chamber (volume V_{AC}) which includes an electric capacitor whose wires are linked to an impedance analyzer (IA). Both vessels are connected by tubes including an expansion valve and a gas circulator which is mandatory for gas mixture experiments, i. e. investigations of coadsorption equilibria. The vessels always should be placed in a thermostat (water, oil, air etc.) and supplied with proper tubing for gas inlet and evacuation. Also thermometers and manometers are needed to measure the temperature (T) and pressure (p) inside the vessels. For coadsorption experiments also a bypass for taking sorptive gas samples and a gas chromatograph (GC) are needed, Fig. 6.24, cp. also Fig. 2.1, [6.25].

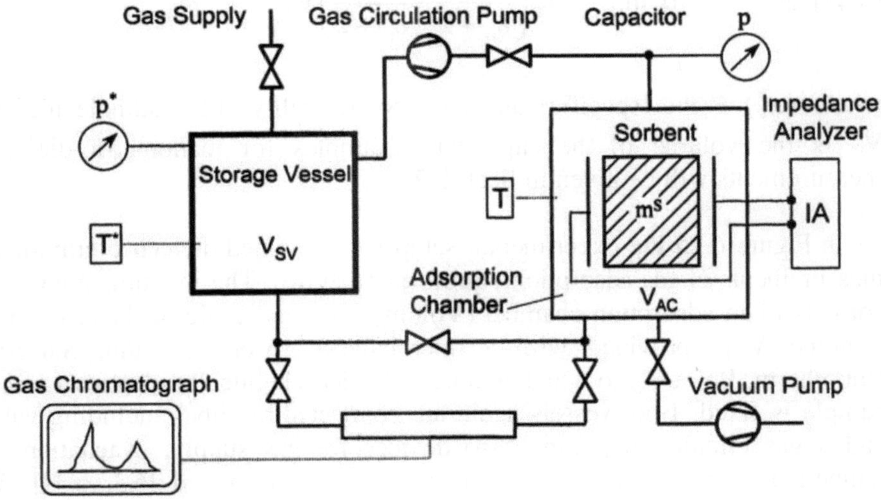

Figure 6.24. Experimental setup for combined dielectric-manometric measurements of pure gas and gas mixture adsorption equilibria. © IFT University of Siegen, 1990.

The capacitor includes a sample of sorbent material (mass m^s) which is "activated" prior to measurements. Hints to proper design and choice of materials for vessels, tubes, sealings etc. have been given in Chap. 2. There some information for the experimental procedure of manometric measurements of gas adsorption equilibria also can be found [6.13].

The quantities to be measured are: the pressure (p) and temperature (T) of the sorption gas inside the adsorption vessel, the frequency dependent capacitance (C(p)) and – if possible – the static capacitance $(C_s = \lim_{p \to 0} C(p))$ of the capacitor in both the evacuated and gas filled state, and the mass of adsorbent (m^s). Also the helium volume (V^s_{He}) as an approximate value of the void volume of the sorbent material must be known, cp. Sect. 2.2. Given these data the following quantities can be calculated:

(2.4) $$\Omega_V = m^a - \rho^f V^{as}$$

(2.5) $$\Omega_V \equiv m^* - \rho^f (V_{SV} + V_{AC})$$

(2.7) $$V^{as} \simeq V^s_{He}$$

The Gibbs excess mass adsorbed $m^a \simeq m^a_{GE}$ is determined from equation (6.47), as

$$\alpha^a_s m^a_{GE} = 3\varepsilon_0 \left(\frac{\varepsilon_{rs} - 1}{\varepsilon_{rs} + 2} - \frac{\varepsilon^s_{rs} - 1}{\varepsilon^s_{rs} + 2} \right) V_C$$

where (α^a_s) is the (specific) dielectric polarizability of the admolecules and V_C is the volume of the capacitor. Examples for manometric-dielectric measurements will be given in Sect. 3.2.

In Figure 6.25 the experimental setup for combined dielectric-gravimetric measurements of gas adsorption equilibria is shown. The instrument basically consists of an adsorption chamber (Volume V_{AC}) and a microbalance chamber (volume V_{MB}) provided with a microbalance (Mettler, Cahn, Sartorius, Rubotherm, Bel, etc.) to which another (physico-chemically identical) sorbent sample is fixed. Both vessels again are connected by tubes including valves and a gas circulation pump. Also devices for gas supply, evacuation and temperature and pressure measurements must be provided [6.3, 6.33]. The sorbent sample in the adsorption chamber is placed within a capacitor whose plates are linked to an impedance analyzer (IA) outside the chamber. The whole installation always should be placed within a thermostat (water, oil, etc.). Automated recording of data supported by appropriate software programs (Lab View etc.) is recommended. Suggestions for the design of an instrument for gravimetric-dielectric measurements, choice of materials for vessels, tubes, etc. can be found in Chaps. 3, 4, cp. [6.3, 6.18, 6.19].

Physical quantities to be measured are: the pressure (p) and temperature (T) of the sorptive gas inside the adsorption and the microbalance vessel, the frequency dependent capacitance C(p) of the capacitor including – if possible – its static limiting value $C_s = \lim_{p \to 0} C(p)$, the masses of the sorbent samples (m^s, m^s_0), and the balance reading (Ω). From these quantities the Gibbs excess mass adsorbed (m^a_{GE}) and the (specific static) polarizability (α^a_s) of the adsorbed molecules can be calculated as follows: Gravimetric measurements lead to the "reduced mass" Ω_G of the sorbate phase. i. e.

$$\Omega_G = m^a - \rho^f V^{as} \tag{3.5}$$

where $(\Omega_G \equiv \Omega)$ is given by Eqs. (3.6), (3.10), (3.30) depending on whether a two beam or a single beam microbalance is used. Using the helium approximation for the void volume (V^{as}) seen by the adsorptive gas molecules

(3.12) $$V^{as} \cong V^{s}_{He},$$

we can calculate the Gibbs excess mass adsorbed from (3.5) as

$$m^{a}_{GE} = \Omega_{G} + \rho^{f} V^{s}_{He}.$$

Hence, from combined impedance measurements the specific dielectric polarizability (α^{a}_{s}) of the admolecules again can be calculated using equation

(6.47) $$\alpha^{a}_{s} m^{a}_{GE} = 3\varepsilon_{0} \left(\frac{\varepsilon_{rs} - 1}{\varepsilon_{rs} + 2} - \frac{\varepsilon^{s}_{rs} - 1}{\varepsilon^{s}_{rs} + 2} \right) V_{C}.$$

Here (V_C) is the volume of the capacitor.

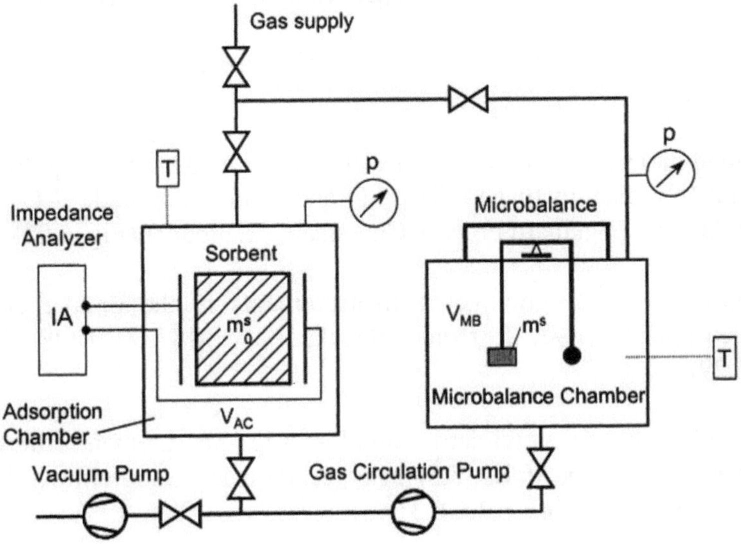

Figure 6.25. Experimental setup for combined dielectric-gravimetric measurements of pure gas adsorption equilibria. © IFT University of Siegen, 1988.

Examples of combined dielectric-gravimetric measurements will be given in Sect. 3.2.

It should be emphasized that in principle impedance measurements also can be added to other experimental methods aiming at an analysis of gas sorption systems. Examples for these are volumetric-gravimetric measurements, Chap. 4, oscillometric measurements, Chap. 5, or ZLC-measurements, Chap. 1. Impedance measurements always will give new and additional information on the state of the sorption system. Though often it does not seem to be easy to interpret these measurements on a molecular basis, impedance spectra always can be used to characterize the system, i. e. for example to check the "quality" or state of industrial sorbent materials.

3.2 Examples

The purpose of this Section is to present several examples of combined dielectric measurements taken of gas adsorption systems. General information on these and especially on the sorbent materials given at the beginning of Sect. 2.4 also should be taken into account. The examples refer to:

1. Combined dielectric volumetric measurements of adsorption equilibria: Methane (CH_4, 5.5) on dealuminated zeolite of Y-type (DAY, Degussa) (Fig. 6.26);
   Hydrogen sulfide (H_2S, 1.8) on molecular sieve MS13X (UOP, Erkrath) (Fig. 6.27).

2. Combined dielectric gravimetric measurements of adsorption equilibria: Nitrogen (N_2, 5.0), methane (CH_4, 5.5) and carbon monoxide (CO, 3.7) on pellets AC-20 (Engelhard Process Chemicals, Nienburg) (Fig. 6.28).

3. Combined dielectric gravimetric measurements of the uptake curve of hydrogen sulfide (H_2S, 1.8) on molecular sieve MS13X (UOP, Erkrath) (Fig. 6.29).

4. Combined dielectric gravimetric oscillometric measurements of sorption equilibria of methane (CH_4, 5.5) in polycarbonate (Goodfellow, UK), cp. Chap. 5, Sect. 3.4, (Figs. 6.30, 6.31).

Ad 1. Dielectric-volumetric measurements of adsorption equilibria, cp. Fig. 6.24.
In Figure 6.26 the Gibbs excess amounts of methane (CH_4, 5.5) on dealuminated zeolite of Y-type (DYA, Degussa, Hanau, Germany) at 298 K for pressures up to 2 MPa is shown. The adsorption isotherm is of Type I (IUPAC). Also data for the "reduced" molar polarizability of the adsorbed methane molecules ($\alpha_m^a(p)/\varepsilon_0$) are given [6.41]. The values of the static permittivities ($\varepsilon_r, \varepsilon_r^s$) needed in Eq. (6.47) to calculate α_s^a and $\alpha_m^a = \alpha_s^a M$ respectively, have been determined by

fitting frequency spectra of $\varepsilon_r(\omega), \varepsilon_r^s(\omega)$ to a generalized Debye model (6.79), (6.80), cp. [6.3].

The horizontal lines in the upper part of the figure indicate the values of (α_m/ε_0) for the compressed liquid state $(\alpha_m^L/\varepsilon_0)$ at T = 112 K and the gaseous state of methane at T = 298 K [6.21, 6.22]. As can be seen from the data, the polarizability of the admolecules is always much smaller than those in the liquid or gaseous state. This indicates strong binding forces between the surface of the adsorbent and the adsorbed molecule (admolecule) so that the interaction of the electric field with the admolecule is fairly small. It also is nearly constant for gas pressures p > 0.5 MPa independent of the increasing amount of CH_4 adsorbed.

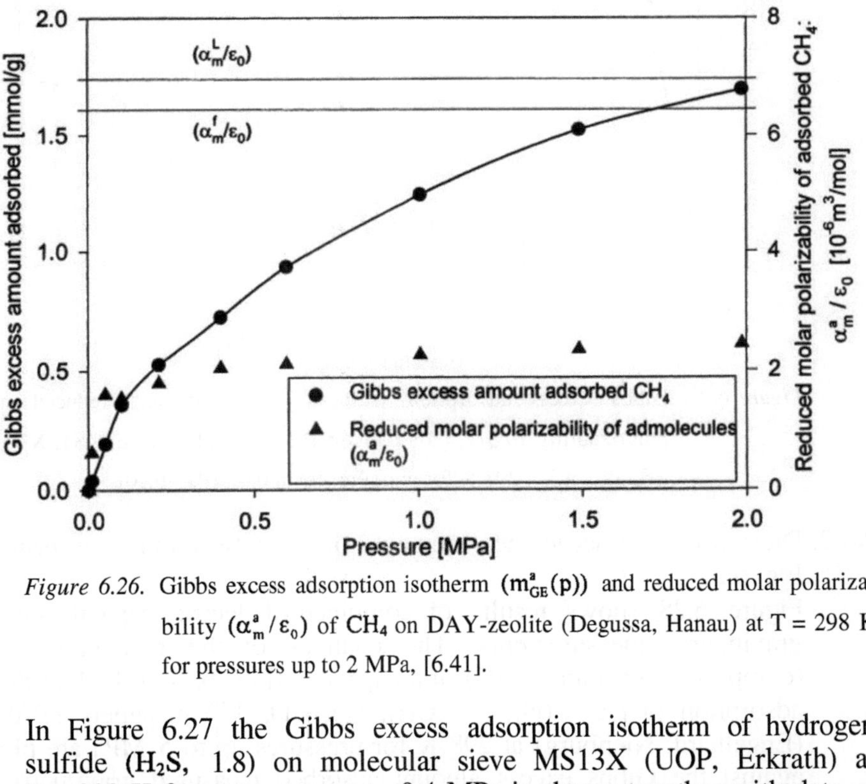

Figure 6.26. Gibbs excess adsorption isotherm $(m_{GE}^a(p))$ and reduced molar polarizability $(\alpha_m^a/\varepsilon_0)$ of CH_4 on DAY-zeolite (Degussa, Hanau) at T = 298 K for pressures up to 2 MPa, [6.41].

In Figure 6.27 the Gibbs excess adsorption isotherm of hydrogen sulfide (H_2S, 1.8) on molecular sieve MS13X (UOP, Erkrath) at T = 298 K for pressures up to 0.4 MPa is shown together with data of the (reduced) molar polarizability of adsorbed H_2S $(\alpha_m^a(p)/\varepsilon_0)$, [6.10]. Again the horizontal lines in the figure indicate the values of (α_m/ε_0) for H_2S in the compressed liquid state $(\alpha_m^L/\varepsilon_0)$ at T = 213 K and the gaseous state for T = 298 K, [6.21, 6.22]. The data show that at low adsorption loads the polarizability of the admolecules is much smaller than those in the gaseous or liquid state. This indicates again strong interaction forces between sorbent and admolecules that seem to freeze or fix the H_2S-molecule to the surface. However, at increasing sorption loads and near saturation, the polarizability of the admolecules increases exceeding even the value corresponding to the gaseous phase. This indicates a decrease of the binding forces between admolecule and sorbent, which also is reflected in a decrease of the heats of adsorption in this system [2.2].

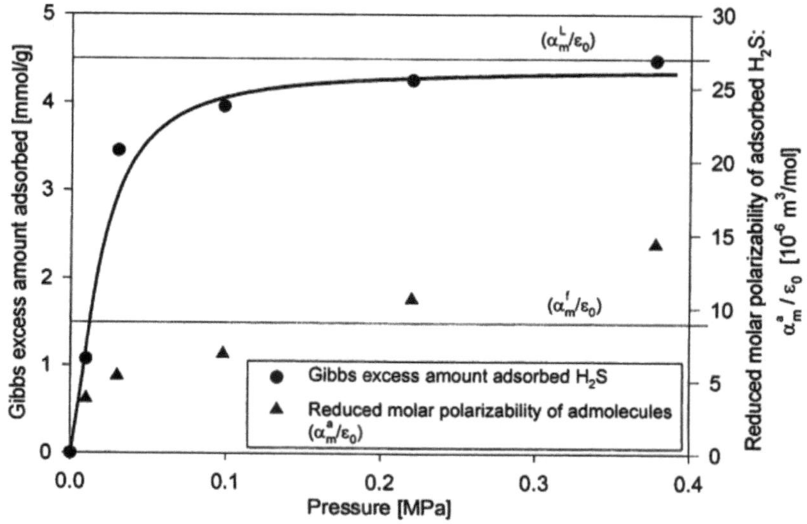

Figure 6.27. Gibbs excess adsorption isotherm $(m^a_{GE}(p))$ and reduced molar polarizability $(\alpha^a_m/\varepsilon_0)$ of H_2S on the molecular sieve MS13X (UOP, Erkrath) at T = 298 K for pressures up to 0.4 MPa, [6.10].

Ad 2. Dielectric-gravimetric measurements of gas adsorption equilibria, cp. Fig. 6.25.

Figure 6.28 shows results of combined dielectric capacitance and gravimetric measurements. The changes of the real part of the (complex) capacitance (ΔC) of a capacitor (Fig. 6.2) at 4 MHz due to adsorption of gases (N_2, 5.0; CH_4, 5.5; CO, 3.7) in pellets TPAC-20 (Engelhard, Nienburg) at 298 K for pressures up to 5 MPa are plotted against the Gibbs excess amount adsorbed measured gravimetrically (Sartorius 4104 S), [6.10]. For non-polar molecules (N_2, CH_4) the changes in the capacitance (ΔC) of the capacitor relative to the empty sorbent show a steep increase for small amounts adsorbed, i. e. at low coverages of the surface of the sorbent. At higher loadings ΔC is growing much slower and nearly linearly with increasing amount of gas adsorbed. This already has been observed by Channen et al. [6.9] doing similar dielectric-manometric measurements for activated carbon sorbents. For polar sorptive gases (CO, μ=0.1 D) these measurements reveal a basically different behavior: the capacitance of the AC-20/CO system (ΔC) relative to its value for the AC-20/vacuum system increases with the square of the amount of (CO) adsorbed. This behavior must be related to dielectric properties of the activated groups of the carbon component in the sorbent AC-20 that, according to manufacturer's information, is a mixture of zeolite molecular sieve and activated carbon (20 % weight) which has been thermally and chemically "activated". This also is indicated by similar measurements (not shown here but presented in [6.10]) of adsorption equilibria of CO on MS5A, showing a much smaller increase of (ΔC) with the amount adsorbed at the same frequency of the electric field applied (4 MHz).

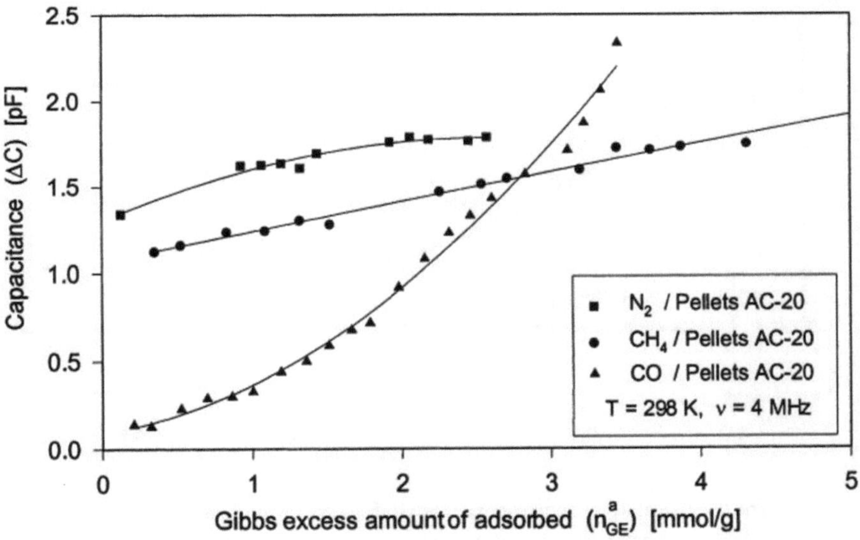

Figure 6.28. Change of the capacitance (ΔC) of systems pellets AC-20/gas relative to the system pellet AC-20/vacuum for gases (N_2, 5.0; CH_4, 5.5; CO, 3.7) as function of the amounts of gases adsorbed (per gram sorbent) at T = 298 K and ν = 4 MHz, corresponding gas pressures ranging up to 4 MPa, [6.10].

Ad 3. Figure 6.29 shows uptake curves of the system hydrogen sulfide (H_2S, 1.8) on molecular sieve MS13X (UOP) taken during adsorption ($p < 1$ Pa $\rightarrow p = 10$ kPa) at T = 298 K [6.10, 6.14]. The upper curve (IA) indicates data of the (real part of the) relative capacitance ($C(t)/C(t \rightarrow \infty)$) at time (t) taken at ν = 250 kHz. the lower curve (MB) marks data of the relative uptake of the Gibbs excess mass adsorbed

$(m^a_{Ge}(t)/m^a_{GE}(t \to \infty)$. The microbalance data are monotonously increasing, assuming the asymptotic value $\lim_{t\to\infty} m^a_{GE}(t) = m^a_{GE}(\infty)$ after 10 minutes, i. e. after this time no change of the mass adsorbed was observed. Contrary to this the impedance analyzer data show a sharp increase during (0 < t < 5 min.) and to pass through a maximum and then to decrease along a typical relaxation curve to their asymptotic value $\lim_{t\to\infty} C(t) = C(\infty)$. This behavior is due to catalytic properties of the molecular sieve indicating a dissociation reaction of the H_2S molecules adsorbed which does not change their mass but their electric dipole moment and hence the permittivity of the system considerably.

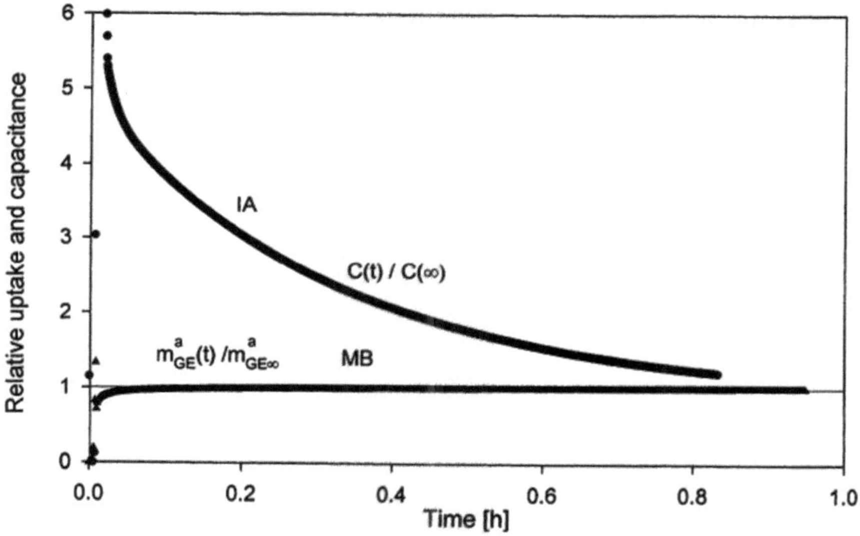

Figure 6.29. Relative uptake curve of (H_2S, 1.8) on molecular sieve MS13X (UOP) during an adsorption process (p < 1 Pa → p = 10 kPa) at T = 298 K. IA: relative changes of the real part of the capacitance (C(t)/C(∞)) of the sorption system measured at ν = 250 kHz with an impedance analyzer (HP).
MB: relative changes of the excess mass adsorbed $(m^a_{Ge}(t)/m^a_{GE}(t \to \infty)$ measured with a magnetic suspension balance (Rubotherm, Bochum), [6.3, 6.10, 6.14].

Ad 4. Dielectric-gravimetric-oscillometric measurements of sorption equilibria of methane in (swelling) polycarbonate.
The uptake of methane (CH_4, 5.5) in polycarbonate (Goodfellow, UK) and the increase of the volume of this sorbent/sorbate system have been determined by combined gravimetric and oscillometric measurements [5.7, 5.8, 6.3], cp. Fig. 5.12.

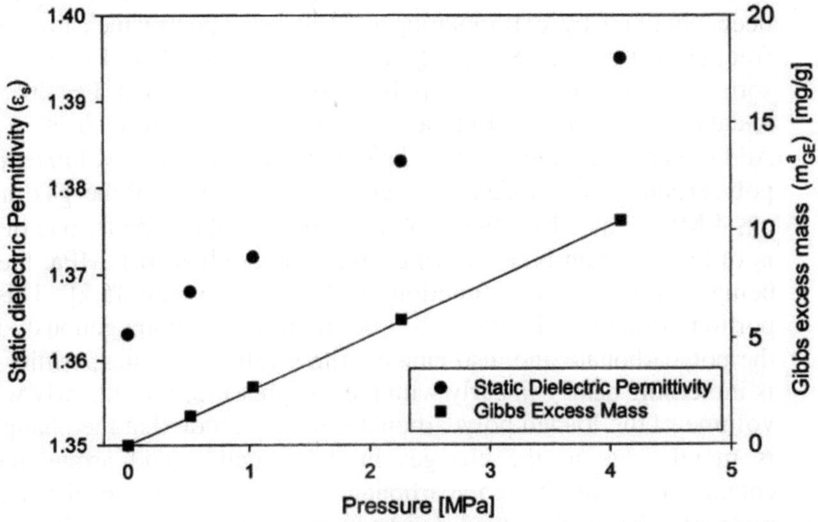

Figure 6.30. Gibbs excess mass of (CH_4, 5.5) sorbed in polycarbonate (Goodfellow, UK) at T = 308 K for gas pressures up to 4 MPa and static dielectric permittivity (ε_s) of the (methane loaded) polymer [5.2, 6.3].

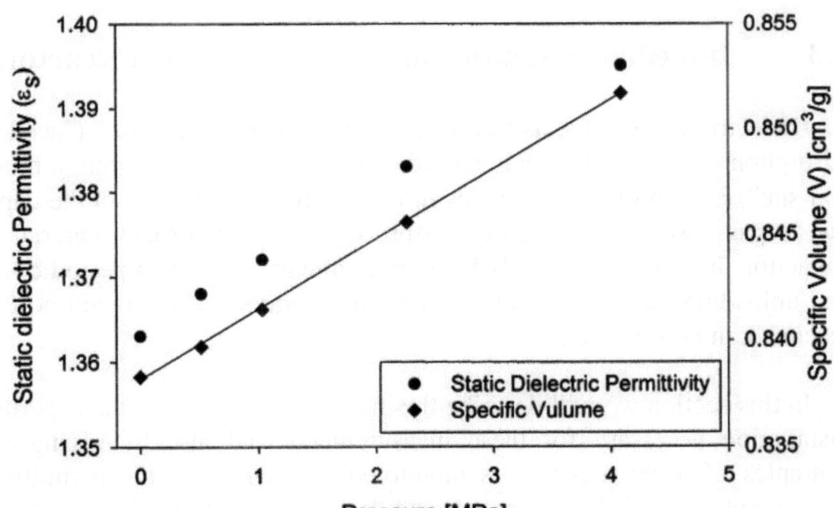

Figure 6.31. Specific volume of polycarbonate (Goodfellow, UK) swelling due to the sorption of methane (CH_4, 5.5) at T = 308 K for pressures up to 4 MPa and static dielectric permittivity (ε_s) of the (methane loaded) polymer [5.2, 6.3]. The volume has been determined by combined oscillometric-gravimetric measurements, cp. Chap. 5. Figures 6.30, 6.31 refer to the same set of combined oscillometric-gravimetric-dielectric experiments [6.3, 6.41].

Simultaneously the static dielectric permittivity (ε_s) of this system has been obtained by extrapolating capacitance spectra measurements to frequency $\nu = 0$, cp. Sect. 2.2 and Figs. 6.18 – 6.21. In Figure 6.30 the sorption isotherm of CH_4 in polycarbonate measured at T = 308 K and ε_s-data is shown as functions of the gas pressure $p \leq 4$ MPa. Additionally in Fig. 6.31 data for the specific volume of the polycarbonate and ε_s-data are depicted as function of the gas pressure $p \leq 4$ MPa. The adsorption isotherm consists of two parts, one of which is of Langmuirian type showing saturation at about 0.1 MPa, the other being a simple linear function of the gas pressure [5.2]. This later portion is due to CH_4 molecules being dissolved homogenously within the polycarbonate and also causing the swelling. As the permittivity ε_s is increasing nearly linearly with the gas pressure, i. e. linearly with the volume of the loaded polycarbonate, we conclude that the change in ε_s is mainly due to the changes in the structure and arrangement or entanglement of the polycarbonate molecules due to the methane molecules being dissolved nearly homogenously in the polymer. That is, the CH_4 molecules adsorbed at low gas pressures in interstitial spaces between the polymeric chains seemingly do not contribute to changes of ε_s.

3.3 Impedance Measurements in Adsorption Reactors

Dielectric measurements also can be performed in capacitors placed inside adsorption columns. Here they provide a means to detect adsorption processes "on site", i. e. to observe changes in the electric capacitance of the capacitor due to gas adsorption or desorption processes on the sorbent material in the capacitor. In this way not only breakthrough curves but also periodic loading and unloading curves of sorptive gases in laboratory or industrial adsorption reactors can be observed.

In this section we will describe this method by presenting the experimental installation necessary for these measurements and also by giving several examples of sorption processes monitored by changes of the permittivity of the sorbent material due to gas ad- and desorption [6.10, 6.15].

The experimental setup for dielectric measurements in sorption columns is sketched in Figure 6.32. It consists mainly of an adsorption column filled with sorbent material, a capacitor inside the column connected to an impedance analyzer (HP 4192 A) and a (traditional) thermal conductivity detector (TCD) (Perkin Elmer, Germany) to detect changes in the concentration of the gas leaving the column.

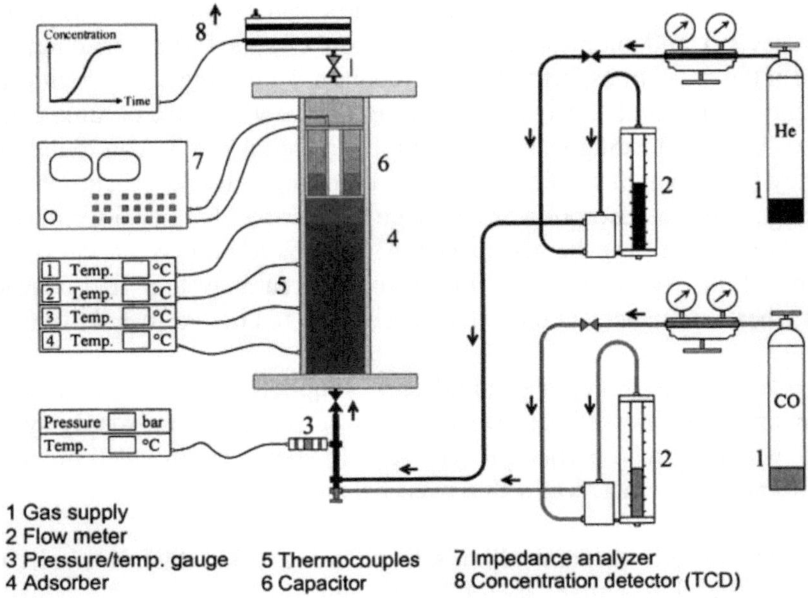

Figure 6.32. Experimental setup for measurements of gas adsorptions and desorption processes within a column [6.3, 6.10].

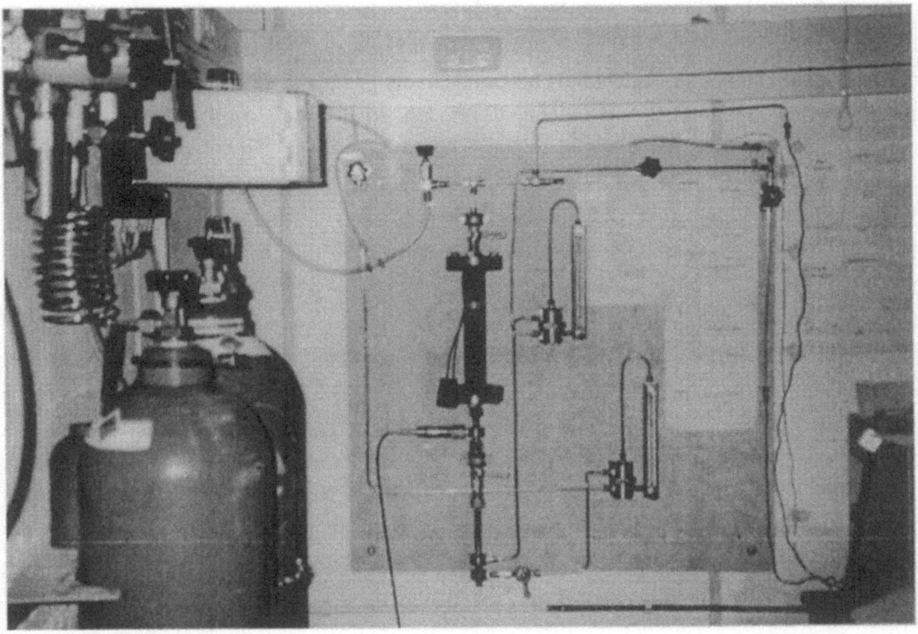

Figure 6.33. Adsorption column including an electrical capacitor for impedance measurements of adsorption and desorption processes, IFT Lab PB-A 0126, 1996.

The electrical capacitance of the capacitor can be measured at various frequencies (ν) and monitored as a function of time. The capacitance is changed if the mass transport within the column is entering the capacitor and adsorption occurs. In this way a breakthrough curve within the column can be observed, i. e. the loading state of the sorbent material can be controlled. The frequency (ν) at which the capacitance is monitored should be chosen such that changes due to adsorption are as large as possible. No general rule for a proper choice of (ν) can be given. Hence it seems to be necessary to take a spectrum of the capacitance of the sorbent/sorbate system and consider regions where resonances in the capacitance seem to emerge [6.3, 6.10, 6.14].

The adsorption column should be supplied with a number of thermocouples to observe heat effects during sorption. Also a supply system for sorptive gases and a carrier gas (He, Ar, N_2) should be provided including well calibrated flow meters to set up mass balances.

A photo of an experimental installation given in Fig. 6.32 which was built and operated at the IFT, University of Siegen during 1994-2001 is shown in Figure 6.33,[*)] [6.3, 6.10].

In Figure 6.34 two breakthrough curves for the system carbon monoxide (CO, 3.7), molecular sieve MS13X (UOP, Erkrath) measured with a thermal conductivity detector (TCD) and an impedance analyzer (IA) are shown. Helium (He, 5.0) has been used as carrier gas. The capacitance was measured at a frequency of $\nu = 10$ MHz. The column was operated at ambient temperature (295 K) and gas pressure p = 0.151 MPa. Flow rates of the gases were: He ... 3.96 ml/s, CO ... 5.59 ml/s. The TCD recorded the breakthrough of the CO after ca. 100 s with the concentration versus time curve showing normal structure. The IA detected the entry of the mass transfer into the capacitor after 65 s, this corresponding well to its position at about 2/3 of the column length and the TCD-breakthrough time of ca. 100 s. The IA-curve representing the real part of the capacitance of the sorbent filled capacitor at 10 MHz shows a maximum after which it falls off to approach a stable asymptotic value. This effect is probably due to the heat generated during the adsorption process which very well can lead to kind of overshooting effect of both the amount of CO adsorbed and hence also the capacitance, both quantities ultimately levelling off due to heat transfer to the sorptive gas. Another phenomenon which also may contribute to the occurrence of a maximum in the IA-curve is that CO molecules first are adsorbed in macro-

*) The instrument meanwhile (2004) has been transferred to the Institute of Non-Classical Chemistry at the University of Leipzig, Leipzig, Germany.

and mesopores but from there diffuse to micropores where they may find adsorption places completely different from the former ones. This can lead to a partial neutralization of primary dipoles of the sorbed atoms and hence to a decrease of the electric capacitance of the sorbent/sorbate system within the capacitor.

After breakthrough of CO has been observed at the TCD, the valve for CO supply was closed and desorption of CO by the carrier gas (He) could be observed. The corresponding desorption curves detected by the TCD and the IA respectively are shown in Figure 6.35. The desorption process is starting at the (lower level) entrance of the adsorption column, the desorption zone reaching the capacitor after 65 s which is regenerated after ca. 100 s, i. e., is free of CO.

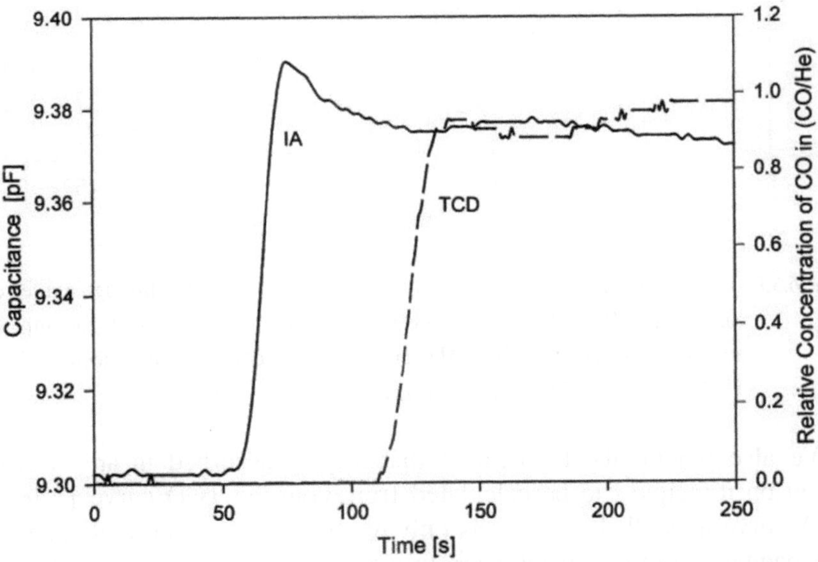

Figure 6.34. Breakthrough curves of an adsorption process of (CO, 3.7) on MS13X (UOP) detected with an impedance analyzer (IA) at ν = 10 MHz and a thermal conductivity detector (TDC) at T = 295 K, p = 0.151 MPa.. The IA curve shows the (real part of the) capacitance of the sorbent/sorbate system inside the capacitor. The TCD curve represents the CO-concentration of the gas flow leaving the column [6.14].

The TCD at the (upper level) end of the column registers the end of the desorption process after ca. 120 s, the last remnant molecules of CO leaving the TCD after 200 s. This clearly demonstates that the actual state of a porous sorbent during an ad- or desorption process easily can be checked and monitored by impedance measurements. For more details and other examples of breakthrough curves in adsorption columns the interested reader is referred to [6.3, 6.10, 6.14] and the literature cited therin.

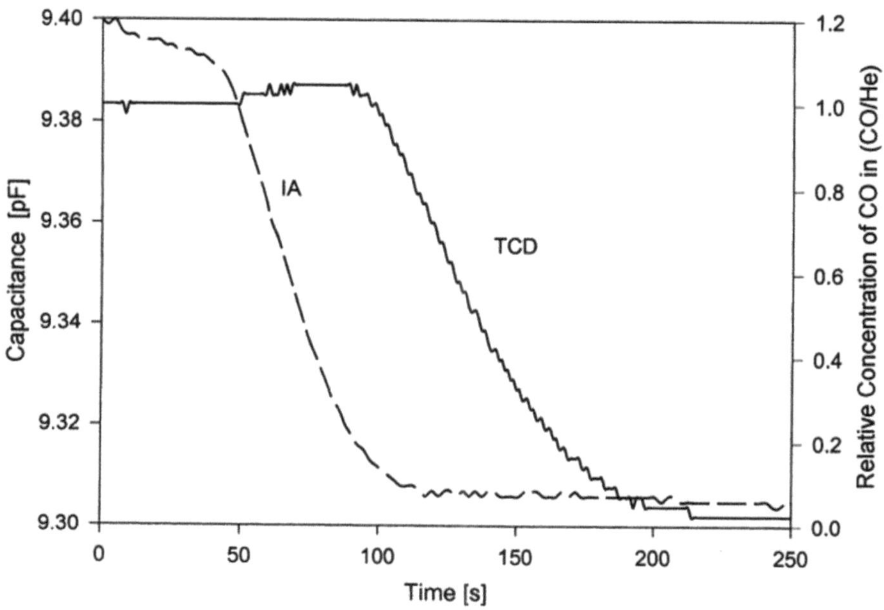

Figure 6.35. Desorption curves of (CO, 3.7) on MS13X (UOP) – or breakthrough curves of the carrier gas (He, 5.0) – in an adsorption column detected with an impedance analyzer (IA) at ν = 10 MHz and a thermal conductivity detector (TCD) at T =295 K, p = 0.151 MPa, cp. also caption of Fig. 6.34, [6.14].

We also would like to mention that masses adsorbed in an adsorption column on principle can be calculated from both the TCD curve [3.48] and the IA curve as well. For this it is only necessary to assume that changes in the capacitance of the capacitor are proportional to the amount of molecules adsorbed which certainly is only an approximate but often useful hypothesis. As an example, the masses adsorbed in saturation states of various sorbent materials for CO and CO_2 as sorptive gases with He and N_2 as carrier gases have been calculated from both TCD- and IA-signals, cp. Figs. 6.34, 6.35. Results are presented in Tables 6.1, [6.14, 6.15]. Although the masses resulting from IA-measurements always are smaller by approximately 5 %, the coincidence of the results is reasonable, recommending again impedance measurements as a means to investigate gas adsorption systems.

Table 6.1: Excess masses adsorbed determined from breakthrough curves, measured by a thermal conductivity detector (TCD) and an impedance analyser (IA) for several sorptive gases and sorbent materials, T = 295 K, p ≈ 0.11 MPa [6.10, 6.45].

Gas	Porous solid	Mass adsorbed [mg/g]	
	Carrier gas: He	TCD	IA
CO	Norit R1	14.64	15.81
CO	MS 5A	2.692	2.667
CO	MS 13X	5.019	4.656
CO	DAY	5.351	5.033
CO	Envisorb b+	4.479	4.583
CO_2	MS 13X	52.455	49.291
	Carrier gas: N_2		
CO_2	Envisorb b+	6.367	6.079

Impedance measurements also can be used to monitor industrial adsorption processes consisting of multi-step cycles like pressure swing adsorption processes (PSA) or vacuum swing adsorption processes (VSA), [3.37, 3.48, 3.42, 3.43].

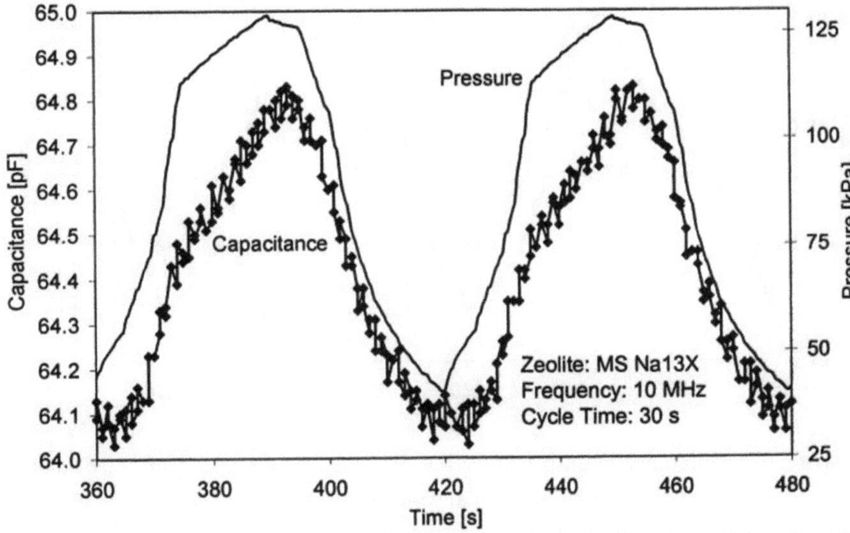

Figure 6.36. Combined measurements of the gas pressure and the (real part of the) dielectric capacitance at ν = **10 MHz** of a capacitor filled with MSNa13X (UOP) inside a commercially operating VSA-process for air separation (Mahler AG) [6.15]. Changes in the gas pressure and the capacitance are due to ad- and desorption of nitrogen (N_2) on the sorbent.

To give examples we present data for the sorptive gas pressure and the dielectric capacitance taken in a commercially operating VSA process for air separation (Messer-Mahler, Stuttgart, Germany), Figs. 6.36, 6.37 [6.15]. Figure 6.36 represents air pressure and capacitance data taken within a fixed bed adsorption column. The pressure curve exhibits the four characteristic steps of a Skarstrom PSA cycle: pressurization, adsorption, depressurization and purge [6.42-6.44]. The corresponding increase and decrease of the capacitance is due to ad- and desorption of nitrogen (N_2). Seemingly these processes are very fast so that the pressure related equilibria states are always realized, i. e. the pressure curve and the capacitance curve are (nearly) in phase. Contrary to this behavior the corresponding curves in Figure 6.37 show a certain displacement of phases of about 4 s. This is due to the fact that these curves were taken at another location within the air separation plant where the incoming air is dried, i. e. changes of the capacitance are due to the ad- and desorption of water. These processes always are accompanied by considerable thermal effects, i. e. heating and cooling of the sorbent, which need some time to level off due to heat transfer from the sorbent to the sorptive air or vice versa [6.15].

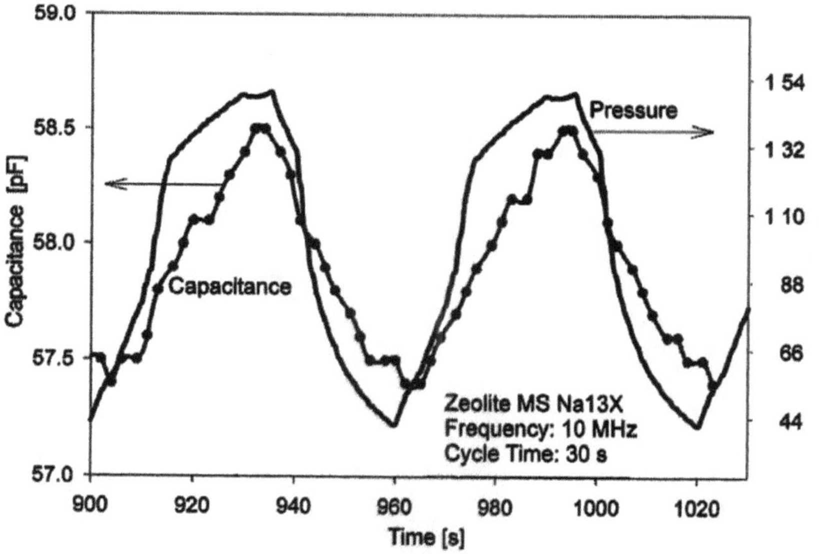

Figure 6.37. Combined measurements of the gas pressure and the (real part of the) dielectric capacitance at ν = 10 MHz of a capacitor filled with MSNa13X (UOP) inside a commercially operating VSA-process for air separation (Mahler AG). Changes in the capacitance are displaced by 4 s against those of the air pressure. This is caused by (non-isothermal) water ad- and desorption on the zeolite [6.15].

4. PROS AND CONS OF IMPEDANCE SPECTROSCOPY

Advantages and disadvantages of impedance spectroscopy (IS) as a method to investigate gas adsorption systems are discussed in brief.

4.1 Advantages

1. Measuring techniques
   Dielectric impedance measurements can be performed easily and quickly even under extreme conditions of pressure and temperature in both laboratory and industrial adsorption columns and systems. They easily can be automated and once installed are of low cost during operation.

2. Versatility
   Impedance measurements can be used for a variety of purposes, namely
   - Characterization of sorbent materials by impedance spectra (10 kHz $< \nu <$ 10 MHz);
   - quality control of industrial sorbents;
   - if calibration data, i. e. combined dielectric-manometric or dielectric-gravimetric already are at hand, impedance measurements allow one to determine the mass adsorbed in a sorbent material;
   - in coadsorption systems including strongly polar and non-polar gas components as for example humid air (H_2O, N_2, O_2) or sour natural gas (CH_4, H_2S, SO_2) the polar components can be "seen" by impedance measurements and thus their mass be determined;
   - in industrial adsorption columns the loading state of the sorbent can be checked "on site". Hence breakthrough of gases can be determined much earlier than by (traditional) TCD- or GC-measurements.

3. Kinetics
   Impedance measurements allow one to detect internal diffusion and even chemical processes in a porous sorbent which cannot be seen by manometric or gravimetric measurements, cp. Fig. 6.29.

4.2 Disadvantages

1. Sorbent material
   Only purely dielectric or weak electrically conducting materials (like activated carbons) should be considered for permittivity measurements. Electrically conducting materials like porous metallic foams often exhibit very low electrical resistances. Thus sample materials will be heated during long measurement periods.
   Also for reproducible measurements normally several grams of a sorbent material are needed. That is, it is not easy to build capacitors which can handle sorbent probes in the mg range. Glues used often in designs of this kind normally penetrate some pores and change the capacitance spectrum of the sorbent. Hence these devices cannot be recommended.

2. Frequency spectrum
   To achieve high sensitivity of impedance measurements it is recommended to consider frequencies (ν) of the alternating electric field applied where the capacitance (C) of the sorbent/sorbate system is changing a lot, as is the case in the vicinity of resonance frequencies of the system. As these frequencies normally are unknown one has to take a spectrum $C = C(\nu)$ in as broad a range of frequencies as possible and to select an appropriate characteristic frequency afterwards. This is laborious and requires an efficient impedance analyzer which could be fairly expensive.

3. Molecular interpretation and Thermodynamics
   The interpretation of impedance measurements, i. e. curves $C = C(\nu)$ in adsorption equilibria, or capacitance versus time curves $C = C(t)$ at constant frequency ($\nu = \text{const}$) for adsorption processes, is not easy and still a field for analytic and simulative investigations. Also the dielectric equation of state (EOS) ($\varepsilon_r = \varepsilon_r(T, p, m^s)$)of an adsorption system does not seem to have been investigated in a systematic way – except some virial like series expansions considered in [3.40]. Correspondingly interrelations between the dielectric EOS and the adsorption isotherm or the heat of adsorption are – though they must exist, cp. [6.16, 6.17] – unknown to the best knowledge of the authors.

5. LIST OF SYMBOLS

Symbols and SI-units of most physical quantities used in this Chapter are given.

A	m^2	area of plates of an electric capacitor, cp. Figs. 6.1, 6.8.
C_s	As/V	static capacitance of an electric capacitor
D	As/m^2	dielectric displacement vector
d	m	distance between the (flat) plates of an electric capacitor
E	V/m	electric field vector
G	m^{-1}	geometric factor of a capacitor, cp. Eq. (6.77)
I(t)	A = C/s	electric current, cp. Fig. 6.8.
Im z	[z]	imaginary part of any complex quantity z
$k = 1.381 \cdot 10^{-23}$	J/K	Boltzmann constant
$N_A = 6.0221 \cdot 10^{+23}$	mol^{-1}	Avogadro or Loschmidt number
$N_T = n_T V_C$		total number of particles (molecules, ions, etc.) of a material system
$n_T = N_T / V_C$		number of particles (molecules, ions) per unit volume
n_i	mol	mole number of component i = 1, 2... of a multi-component system
P	Cm/m^3	electric polarization, i. e. electric dipole moment per unit volume of a dielectric material
p	Pa	pressure in a sorptive gas
P_e	Asm	macroscopic dipole moment of a material system between the plates of a capacitor
p_e	Cm, D 1D/Cm= $= 3.338 \cdot 10^{-30}$	electric dipole moment of a single particle (molecule, ion, etc.)
Q(t)	C = As	electric charge on the plates of a capacitor

$Q_p = DF$	$C = As$	electric charge of the polarization on the test plates of area (F) in a slit perpendicular to an electric field (D, E), cp. Fig. 6.8
q	$C = As$	electric charge of an electron or ion
Rez	[z]	real part of any complex quantity z
r_a	m	radius of the outer cylinder of a cylindrical capacitor, cp. Fig. 6.2
r_i	m	radius of the inner cylinder of a cylindrical capacitor, cp. Fig. 6.2
T	K	absolute temperature
U(t)	V	electric potential difference or electromotive force between the plates of a capacitor, cp. Fig. 6.8
U_s	V	static electric potential difference or electromotive force between the plates of a capacitor, electric voltage
$V_c = Ad$	m^3	volume of a plate capacitor
$Y(p)=(Z(p))^{-1}$	$\Omega^{-1} = A/V$	admittance function of a dielectric system or an electric model network
$Z(p)=(Y(p))^{-1}$	$\Omega = V/A$	impedance function of a dielectric system or an electric model network
$\alpha = p_e / E$	Asm/(V/m) =(As/Vm)m^3	molecular polarizability
α_{ind}	(As/Vm)m^3	displacement or induced polarizability of a single molecule
α_{ori}	(As/Vm)m^3	oriental polarizability of a single polar molecule with permanent dipole moment (P_e)
$\alpha_m = \alpha N_A$	(As/Vm)m^3/mol	molar polarizability, i. e. polarizability of 1 mole of particles
α_m^a	(As/Vm)m^3/mol	molar polarizability, i. e. polarizability of 1 mole of adsorbed molecules (admolecules)
$\alpha_s = \alpha_m/M$	(As/Vm)m^3/g	specific polarizability, i. e. polarizability of 1 g of a material

α_s^a	(As/Vm)m^3/g	specific polarizability, i. e. polarizability of 1 g of the adsorbed phase (adsorbate)
$\varepsilon = \varepsilon_0 \varepsilon_r$	(As/Vm)	absolute dielectric permittivity of a material system
$\varepsilon_0 = 8.8542 \cdot 10^{-12}$	(As/Vm)	electric field constant or dielectric permittivity of the vacuum
ε_r	1	relative dielectric permittivity of a material
ε_{rs}	1	static dielectric permittivity of a material
$\varepsilon_{r\infty} = n^2$	1	optical dielectric permittivity of a material
ε_r^f	1	relative dielectric permittivity of a sorptive gas
ε_r^s	1	relative dielectric permittivity of a sorbent material in vacuum
μ	Cm, D	permanent dipole moment of a polar molecule
ρ	kg m^{-3}	mass density
ν	s^{-1}	frequency of an alternating electric current
σ_x^2	[x^2]	mean square deviation or dispersion of x, cp. Eqs. (2.13, 2.14)
$\omega = 2\pi\nu$	s^{-1}	angular frequency of an alternating current
$\Omega_{DE} = P_e/E$	Asm/(V/m) =(As/Vm)m^3	macroscopic dielectric polarizability of a material

REFERENCES

[6.1] **Jackson J. D.**
Classical Electrodynamics
John Wiley & Sons, New York, 2nd Ed., 1975

Klassische Elektrodynamik,
W. de Gruyter, Berlin-New-York, 1993.

[6.2] **Böttcher C. J. F.**
Theory of Electric Polarization,
Vols. 1, 2, Elsevier, 2nd Ed. Amsterdam, 1973.

[6.3] **Staudt R.**
Gasadsorption, Ingenieurwissenschaftliche Grundlagen, Messmethoden zur Bestimmung und Modelle zur Beschreibung von Adsorptionsgleichgewichten reiner Gase und Gasgemische an porösen Stoffen,
Habilitationsschrift, IFT, University of Siegen, Siegen, 2000.

[6.4] **Doll Th., Editor**
Advanced Gas Sensing – The Electroadsorptive Effect and Related Techniques,
Kluwer, Dordrecht etc., 2003.

[6.5] **Grimnes S., Martinsen O. G.**
Bioimpedance and Bioelectricity Basics,
Academic Press, New York etc., 2000.

[6.6] **Macdonald J. R.**
Impedance Spectroscopy, Emphasizing solid materials and systems,
J. Wiley & Sons, New York etc., 1987.

[6.7] **Rost A.**
Messung dielektrischer Stoffeigenschaften,
Vieweg, Braunschweig, 1978.

[6.8] **Dohrmann St.**
Charakterisierung von porösen Stoffen mit Hilfe der Impedanzspektroskopie,
Diploma-Thesis, IFT, University of Siegen, Siegen, 1997.

[6.9] **Channen E. W., McIntosh R.**
Investigation of the Physically Adsorbed State by Means of Dielectric Measurements,
Canad. J. Chem., 33 (1955), p. 172.

[6.10] **Staudt R., Rave H., Keller J. U.**
Impedance Spectroscopic Measurements of Pure Gas Adsorption Equilibria on Zeolites,
Adsorption, 5 (1999), p. 159-167.

[6.11] **Geis S., Müller B., Fricke J.**
The Response Function Method as a Novel Technique to Determine the Dielectric Permittivity of Highly Porous Materials,
Studies in Suface Science and Catalysis, Vol. 128, p. 545-555,
K. Unger et al., Eds., Elsevier Science B. V., Amsterdam etc., 2000.

[6.12] **Cauer W.**
Theorie der linearen Wechselstromschaltungen,
Springer, Berlin etc., 1954.

[6.13] **Staudt R., Saller G., Tomalla M., Keller J. U.**
Electro-gravimetric measurements of binary coadsorption equilibria,
Characterization of Porous Solids III (COPS III), Studies in Surface Science and Catalysis, Vol. 87, p. 345-351, Elsevier Science B. V., Amsterdam etc., 1994.

[6.14] **Staudt R., Kramer S., Dreisbach F., Keller J. U.**
Impedance Spectroscopic Measurements of Breakthrough Curves,
Fundamental of Adsorption 6, Proceedings of FoA 6 Conference, May 1998, Giens, p. 153-158, Meunier, Ed., Elsevier, Paris, 1998.

[6.15] **Staudt R., Ohm M., Löhr M., Keller J. U.**
Impedanzspektroskopische Untersuchungen von Adsorptionsgleichgewichten zur Optimierung von technischen Adsorbern,
Chemie Ingenieur Technik (CIT) 10 (2002), p. 1410-1413.

[6.16] **Kestin J.**
A Course in Thermodynamics, Vols. I, II, Blaisdell Publ. Comp., London, 1968.

[6.17] **Keller J. U.**
Thermodynamik der irreversiblen Prozesse mit Aufgaben, Rechenweg und Lösungen, Teil I: Thermodynamik und Grundbegriffe, Kap. A 14,
W. de Gruyter, Berlin – New York, 1977.

[6.18] **Michels W., Keller J. U.**
Elektrogravimetrische Untersuchung der Adsorption von reinem Stickstoff und Stickstoff-Wasser-Gemischen bei Umgebungstemperatur im Druckbereich 0,1-12 MPa an Molekularsieben und Aktivkohle,
Diploma-Thesis, IFT, University of Siegen, Siegen, 1988.

[6.19] **Fischer M., Keller J. U.**
Elektrogravimetrische Untersuchung von Adsorptionsgleichgewichten von reinem Argon und Argon-Wasser-Gemischen an Molekularsieben und Aktivkohle,
Diploma-Thesis, IFT, University of Siegen, Siegen, 1988.

[6.20] **Gray D. E., Editor**
American Institute of Physics Handbook,
Mc Graw Hill, New York etc., 1990.

[6.21] **Landolt-Börnstein**
Zahlenwerte und Funktionen der Physik etc., Springer, Berlin etc., 1980 etc.

[6.22] **Frohlich H.**
Theory of Dielectrics, Dielectric Constant and Dielectric Loss,
Oxford Science Publ., Oxford, UK, Reprint 1986.

[6.23] **Deutsches Institut für Normung e. V. (DIN)**
Einheiten und Begriffe für physikalische Größen (AEF-Taschenbuch 1)
DIN-Taschenbuch 22, Beuth-Verlag, Berlin etc., 7. Aufl., 1990.

[6.24] **Kohlrausch F.**
Praktische Physik, D. Hahn, S. Wagner, Eds.,
Vol. 2, 23rd Ed., B. G. Teubner, Stuttgart, 1985.

[6.25] **Keller J. U., Zimmermann W., Schein A.**
Determination of Absolute Gas Adsorption Isotherms by Combined Calorimetric and Dielectric Measurements,
Adsorption, 9 (2003), p. 177-188.

[6.26] **Ohm M.**
Impedanzspektroskopische Untersuchungen von Adsorptionsgleichgewichten reiner Gase, Gasgemischen und feuchter Luft an porösen Stoffen,
Diploma-Thesis, IFT, University of Siegen, Siegen, 2000.

[6.27 a] **König H., Meixner J.**
Linear Passive Transformationen,
Math. Nachrichten, 19 (1958), 265.

[6.27 b] **König H.**
Archiv für Mathematik,
10 (1959), 447.

[6.27 c] **Meixner J.,**
J. Appl. Mech., 33 (1966), 481.

[6.28] **Keller J. (U.)**
Über Linear Passive Transformationen Stochastischer Prozesse I, II
Zeitschrift für Naturforschung, 23 a (1968), p. 1430; 24 a (1969), p. 716.

[6.29] **Coelho R.**
Physics of Dielectrics for the Engineer,
Elsevier Science Publishers B. V., Amsterdam, 1979.

[6.30] **Meixner J.**
Network Theory in its Relation to Thermodynamics,
Proc. Symp. on Generalized Networks, New York, 1966; Vol. XVI of the Microwave Research Institute Symposia Series, Polytechnic Press of the Polytechnic Institute of Brooklyn, N. Y., 1966.

[6.31] **Mandelbrot B. B.**
The Fractal Geometry of Nature,
W. H. Freeman & Co., New York, 1977 etc.

[6.32] **Bunde A., Havlin Sh. (Eds.)**
Fractals and Disordered Systems, Springer, Berlin etc., 1991.

[6.33] **Staudt R.**
„Analytische und experimentelle Untersuchungen von Adsorptionsgleichgewichten von reinen Gasen und Gasgemischen an Aktivkohlen und Zeolithen.",
PHD-Thesis, IFT, University of Siegen, Siegen, 1994.

[6.34] **Kramer St.**
Messung und Korrelation von Durchbruchskurven an einer PSA-Anlage mit Hilfe der Impedanzspektroskopie
Diploma-Thesis, IFT, University of Siegen, Siegen, 1998.

[6.35] **Keller J. U., Schmidt O., Michels W., Fischer M., Seelbach M., Staudt R.**
Influence of Water on Adsorption Kinetics and Equilibria of Ar, N_2, and CO on Zeolite and Molecular Sieve,
Proceedings Conf. "Fundamentals of Adsorption 7", Nagasaki, May 2001, K. Kaneko et al., Eds., p. 594-600, IAS, IK International, Chiba, 2002.

[6.36] **Kienle H. von, Bäder E.**
Aktivkohle und ihre industrielle Anwendung
F. Enke Verlag, Stuttgart, 1980.

[6.37] **Stoeckli H. F.**
Microporous Carbons and their Characterization: The Present State of the Art,
Carbon, 28 (1990), p. 1-6.

[6.38] **Mersmann A.**
Adsorption, 3.2 Carbon Adsorbents, article in ULLMANN'S Encyclopedia of Industrial Chemistry, Release 2001, 6th Ed., Wiley-VCH, Weinheim etc., 2001.

[6.39] **Yang R. T.**
Adsorbents, Fundamentals and Applications,
Wiley-Interscience, Hoboken, New Jersey, 2003.

[6.40] **Dorf R. C., Svoboda J. A.**
Introduction to Electric Circuits,
Wiley Interscience, 6th Ed., Chichester, London, New York etc., 2004.

[6.41] **Schwarzer R., Rave H., Staudt R., Keller J. U.**
Messung von Sorptionsgleichgewichten von CO_2 an Polymeren, part of Diploma-Thesis, IFT, University of Siegen, Siegen, 1999.

[6.42] **Ruthven D. M.**
Principles of Adsorption and Adsorption Processes,
J. Wiley & Sons, New York etc., 1984.

[6.43] **Bathen D., Breitbach M.**
Adsorptionstechnik, VDI-Buch, Springer, Berlin, New York etc., 2001.

[6.44] **Ruthven D. M., Shamsuzzaman F., Knaebel K. S.**
Pressure Swing Adsorption, VCH Publishers, New York, Weinheim, 1994.

[6.45] **Gummersbach M., Staudt R., Keller J. U.**
Impedanzspektroskopische Untersuchungen von porösen Stoffen und deren Adsorbat, Diploma-Thesis, IFT, University of Siegen, Siegen, 1999.

This page intentionally left blank

Chapter 7

ADSORPTION ISOTHERMS

Abstract A short overview of adsorption isotherms often used (i) to correlate adsorption equilibria data for sorbent materials characterization and (ii) for design of industrial gas adsorption processes is given. The basic types of adsorption isotherms observed experimentally are discussed in brief. The Langmuir adsorption isotherm (LAI) and certain of its extensions to (energetically) heterogeneous sorbent materials and to admolecules with interactions are presented (Sect. 2). Several empirical isotherms used for micro- and mesoporous materials showing pore condensation are discussed in Sect. 3 briefly. In Sect. 4 an outline of thermodynamics of adsorbate phases of fractal dimension is given. Several isotherms are presented which are generalizations of well-known isotherms (Langmuir, BET, etc.) to multicomponent adsorbates on heterogeneous surfaces. These isotherms are solutions of Maxwell's relations of the underlying Gibbs fundamental equation of the adsorbed phase. List of symbols. References.

1. INTRODUCTION

The purpose of this chapter is to give a comprehensible overview of a few frequently used adsorption isotherms (AIs) for single- and multicomponent gas adsorption systems. The adsorption isotherm (AI) is, in the sense of thermodynamics, the thermal equation of state [7.81] for the adsorbed phase, i. e. a function

$$m = m\,(p, T, m^s)$$

for single component systems, or a set of functions

$$m_i = m_i\,(p_1 \ldots p_N, T, m^s), \quad i = 1 \ldots N,$$

for a multicomponent system. Here (m) and ($m_1 \ldots m_N$) indicate the mass(es) of components (i = 1...N) adsorbed, $p_i = y_i p$ is the partial pressure of component i in the gas phase with molar concentration (y_i) and T is the temperature of the system. The parameter (m^s) indicates the mass of the sorbent material.

As many isotherms, especially for single component systems, already have been developed during a period of nearly 100 years, it is not possible within this monography to present all of them. Instead we will restrict the discussion to those which proved to be useful for engineering applications or at least have the potential to be so, especially in higher pressure regimes ($p \simeq 10$ MPa) and in a broad temperature interval (200 K < T < 400 K). Good overviews of adsorption isotherms can be found in [7.1-7.4]. Hints to additional literature will be given whenever it is appropriate.

Adsorption isotherms (AI) are used for (i) characterization of porous solids and (ii) design of industrial adsorption processes. Here they are needed not only for single- but also for multicomponent gas mixture systems. Due to the complexity of interactions of admolecules (a) with the atoms and molecules of the sorbent material (s) (a-s-interactions) [7.5-7.7], it is not possible today to calculate AIs either "ab initio" by statistical molecular methods (DFT or other methods from Statistical Mechanics) or by phenomenological methods based on a few macroscopic or microscopic data of the sorptive-sorbent system. Even ideal molecular models of surfaces and pores of sorbent materials need not only model functions for the (a-s)-interactions, but in principle will lead to a so-called calibration problem [7.8] which only can be solved by introducing experimental data regarding absolute amounts or masses of gas adsorbed at well defined conditions [7.9]. As these are fairly cumbersome to measure and normally are not available at all, cp. remarks in Chap. 1, it is necessary to measure gas adsorption equilibria data, especially for multicomponent systems, i. e. so-called coadsorption equilibria data. These often exhibit or at least indicate the physical mechanism of the adsorption process observed and thus allow one to make a proper choice for an analytical AI to correlate the data and then to try to extrapolate them to other pressures and temperatures needed for process design. As in most industrial sorbent materials both micro- and mesopores occur to a more or lesser extent, one often can find a combination of basically different adsorption mechanisms like surface coverage and pore condensation to occur, thus leading to fairly complex AIs the structure of which cannot be foreseen without any experimental information.

Adsorption equilibria data normally are graphically presented as

Isotherms : m, p-diagrams at T = const,
Isobars : m, T-diagrams at p = const,
Isosteres : In (p/p_0), T-diagrams at m = const.

Typical examples for these curves often observed for microporous materials are sketched in Table 7.1. In addition adsorption enthalpies are often needed for process design. In view of space limitations we do not consider them here but refer to the literature [7.2, 7.5, 7.10, 7.11].

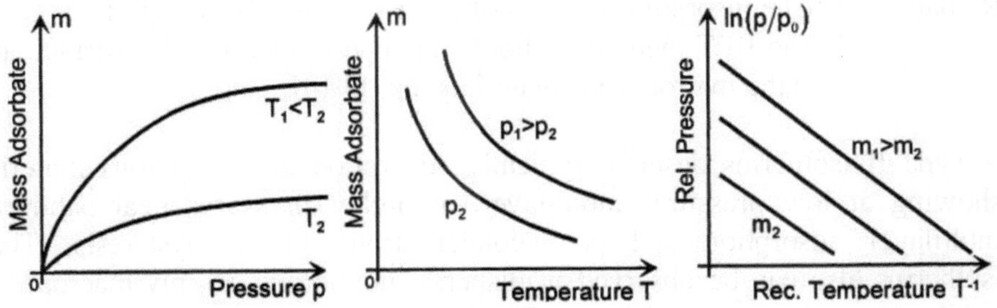

Table 7.1. Representation of gas adsorption equilibria data as isotherms (T = const), isobars (p = const), and isosteres (m = const).

The experimentally observed adsorption isotherms (AI) can be classified according to IUPAC-recommendations in 6 different types I-VI [7.1-7.3, 7.5, 7.12, 7.13]. Recently some subclasses of these have been introduced [7.5], but are not considered here. A scheme of these types of AIs is given in Tab. 7.2, [7.2, 7.5, p 19].

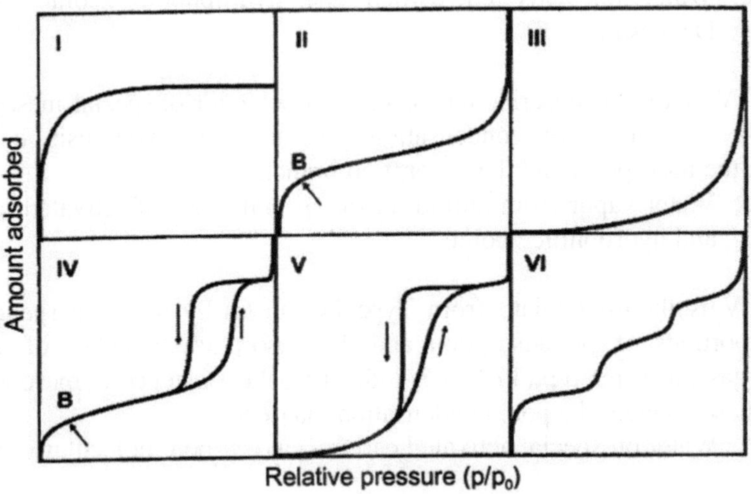

Table 7.2. Main types of gas physisorption isotherms (IUPAC, 1985).

Type I isotherms can be described by the Langmuir equation, cp. Sect. 2.1. They are characterized by a horizontal plateau, i. e. the asymptotic value the mass adsorbed approaches and maintains for even very high gas pressures.

These isotherms are characteristic for microporous materials showing micropore filling but no multilayer adsorption.

Example: Water on zeolite or inorganic (i. e. polar) molecular sieves. Organic vapors (C_nH_{2n+2}) on zeolites and molecular sieves.

Remarks: 1. Chemisorption systems often also show AIs of Type I.

2. The BET method cannot be applied to measure the surface area of the microporous materials, cp. [7.14].

Type II isotherms describe typically adsorption in mesoporous materials showing at low pressures monolayer, at higher pressures near saturation multilayer adsorption and pore condensation but no hysteresis. These isotherms also can be observed in disperse, nonporous or only macroporous solids. (pore diameter > 50 nm). They often can be described by the BET equation or its generalizations, cp. Sects. 3.4, 4.2.

Example: Vapors of polar media (H_2O, $CCl_iF_kH_{4-i-k}$) on molecular sieves, vapors of non-polar organic substances on slightly microporous but mainly mesoporous activated carbons.

Type III isotherms occur in systems where the adsorbate-sorbent (a-s) interaction is small compared to the adsorbate-adsorbate (a-a) interaction, i. e. strongly associating admolecules.

Example: Water on activated carbon and hydrophobic zeolites (DAY, Degussa).

Type IV isotherms describe the adsorption behavior of special mesoporous materials showing pore condensation together with hysteresis behavior between the adsorption and the desorption branch.

Example: Water vapor from humid air on special types of activated carbons and hydrophilic zeolites.

Type V isotherms deviate from Type IV curves by nearly perpendicular middle portions of the adsorption and the desorption branches often near relative gas pressures $p/p_s(T) \cong 0.5$, indicating the existence of mesopores in which phase change like pore condensation may occur.

Example: Water on special activated carbons and carbon molecular sieves.

Type VI isotherms present stepwise multilayer adsorbates, the layers becoming more pronounced at low temperatures.

Example: Nonpolar, spherical molecules (noble gases) on planar graphite surfaces, butanol (C_4H_9OH) on aluminum silicate.

A much more detailed analysis of these isotherms and their molecular interpretation is given in [7.1, 7.3, 7.5], cp. also [7.62].

In this chapter we are going to discuss the Langmuir adsorption isotherm (LAI) and several of its generalizations for single- and multicomponent systems in Sect. 2. This isotherm has proved to be most useful to describe adsorption in microporous materials. In Sect. 3 we will provide some information on a few empirical adsorption isotherms used to describe gas adsorption in micro- and mesoporous materials, i. e. showing pore condensation leading to an unlimited amount of mass in the adsorbed phase.

In Sect. 4 we present several adsorption isotherms which are solutions of the Maxwell relations of the Gibbs fundamental equation of the multicomponent adsorbate [7.15]. These isotherms are thermodynamically consistent generalizations of several of the empirical isotherms presented in Sect. 3 to (energetically) heterogenous sorbent materials with surfaces of fractal dimension. In Sect. 5 some general recommendations for use of AIs in industrial adsorption processes are given.

2. SIMPLE MOLECULAR ISOTHERMS

2.1 Langmuir Adsorption Isotherm

The classical Langmuir adsorption isotherm and several of its generalizations are presented. The physical background is enlightened and hints for engineering applications are given. The generalizations refer to

a) real gas adsorptives,
b) energetically heterogeneous sorbent materials,
c) admolecules with interactions.

For more information the reader is referred to the vast and still growing literature of adsorption, experiments, theory, numerical simulation and engineering applications [7.1-7.5].

2.1.1 Classical Form

The classical form of the Langmuir adsorption isotherm (LAI) already published in 1916, [7.16] is

$$m(p,T,m^s) = m_\infty(T,m^s)\frac{b(T)p}{1+b(T)p} \qquad 7.1$$

Parameters:

$m = Mn$	Mass of adsorbate
p	pressure
T	Absolute temperature
m^s	mass of sorbent material
$m_\infty = \lim_{p\to\infty} m$	limiting value of mass adsorbed at T = const, $p \to \infty$. Also mass of a monolayer adsorbate covering completely the surface of the sorbent [7.3].

$$m_\infty(T,m^s) = \mu_\infty(T_0)\left(\frac{T_0}{T}\right)^{k_s} m^s, \quad k_s > 0 \qquad 7.2$$

μ_∞	limiting mass ($p \to \infty$) of adsorbate on sorbent of mass (m^s) at reference temperature (T_0).

$$b(T) = \frac{1}{p_0}\sqrt{\frac{T_0}{T}}\exp\left(\frac{q}{RT}\right) \quad \text{...Langmuir parameter}$$

also reciprocal of half-loading pressure, cp. (7.1), [7.3, 7.76]

$$b^{-1} = p(T,\ m = m_\infty/2,\ m^s), \qquad 7.3$$

$R=\mathbf{R}/M$	specific gas constant.

Enthalpy of adsorption

$$\Delta H = H^a - H^f = -qm < 0 \qquad 7.4$$

$q = \text{const}$	specific heat of adsorption assumed to be constant [7.5, 7.33]

Enthalpy of desorption:

$$H^f - H^a = -\Delta H = qm > 0. \qquad 7.4a$$

Physical prerequisites of the LAI (7.1); cp. also Fig. 7.1.

1. Adsorption occurs on a fixed number of sites.
2. Each site can only take one admolecule but no more.
3. All sites are energetically equivalent (q = const), cp. (7.4)
4. Interactions between adsorbed molecules are neglected as they are assumed to be small compared to the sorbate / sorbent interactions characterized by the (specific) adsorption energy (q) [7.18, 7.64].

The energetic situation of the admolecules in a Langmuir adsorbate is sketched in Figure 7.1. A graphic representation of the LAI, Eq. (7.1) is given in Fig. 7.2

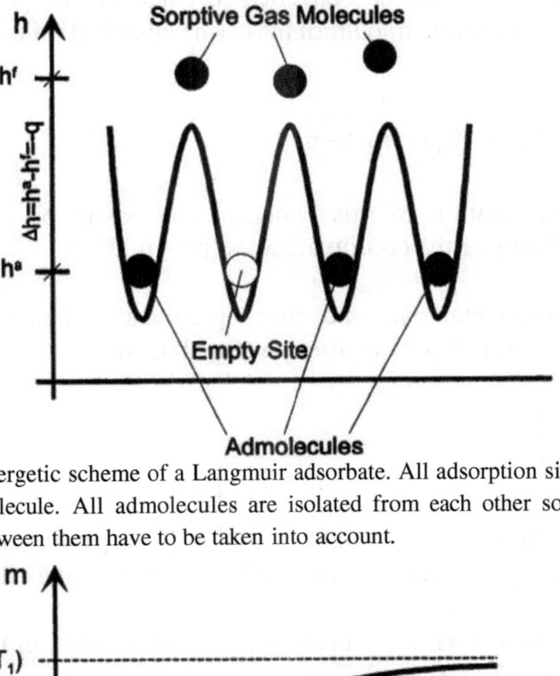

Figure 7.1. Energetic scheme of a Langmuir adsorbate. All adsorption sites can only take one molecule. All admolecules are isolated from each other so that no interactions between them have to be taken into account.

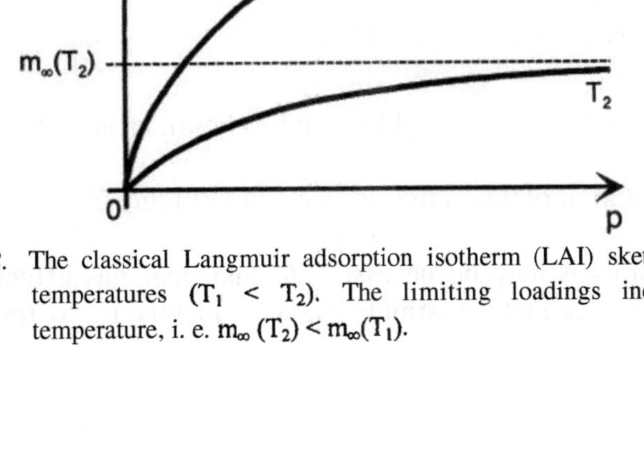

Figure 7.2. The classical Langmuir adsorption isotherm (LAI) sketched for two different temperatures ($T_1 < T_2$). The limiting loadings increase with decreasing temperature, i. e. $m_\infty(T_2) < m_\infty(T_1)$.

The classical LAI (7.1) seems to be the limiting form of all gas adsorption isotherms for high temperatures. This simply can be explained by the fact that for $T \to \infty$ the average (specific) thermal energy (RT) of the ideal sorptive gas is much larger than the adsorption energy (q), i. e. $RT >> q$. Under such conditions the sorbent surface appears to be nearly homogeneous for all admolecules. Also the interaction energy of admolecules represented by the heat of evaporation / condensation ($r < q$) can be neglected. Hence the prerequisites for the LAI given above hold and Langmuirian behavior of the AI should be expected.

The LAI (7.1) easily can be converted to calculate the gas pressure (p) needed to maintain a certain amount of mass (m) adsorbed on (m^s):

$$p(m,T,m^s) = \frac{1}{b(T)} \cdot \frac{m}{m_\infty(T,m^s) - m} \qquad 7.5$$

A graphical representation of this function can easily be deduced from Fig. 7.1 by interchanging the coordinate axes ($p \leftrightarrow m$).

To represent experimental data and check graphically their consistency, a linearized in (p) version of (7.1) is often useful. It reads

$$\frac{p}{m} = \frac{1}{bm_\infty} + \frac{p}{m_\infty} \qquad 7.6$$

Scattering of data around this straight line in a (p/m, p)-diagram often indicates experimental inconsistencies, especially at low pressures ($p \to 0$, $m \to 0$).

The isosteres of the LAI, i. e. curves $p = p$ (T, m = const, m^s) easily can be calculated from eqs. (7.1, 7.2) as

$$\ln\left(\frac{p}{p_0}\sqrt{\frac{T_0}{T}}\right) = -\frac{q}{RT_0}\left(\frac{T_0}{T}\right) + \ln\left(\frac{m/m_\infty}{1-(m/m_\infty)}\right) \qquad 7.7$$

Here (T_0) is a reference temperature which can be chosen arbitrarily.

A graphical representation of this function is given in Figure 7.3.

At higher pressures it may be necessary to take real gas effects of the sorptive medium into account. A simple way to do this is, to replace the pressure (p) in Eq. (7.1) by the fugacity f= f(T, P) of the gas [7.17]. We then get

$$m(p,T,m^s) = m_\infty(T,m^s)\frac{b(r)f(p,T)}{1+b(r)f(p,T)} \qquad 7.8$$

Adsorption isotherms of this type have proved to be useful to describe adsorption of small molecule gases like (CH_4, N_2, O_2) at ambient temperature and pressures up to 50 MPa, [7.80]. The enthalpy of adsorption then is, cp. Eq. (7.4):

$$\Delta H = H^a - H^f = -\left(q - \frac{RT^2}{f}\left(\frac{\partial f}{\partial T}\right)_p\right)m \qquad 7.8a$$

We finally want to draw reader's attention to the energetic situation of the classical Langmuir adsorbate described by Eq. (7.1).

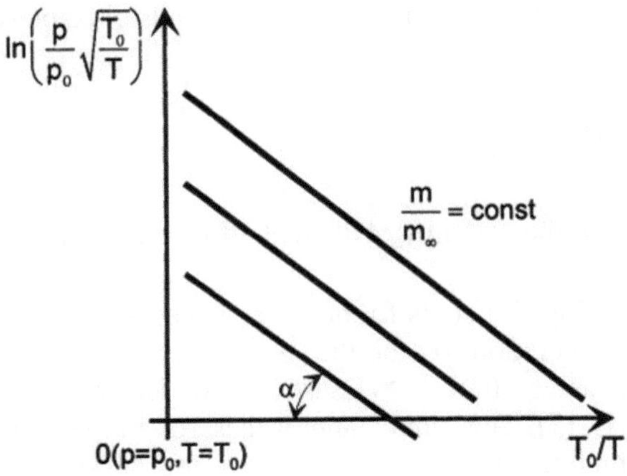

Figure 7.3. Isosteres p = p (T, m = const, m^s) of the classical Langmuir adsorption isotherm, Eq. (7.7).

It is characterized by a single constant (q) which is – according to (7.4) – the specific change, i. e. increase of the enthalpy of the sorptive gas due to desorption for an adsorbate mass $0 \leq m \leq m_\infty$.

$$\Delta h_{des} = \frac{\Delta H}{m} = +q > 0 \qquad 7.9$$

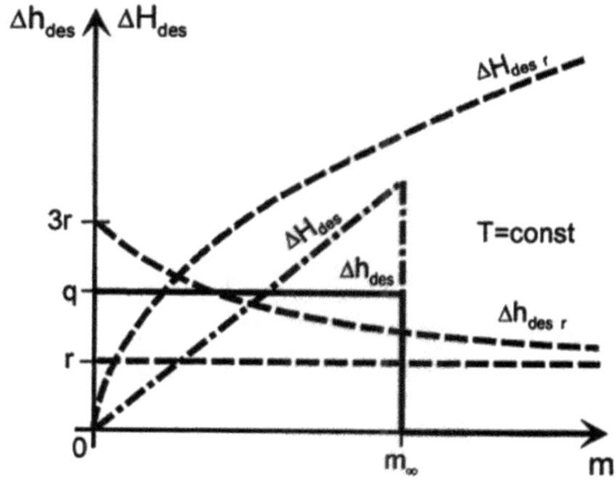

Figure 7.4. Specific enthalpy of desorption of a Langmuir adsorbate (Δh_{des}) and a real adsorbate (Δh_{desr}). The integral enthalpies of desorption $\left(\Delta H_{des}, \Delta H_{des,r}\right)$ also have been sketched qualitatively.

In Figure 7.4 this curve is sketched together with the differential heat of desorption of a real, i. e. energetically heterogeneous adsorbed phase

$$\Delta h_{des,r} = \left(\partial (H^h - H^a_r)/\partial m\right)_T \qquad 7.9a$$

Here (H^a_r) is indicating the enthalpy of the real adsorbed phase of mass (m).

The curve $\Delta h_{desr}(m)$ in Fig. 7.4 starts for m = 0 at values which are about three times the heat of evaporation (r) of the liquid sorptive medium at the temperature (T) considered. (For $T > T_c$ use the Riedel-line approximation, cp. Eq. (7.23) and [7.17]).This curve approaches for $m \rightarrow \infty$ the heat of evaporation (r) asymptotically. A comparison of the curves $\Delta h_{des}(m)$ and $\Delta h_{desr}(m)$ in Figure 7.4 clearly shows the limitations of the Langmuir adsorption model leading to the isotherm (7.1), but also its main advantage, namely its simplicity.

Langmuir originally based the derivation of the adsorption isotherm (7.1) on kinetic arguments which for sake of completeness are mentioned here in brief [7.2 – 7.4, 7.7, 7.15, 7.19, 7.20]:

1. The number of molecules (dn^+) which are adsorbed on a surface including a fixed maximum number ($n_\infty = m_\infty/M$)of adsorption sites within the

differential time interval (dt) is proportional to the number of empty sites (n_∞ - n) and the pressure (p) of the adsorptive

$$dn^+ = k^+ p(n_\infty - n)dt, \tag{7.10}$$

k^+ being a positive constant.

2. The number of molecules (dn^-) which desorb from the surface within (dt) is proportional to the number (n) of molecules already adsorbed:

$$dn^- = k^- ndt, \tag{7.11}$$

k^- being a positive constant.

Hence we have for the net gain of molecules adsorbed within (dt)

$$dn = dn^+ - dn^- \tag{7.12}$$

from (7.10, 7.11).

$$\frac{dn}{dt} = k^+ pn_\infty - (k^+ p + k^-)n \tag{7.13}$$

This ordinary differential equation (ODE) has the solution

$$n(t) = n_0 e^{-at} + n_\infty \frac{bp}{1+bp}(1 - e^{-at}) \tag{7.14}$$

with

$$a = k^+ p + k^-, \quad b = \frac{k^+}{k^-} \tag{7.15}$$

and $n_0 = n\,(0)$, cp. Fig. 7.5. For $t \to \infty$ or, likewise (dn/dt) = 0, we get either from (7.13) or (7.14) after multiplication with the mole mass (M) of the sorptive gas the equilibrium value of the sorbate mass

$$m_{eq} = m_\infty \frac{bp}{1+bp}$$

which is the LAI, Eq. (7.1).

For more information on the kinetics of adsorption processes the reader is referred to the literature, especially to the review articles by W. Rudzinski and Panczyk [7.19], [7.20] and the monograph by D. D. Do [7.3].

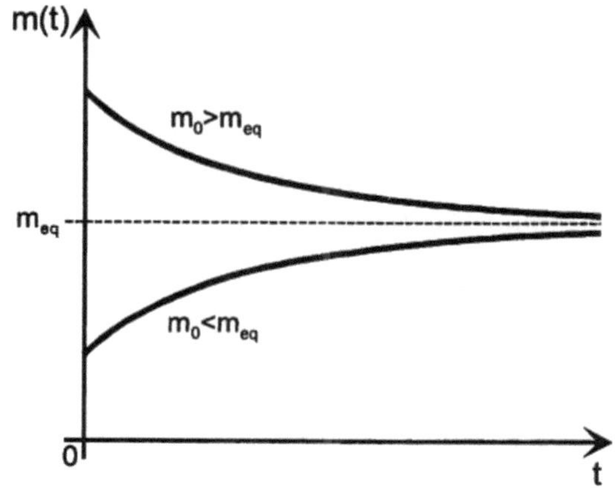

Figure 7.5. Kinetics of a Langmuir adsorption process with initial adsorbate mass $m_0 = n_0 M$ and asymptotic mass $m(\infty) = m_\infty = n_\infty M$. The time of relaxation of the process is given by $\tau = 1/a\ 0\ 1/(k^+p+k^-)$.

The Langmuir model of adsorption easily can be extended to multi-component adsorption processes [7.2 Chap. 4.6, 7.3 Chap. 5, 7.29, 7.75, 7.76]. The resulting adsorption isotherm is

$$m_i(p_i \ldots p_N, T, m^s) = m_{i\infty}(T, m^s) \frac{b_i p_i}{1 + \sum_{k=1}^{N} b_k p_k}, \quad i = 1 \ldots N \tag{7.16}$$

Parameters:

$m_i = n_i M_i$ ... mass of component i = 1...N of the adsorbed phase

$p_i = y_i p$ ... partial pressure of component (i) with molar fraction (y_i) in the sorptive gas mixture at pressure (p).

$m_{i\infty} = \lim_{\substack{p_i \to \infty \\ p_k = 0}} m_i$ ... limiting value of pure component (i) adsorbed at T = const for $p_i = p \to \infty$. Also mass of a monolayer of component (i) covering completely the surface of the sorbent material.

$$b_i(T) = \frac{1}{p_{0i}} \sqrt{\frac{T_{0i}}{T}} \exp\left(\frac{q_i}{R_i T}\right), \quad i = 1 \ldots N \tag{7.17}$$

Langmuir parameter, also reciprocal of half-loading partial pressure, cp. (7.16):

$b_i^{-1}(T) \quad = p_i\,(T, m_i = m_{i\infty} / 2_1\, m^s),\ 1...N.$
$R_i \qquad = R/M_i$...specific gas constant
q_i...specific enthalpy of desorption of pure component (i).

The adsorption isotherms (7.16) easily can be converted to give the partial gas pressures ($p_1...p_N$) necessary for maintaining a given set of masses ($m_1...m_N$) in the adsorbed state:

$$p_i(m_1...m_N, T, m^s) = \frac{1}{b_i(T)} \bullet \frac{m_{i\infty}(T, m^s)}{1 - \sum_{k^{-1}}^{N}(m_k / m_k(T, m^s))}, \quad i = 1...N \qquad 7.18$$

For small pressure values ($p_i \rightarrow 0$, i = 1...N) the isotherms (7.16) can be linearized. Then we have for the mole numbers of the adsorbed components

$$n_i = n_{i\infty} b_i p_i + O(p_k^2) \qquad 7.19$$

with the Henry constants defined by

$$H_i = n_{i\infty} b_i, \quad i = 1...N \qquad 7.20$$

Hence we get for the so-called separation factors of components (i, k) in the gaseous and the adsorbed phase defined by

$$S_{ik} = (x_i / y_i)/(x_k / y_k), \quad i, k = 1...N$$
$$= \frac{n_i p_k}{n_k p_i} \qquad 7.21$$

in view of (7.19, 7.20)

$$S_{ik} = \frac{H_i}{H_k}, \quad i, k = 1...N \qquad 7.22$$

The symbols on the r.h.s of (7.21) indicate respectively the molar concentrations of components (i, k) in the adsorbate (x_i, x_k) and the sorptive phase (y_i, y_k). The practical value of cp. (7.22) is that low pressure measurements of the binary coadsorption equilibria, which deliver the Henry's constants (H_i, H_k) allow one to calculate – approximately – the selectivity (S_{ik}). If $S_{ik} > 10$ the sorbent in question can practically used to purify a gas mixture (i, k) from component (i) by adsorption equilibria effects. However, it should be noted that often kinetic effects are also important in industrial adsorption processes. Hence, appropriate experiments and measurements of kinetic parameters, cp. Eq. (7.14) and Fig. 7.5, should be performed prior to a final decision on the sorbent to be chosen for the adsorption separation process in question [7.2].

The physical assumptions on which the LAI is based often hold for chemisorption processes [7.21]. For physisorption processes of gases they normally hold at low pressures and/or low coverages of the surface ($(m/m_\infty) < 0{,}5$) only. The term "low pressure" can be specified as

$$\begin{aligned} p &< p_s(T) && \ldots T < T_c \\ p &< p_R = p_c + (\partial p/\partial T)_c\,(T - T_c) && \ldots T > T_c \end{aligned} \qquad 7.23$$

Here the index (c) indicates "critical state" of the sorptive gas. The pressure (p_R) is known as Riedel pressure, an equivalent for the saturation pressure for supercritical temperatures of the gas [7.17]

The LAI often successfully has been applied to describe physisorption of supercritical gases including small non-polar molecules like (He, Ar, H_2, N_2, O_2, CH_4) with small heats of adsorption ($|\Delta h| < 30 kJ/mol$) on energetically nearly homogenous sorbent materials. Examples for these are graphitized carbon black [7.5], highly activated carbons (NORIT R1) and also porous polymers (POLYSORB, Sweden). More detailed information is provided in the extensive literature which is available today [7.1, 7.5, 7.21, 7.22].

2.1.2 Heterogeneous Surfaces

As the classical LAI only holds for energetically homogenous surfaces, but most technical sorbent materials are energetically inhomogeneous, as early as 1930 work began to extend the LAI to heterogeneous porous sorbents [7.1, 7.3, 7.5, 7.6, 7.7, 7.22]. A simple way to do this is sketched in Figure 7.6. The sorbent is assumed to consist of two types of pores: macropores offering "primary adsorption sites" (p) to gas molecules, and micropores which cannot directly adsorb molecules form the gas phase but are filled by diffusion processes of primary adsorbed molecules reaching eventually the open end of such a pore and there are adsorbed on "secondary sites" (q).

The adsorption isotherm referring to this situation is

$$m(p,T,m^s) = m^p(p,T,m^s) + m^q(p,T,m^s)$$
$$= m^p_\infty(T,m^s)\frac{b^p(T)p}{1+b^p(T)p} + m^q_\infty(T,m^s)\frac{b^q(T)p}{1+b^q(T)p} \quad 7.24$$

It can be derived from statistical arguments similar to those leading to the classical LAI, Eq. (7.1), [7.23, 7.24].

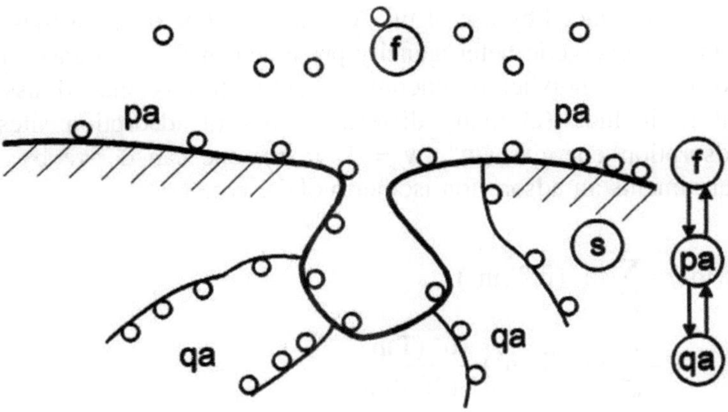

Figure 7.6. Sorbent material with two types of adsorption sites: macropores bearing primary, i. e. from the sorptive gas phase directly adsorbed molecules (pa), and micropores including secondary adsorbed molecules (qa) coming from the primary adsorbate (pa) by diffusion of the molecules.

Parameters:

$m = Mn$	... mass of adsorbate
m^s	...mass of sorbent material
$m^r_\infty = \lim_{p\to\infty} m^r$	...limiting value of mass adsorbed for $p \to \infty$ on adsorption site of type ($r = p, q$). Also mass of a monolayer adsorbate of type (r).
$b^r(T) = \frac{1}{p_0}\sqrt{\frac{T_0}{T}}\exp\left(\frac{q^r}{RT}\right)$	...Langmuir parameter for adsorption sites of type ($r = p, q$).
q^r, $r = p, q$	...specific heat of desorption from adsorption site of type (r).
$R = \mathbf{R}/M$	...specific gas constant.

Enthalpy of adsorption:

$$\Delta H = H^{a} - H^{f} = -(q^{p}m^{p} + q^{q}m^{q}) < 0 \qquad 7.25$$

As an example for the AI (7.24) data of nitrogen (N_2, 5.0) adsorption on activated carbon Norit ACR1 have been correlated. Results are sketched in Figure 7.7, [7.24]. The curves refer to the primary adsorption and the total adsorption i. e. the sum of the primary and the secondary adsorbed molecules. Compared to the classical LAI the dispersion between measured and correlated data is reduced by use of the AI (7.24) by one order of magnitude. The basic idea of energetic heterogeneity presented in Fig. 7.6 easily can be extended to multi- or polyheterogeneous sorbent materials. Indeed, assuming the sorbent to include (z) many different types of adsorption sites with specific adsorption energies ($-q^{w}$, w = 1 ... z), one can derive by simple statistical arguments an adsorption isotherm of the type

$$\begin{aligned} m(p,T,m^{s}) &= \sum_{w=1}^{z} m^{w}(p,T,m^{s}) \\ &= \sum_{w=1}^{z} m^{w}(T,m^{s}) \frac{b^{w}(T)p}{1+b^{w}(T)p} \end{aligned} \qquad 7.26$$

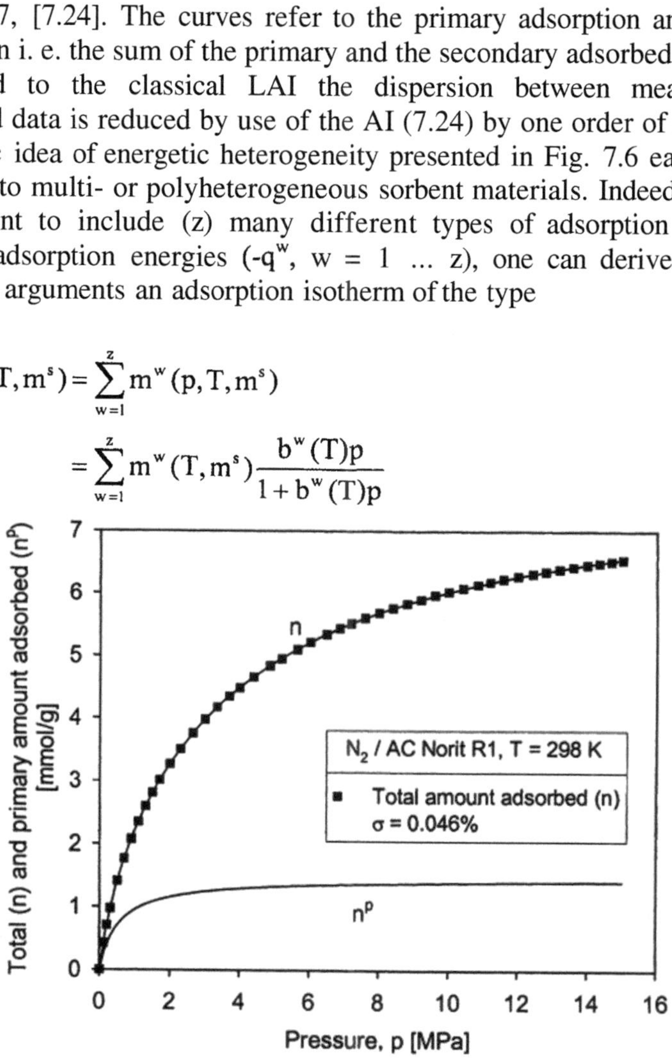

Figure 7.7. Adsorption isotherm of nitrogen (N_2, 5.0) on Norit R1 at T = 298 K. Data have been correlated by Eq. (7.24). The partial amount of N_2 adsorbed on primary sites ($n^{p} = m^{p}/M_{N2}$) is presented by the lower curve. The upper curve indicates the total amount adsorbed ($n = (n^{p} + n^{q})$), [7.24].

The parameters in the AI (7.26) ($m_\infty^w, b^w, w = 1...z$) have the same meaning as indicated after the AI (7.24). The enthalpy of adsorption is similarly to (7.25)

$$\Delta H = H^a - H^f = -\sum_{w=1}^{z} q^w m^w < 0 \qquad 7.27$$

The total mass which can be adsorbed for $p \to \infty$ is

$$m_\infty(T, m^s) = \sum_{w=1}^{z} m_\infty^w(T, m^s) \qquad 7.28$$

It is interesting to consider the limit ($z \to \infty$) of the AI (7.26). Introducing the (integral distribution of the) pore spectrum $V = V(r, T, m^s)$ of the sorbent by the relation

$$dm_\infty(r, T, m^s) = \rho^L(T) dV(r, T, m^s), \qquad 7.29$$

we get from Eq. (7.26) the isotherm

$$m(p, T, m^s) = \rho^L(T) \int_{r=0}^{\infty} \frac{b(r,T)p}{1 + b(r,T)p} \left(\frac{\partial V(r, T, m^s)}{\partial r} \right) dr \qquad 7.30$$

In (7.29, 7.30) $\rho^L(T)$ is for $T < T_c$ the density of the sorptive medium in the liquid boiling state or, for $T > T_c$, the Riedel density $\rho^L(T) = \rho_c + (\partial \rho / \partial T)_c (T - T_c)$, the index "c" always indicating the thermodynamic critical state of the medium. For small pressures ($p \to 0$) (7.30) reduces to:

$$m(p, T, m^s) = H(T, m^s) p \qquad 7.31$$

with the Henry constant given by

$$H(T, m^s) = \rho^L(T) \int_0^{\infty} b(r, T) \left(\frac{\partial V(r, T, m^s)}{\partial r} \right) dr \qquad 7.32$$

The enthalpy of adsorption is

$$\Delta H = H^a - H^f = -\rho^L(T)\int_0^\infty q(r,T)\frac{b(r,T)p}{1+b(r,T)p}\left(\frac{\partial V(r,T,m^s)}{\partial r}\right)dr \qquad 7.33$$

Isotherms of type (7.30 have proved to be useful to describe adsorption of small test molecules (He, Ar, N_2) on impregnated activated carbons, the (often organic acid) impregnation providing a means to change the pore spectrum of the carbon in a prescribed way thus providing specified adsorption sites for biochemical and / or pharmaceutical molecules [7.25].

The adsorption isotherms (7.24, 7.26, 7.30) easily can be generalized to multi-component adsorption equilibria. In view of space limitations and lack of elucidating technical examples we here restrict the discussion to a mention of the coadsorption isotherm corresponding to the dual site isotherm (7.24) only, cp. also Sect. 2.1.1. The mass of component (i) adsorbed on a two site adsorbent is given by

$$\begin{aligned} m_i(p_1...p_N,T,m^s) &= m_i^p(p_1...p_N,T,m^s) + m_i^q(p_1...p_N,T,m^s) \\ &= m_{i\infty}^p(T,m^s)\frac{b_i^p(T)p_i}{1+\sum_{k=1}^N b_k^p(T)p_k} \\ &\quad + m_{i\infty}^q(T,m^s)\frac{b_i^q(T)p_i}{1+\sum_{k=1}^N b_k^q(T)p_k} \end{aligned} \qquad 7.34$$

Parameters:

p_i, i= 1...N	... partial pressure of component (i) in the sorptive gas mixture
$m_{i\infty}^r(T,m^s), \quad r = p,q$	... limiting loading of pure component i = 1...N on the adsorption sites of type (r = p, q).
$b_i^r(T) = \frac{1}{p_0}\sqrt{\frac{T_0}{T}}\, e^{\frac{q_i^r}{R_i T}}$	... Langmuir parameter of component (i) on adsorption sites of type (r = p, q).
q^r, r = p, q	... specific heat of desorption from adsorption sites of type (r).
q_i^r, i = 1...N	... specific heat of desorption of pure component (i) from adsorption sites of type (r).
$R = \mathbf{R}/M_i$	... specific gas constant of component (i) with mole mass (M_i).

The adsorption enthalpy corresponding to (7.34) is

$$\Delta H = H^a - H^f = -\sum_{i=1}^{N}(q_i^p m_i^p + q_i^q m_i^q) \qquad 7.35$$

Coadsorption isotherms of type (7.34) have proved to be useful for the correlations of binary and ternary coadsorption data of natural gas mixtures (CH_4, N_2, CO_2, CO, H_2) on activated carbons. Details are given in [7.26].

2.1.3 Admolecules with Interactions

Admolecules often are not completely isolated from each other as indicated in Fig. 7.1, but are interacting, especially at higher surface coverages or adsorption loadings [7.6, 7.7]. Hence there have been several attempts in the literature to take these interactions into account, i. e. to extend the classical LAI to interacting admolecules on both homogeneous and heterogeneous sorbent surfaces. We here mention only the early and fundamental work of Fowler and Guggenheim (1939) [7.27], which however was not very successful in practical applications, and the work of Rudzinski & Everett (1992) [7.22], which unfortunately is very complex and not easily applicable to engineering systems and processes.

Based on a simple statistical model the classical LAI can be generalized to situations where interactions between adsorbed molecules (admolecules) on neighboring sites no longer can be neglected but explicitly have to be taken into account [7.28]. This is done by introducing the concept of "neighborhood" of admolecules in the statistical derivation of the LAI and specifying it by a simple phenomenological model. The resulting isotherm can be either of Type I, IV, or V (IUPAC) depending on numerical values of the interaction parameters. The energetic situation of the admolecules with interactions is sketched in Figure 7.8. The adsorption enthalpy consists of two parts, namely that for adsorption of an isolated admolecule (q^a) and another part reflecting the interaction between the admolecules (q^{ai}) as it occurs in densely packed micropores and – in principal – also in meso- and macropores.

$$\Delta H = H^a - H^f = -(q^a + q^{ai})m < 0 \qquad 7.36$$

At low coverages approximate values are: $q^a \cong 2.5$ r, $q^{ai} \cong$ r with (r) being the heat of evaporation of the sorptive medium.

The resulting adsorption isotherm is [7.28]:

$$\theta \equiv \frac{m}{m_\infty} = \frac{F(\theta)p^\alpha}{1+F(\theta)p^\alpha} \qquad 7.37$$

$$F(\theta,p) = \frac{b^\alpha + c(1-\theta)^k}{1+d\theta^k}, k>0 \qquad 7.38$$

with

$$\lim_{p\to\infty} m = m_\infty \qquad 7.39$$

being the limiting monolayer mass adsorbed at high gas pressures.
For small pressures ($p \to 0$) we have

$$\theta = F(0)p^\alpha = (bp)^\alpha$$
$$m = m_\infty (bp)^\alpha \qquad 7.40$$

Relations (7.37, 7.38) provide an implicit equation to calculate the mass adsorbed (m) from given values of the parameters (b, c, d, k, α, m_∞) and sorptive gas pressure (p). Resulting curves can be of IUPAC Type I, IV, V depending on numerical values of parameters $b > 0$, $c \gtrless 0$, $d \gtrless 0$, $k > 0$, $\alpha > 0$. Here c, d describe admolecular interactions, b is the Langmuir parameter (7.2) and k, α describe the topology of the "neighborhood" of the admolecule. Hence these parameters are related to the pore structure of the sorbent material.

As an example 3 sets of experimental data [7.29] referring to adsorption equilibria of H_2S, CO_2 and C_3H_8 on mordenite (H) at T = 303 for pressures up to 0.2 MPa have been correlated with both the generalized LAI (7.37, 7.38) and the simple LAI, i. e. neglecting admolecular interactions and assuming in (7.38) a = c = 0, but $\alpha \neq 1$. Results are presented graphically in Figures 7.9, 7.10. As can be seen, neglecting interactions between admolecules leads to systematic deviations between measured and simply correlated data, especially at higher pressures, i. e. adsorption loadings, Fig. 7.9. Correlation of data however is possible by qualitatively taking the interaction of admolecules into account, i. e. by correlating the experimental data by (7.37, 7.38), Fig. 7.10.

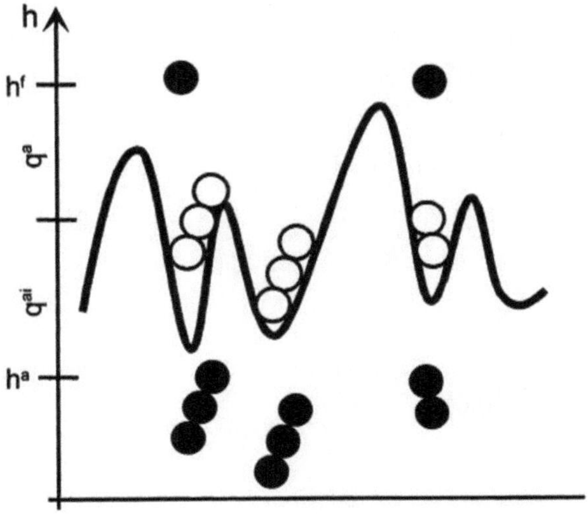

Figure 7.8. Energetic scheme of a Langmuir adsorbate with admolecular interactions. At low coverages the adsorption energy ($q^a \cong 2.5\ r$) is much larger than the interaction energy of the admolecules ($q^{ai} \cong r$) with (r) being the heat of evaporation of the sorptive medium.

To elucidate this difference we mention the mean square deviation (σ) between measured (exp) and correlated (cal) data for both AIS:

$$\sigma^2 = \frac{1}{N}\sum_{i=1}^{N}\left(\frac{n_{ical} - n_{iexp}}{n_{iexp}}\right)^2$$

Numerical values of (σ) in (%) are given in the scheme below:

Dispersions (σ/%) of data correlations of adsorption equilibria of H_2S, CO_2, C_3H_8 on microporous mordenite (H) at T = 303 K for pressures up to 0.2 MPa.

	H_2S	CO_2	C_3H_8
Generalized LAI (7.37)	0.25	0.40	0.75
Classical LAI ($\alpha \neq 1$)	2.63	3.49	20.7

More details about the statistical derivation and range of application of the AI (7.37, 7.38) are given in the literature [7.29, 7.30]. For the sake of completeness we finally present the generalization of the AI (7.37, 7.38) to multicomponent systems ($N \geq 1$):

$$\theta_i = \frac{m_i}{m_{i\infty}} = \frac{F_i p_i^{\alpha i}}{1 + \sum_k^N F_k p_k^{\alpha k}} \leq 1, \quad i = 1 \ldots N \qquad 7.41$$

$$F_i(\theta_1 \ldots \theta_N) = \frac{b_i^{\alpha i} + \sum_l^N c_l \theta_l^{k_l}}{1 + \sum_l^N d_l \theta_l^{k_l}}, \quad i = 1 \ldots N \qquad 7.42$$

Here

$$\lim_{p_i \to \infty} \theta_i = 1, \text{ or } \lim_{p_i \to \infty} m_i = m_{i\infty}, \qquad 7.43$$

i. e. $m_{i\infty}$ is the monolayer limiting mass of component ($i = 1 \ldots N$) to be adsorbed at high pressures. For small pressures we have similarly to (7.40)

$$p_i \to 0 \qquad m_i = m_{i\infty} F_{i\,i=0} p^{\alpha i} \Big/ \left(1 + \sum_{k \neq i}^N F_{k\,i=0} p_k^{\alpha k}\right) \qquad 7.44$$

$$F_{k\,i=0} \equiv F_k(\theta_1 \ldots \theta_{i-1}, \theta_i = 0, \theta_{i+1} \ldots \theta_N), \; i, k = 1 \ldots N \qquad 7.45$$

Conversely to (7.41) we also have

$$p_i^{\alpha i} = \frac{\theta_i}{1 - \sum_l^N \theta_l} \bullet \frac{1}{F_i}, \quad i = 1 \ldots N \qquad 7.46$$

this relation providing an expression allowing one to calculate the partial pressure (p_i) of component (i) necessary to keep given amounts of all components ($\theta_1 \ldots \theta_N$) adsorbed.

In Eq. (7.42) (b_i) is the Langmuir parameter of component ($i = 1 \ldots N$), cp. (7.17). Also parameters c_l, d_l, $l = 1 \ldots N$ describe interactions between admolecules of type (l), and (α_i, k_i, $i = 1 \ldots N$) are related to the surface structure, i. e. the topology of the sorbent material [7.31]. The main physical assumptions leading to the AI (7.41, 7.42) are:

- monolayer adsorbate,
- constant adsorption enthalpies $(q_i, i = 1...N)$,
- densely packed admolecules so that interaction energies have to be taken into account.

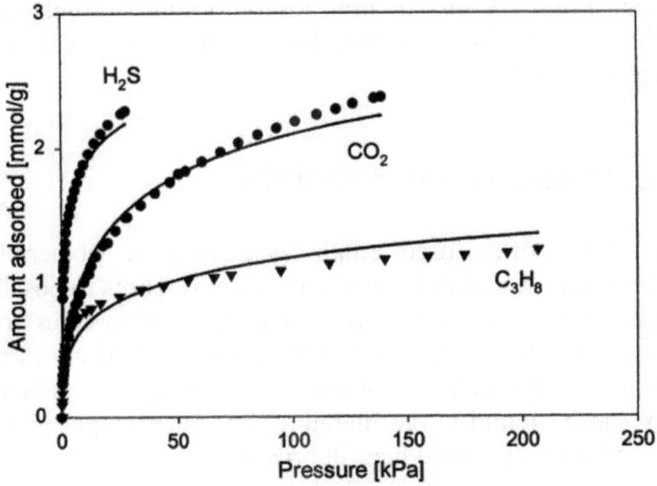

Figure 7.9. Correlation of adsorption data of H_2S, CO_2, C_3H_8 on mordenite (H) at T = 303 K for pressures up to 200 kPa using the classical LAI included in (7.37, 7.38) with $\alpha \neq 1$, a = c = 0, [7.29, 7.24].

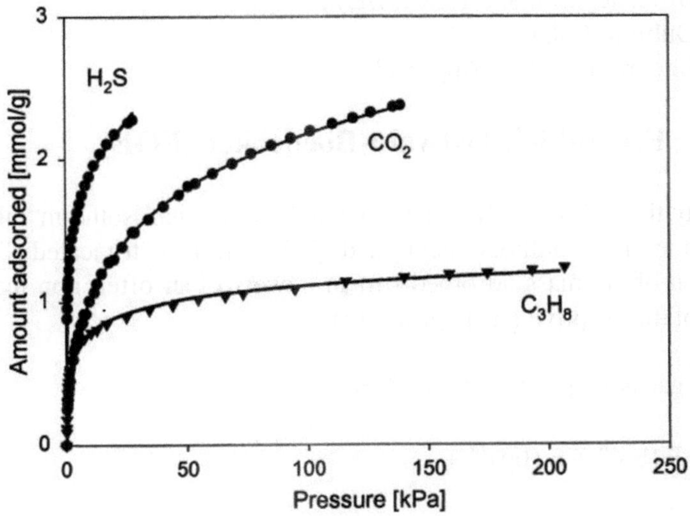

Figure 7.10. Correlation of adsorption data of H_2S, CO_2, C_3H_8 on mordenite (H) at T = 303 K for pressures up to 200 kPa using the generalized LAI (7.37, 7.38) taking interactions between admolecules into account [7.28].

The coadsorption isotherms (7.41, 7.42) for $N \geq 1$ provide many implicit equations to calculate adsorption loadings $\theta_i = m_i / m_{i\infty}$, $i = 1...N$ if all the parameters $b_i > 0, c_i \lesseqgtr 0, d_i \lesseqgtr 0, \alpha_i > 0, k_i > 0$, $i = 1...N$, are known. These normally have to be determined by correlating experimental coadsorption equilibria data using a Gauss minimization procedure. A detailed numerical analysis of eqs. (7.41, 7.42) is still lacking. Hence it is left as a challenge to young research students.

3. EMPIRICAL ISOTHERMS

Several of the often used empirical adsorption isotherms (AIs) are presented for single- and multicomponent systems. Hints to possible physical interpretation of parameters are given. Examples of systems to which the isotherms successfully may be applied are mentioned. Many more empirical AIs referring to anorganic, organic, and biological sorbent materials respectively can be found in the literature, cp. for example [7.3, 7.26, 7.32, 7.33]. The isotherms discussed here in brief are:

1. Freundlich-Ostwald-Boedecker
2. Virial Expansions*)
3. Toth
4. Brunauer-Emmett-Teller (BET)
5. Dubinin-Polanyi
6. Integral Equation Approach

3.1 Freundlich-Ostwald-Boedecker (FOB)

Today this AI most often is known as „Freundlich-isotherm" although it is due to the three authors mentioned [7.34]. It is a truncated Taylor-series expansion of the mass adsorbed (m) in powers of an often non-integer power ($\alpha > 0$) of the sorptive gas pressure (p):

Pure gas adsorption systems ($N = 1$):

$$m = F_1 (bp)^{\alpha} + F_2 (bp)^{2\alpha} + ... \qquad 7.50$$

*) These expansions are not purely empirical but can bederived within the framework of statistical mechanics [7.43, 7.74].

Parameters:

$F_i = F_i(m^s, T), i = 1, 2, 3$ ... Taylor-coefficients

$$b(T) = \frac{1}{p_0}\sqrt{\frac{T_0}{T}}\exp\left(\frac{q}{RT}\right)$$ ... Langmuir parameter, cp. (7.3).

In the low pressure region, i. e. for $p << p_s(T)$ we have

$$p \to 0 \quad m = F_1 (bp)^\alpha \qquad 7.51$$

Henry coefficient:

$$H \equiv \left(\frac{\partial m}{\partial p}\right)_{0,T} = \alpha F_1 (bp)^{\alpha-1} = \begin{cases} 0 \ \ldots \alpha > 1 \\ F_1 b \ldots \alpha = 1 \\ \infty \ \ldots \alpha < 1 \end{cases} \qquad 7.52$$

The limiting cases H = 0 and H → ∞ are realized – for example – for water vapor adsorption on activated carbon (AC) and hydrophilic zeolites respectively.

Multicomponent systems (N>1):

$$m_i = \sum_{k=1}^{N} F_{ik}(b_k p_k)^{\alpha_k} + \sum_{k,l=1}^{N} F_{ikl}(b_k p_k)^{\alpha_k}(b_l p_l)^{\alpha_l} + \ldots \quad i = 1 \ldots N \qquad 7.53$$

$$p_i \to 0: m_i = \sum_{k \neq i}^{N} F_{ik}(b_k p_k)^{\alpha_k} + \sum_{\substack{k \neq i \\ l \neq i}}^{N} F_{ikl}(b_k p_k)^{\alpha_k}(b_l p_l)^{\alpha_l} + \ldots, \qquad 7.54$$

$$i = 1 \ldots N, \quad p_{k=\text{const}}, k \neq i$$

Parameters:

$F_{ik} = F_{ik}(m^s, T), F_{ikl}(m^s, T)$ ... Taylor-coefficients

$$b_i = \frac{1}{p_{0i}}\sqrt{\frac{T_{0i}}{T}}\exp\left(\frac{q_i}{R_i T}\right)$$ ... Langmuir parameter, cp. (7.17).

The exponents (α_i, i = 1...N) are related to energetic inhomogeneities and also to the fractal nature of the surface of the sorbent material [7.15, 7.35-7.37]. Numerical values are often in the range $0.2 < \alpha_i \leq 1$.

3.2 Virial Expansions

Certain functions of the number of moles (n) adsorbed on the sorbent mass (m^s) at sorptive gas pressure (p) and temperature (T) are considered as series expansions of either n or p, [7.2, 7.24, 7.38]:

$$\ln\left(\frac{n}{n_0 bp}\right) = \begin{cases} B(m^s,T)n + C(m^s,T)n^2 + ... \\ B'(m^s,T)p + C'(m^s,T)p^2 + ... \end{cases} \qquad 7.55$$

Parameters:

n_0	... reference loading of a monolayer adsorbate
$b = \frac{1}{p_0}\sqrt{\frac{T_0}{T}}\exp\left(\frac{q}{RT}\right)$	... Langmuir parameter, cp. (7.3).
B, B′,C, C′...	... Virial coefficients of the adsorption system.

Equations (7.55) are also used as

$$n = n_0 bp \begin{cases} \exp(Bn + Cn^2 + ...) \\ \exp(B'p + C'p^2 + ...) \end{cases} \qquad 7.56$$

Low pressure / loading limit ($p \to 0$, $n \to 0$)

$$n = n_0 bp \begin{cases} 1 + Bn + ... \\ 1 + B'p + ... \end{cases} \qquad 7.57$$

Henry coefficient

$$H = \left(\frac{\partial n}{\partial p}\right)_{0,T} = bn_0 \qquad 7.58$$

The series expansions (7.55, 7.56) are useful if Henry coefficients have to be determined from high pressure adsorption data by extrapolation to the limit ($p \to 0$). This may be necessary as low pressure adsorption measurements often are tedious, cumbersome and time consuming. Virial expansions also are used in developing microscopic models for adsorption equilibria on heterogeneous surfaces [7.38].

Multicomponent systems (N > 1):

$$\ln\left(\frac{n_i}{n_{i0} b_i p_i}\right) = \begin{cases} \sum_{k=1}^{N} B_{ik}(m^s, T) n_k + \sum_{k,l=1}^{N} C_{ikl}(m^s, T) n_k n_l + \ldots \\ \sum_{k=1}^{N} B'_{ik}(m^s, T) p_k + \sum_{k=1}^{N} C'_{ikl}(m^s, T) p_k p_l + \ldots \end{cases} \quad 7.59$$

$$i = 1 \ldots N$$

Parameters:

$p_i = y_i p$ ... partial pressure of component (i) in the sorptive gas mixture

n_{i0} ... reference loading of component (i)

$b_i = \frac{1}{p_{0i}} \sqrt{\frac{T}{T_0}} \exp\left(\frac{q_i}{R_i T}\right)$ ... Langmuir parameter, cp. (7.17).

$B_k, B'_k, C_{kl}, C'_{kl}$ ... Virial coefficients of the coadsorption system.
$k,l = 1 \ldots N$

Low pressure limit of (7.59):

$p_i \rightarrow 0$, $p_k = \text{const}$, $k = 1 \ldots i\text{-}1, i+1 \ldots N$

$$n_i = n_{i0} b_i p_i \exp\left\{ \sum_{k \neq i}^{N} B'_{ik} p_k + \sum_{k,l \neq i}^{N} C'_{ikl} p_k p_l + \ldots \right\} \quad 7.60$$

Henry coefficient of component (i) in the coadsorbate:

$$H_i = n_{i0} b_i \exp\left\{ \sum_{k \neq i}^{N} B'_{ik} p_k + \ldots \right\}, \quad i = 1 \ldots N \quad 7.61$$

The exponential function in this expression mirrors the influence of the other components ($k \neq i$) on the adsorption of component (i). The series expansions (7.55, 7.59) can be derived from equivalent virial expansions of the thermal equation of state (EOS) of the single- or multicomponent adsorbate by standard thermodynamic methods. Details are given in [7.2, 7.3]. Real gas effects of the sorptive gas mixture can be taken into account by replacing the partial pressures (p_i, $i = 1 \ldots N$) in (7.60) by the (mixture) fugacities ($f_i = f_i(p_1 \ldots p_N, T)$) of the system [7.17].

3.3 Toth's Isotherm

In 1971 J. Toth [7.39] proposed a generalization of the classical Langmuir isotherm (7.1) as

$$(\theta)^t \equiv \left(\frac{n}{n_\infty}\right)^t = \frac{(bp)^t}{1+(bp)^t} \qquad 7.62$$

$$= (bp)^t(1-(bp)^t + (bp)^{2t} - ...$$

Parameters:

$n_\infty = \lim_{p\to\infty} n$ ... limiting amount of gas adsorbed at $T = \text{const}$ for $p \to \infty$.

$\theta = \frac{n}{n_\infty} \le 1$ ... fractional loading adsorbed.

$b = \frac{1}{p_0}\sqrt{\frac{T_0}{T}}\exp\left(\frac{q}{RT}\right)$ ... Langmuir parameter.

$b^{-1} = p(T, n = n_\infty/2, m^s)$ ... half-loading pressure

$0 \le t \le 1$ ... Toth parameter describing approximately energetic heterogeneity and fractality of the surface of the sorbent.

Low pressure limit and Henry coefficient

$p \to 0$

$$n = n_\infty bp \qquad 7.63$$

$$H = bn_\infty \qquad 7.64$$

The isotherm (7.62) was originally derived from the Langmuir AI (7.1) which can be written as

$$\delta \equiv \frac{\theta}{p}\left(\frac{dp}{d\theta}\right)_T = 1 + bp \qquad 7.65$$

which gives a linear relation in the (δ, p)-plane. To make up for deviations of experimental δ-data from this straight line, which were caused by heterogeneities of the sorbent materials analyzed, Toth proposed instead of (7.65) the relation

$$\frac{\theta}{p}\left(\frac{dp}{d\theta}\right)_T = 1 + (bp)^t, \qquad 0 < t \leq 1 \tag{7.66}$$

which after integration leads to (7.62).

Multicomponent systems (N>1):

$$\theta_i^t \equiv \left(\frac{n_i}{n_{i\infty}}\right)^t = \frac{(b_i p_i)^t}{1 + \sum_{k=1}^{N} (b_k p_k)^t}, \qquad i = 1...N \tag{7.67}$$

p_i ... partial pressure of component i in the sorptive gas mixture.

In the low pressure limit ($p_i \rightarrow 0$, p_k = const, $k \neq i$), (7.67) leads to the Henry coefficient of component (i) in the coadsorbate

$$H_i = \frac{b_i n_{i\infty}}{\left(1 + \sum_{k \neq i}^{N} (b_k p_k)^t\right)^{1/t}}, \qquad i = 1...N. \tag{7.68}$$

In this expression the denominator on the r.h.s. describes the impact of the other components ($k \neq i$) on the adsorption of component i in the low pressure region ($p_i \rightarrow 0$).

Today Toth's AI (7.62) is quite often used to describe adsorption of a variety of gases and vapors of organic compounds on zeolites and other inorganic sorbent materials [7.3, 7.33, 7.40-7.42].

3.4 Brunauer-Emmett-Teller Isotherm (BET)

Extending the classical Langmuir adsorption isotherm (7.1) from monolayer to ideal multilayer adsorption and considering the limiting case of infinite many layers, Brunauer, Emmett, and Teller derived in 1938 the AI [7.1-7.5, 7.42]

$$\frac{n}{n_l} = \frac{Cp/p_s}{(1-p/p_s)(1+(C-1)p/p_s)} \tag{7.70}$$

which has IUPAC-Type II character [7.5]. Its linear form in (p) is

$$\frac{p}{n(p_s - p)} = \frac{1}{Cn_l} + \frac{C-1}{Cn_l} \bullet \frac{p}{p_s} \qquad 7.71$$

Parameters:

n	...	Amount of gas adsorbed on sorbent of mass (m^s) at temperature ($T < T_C$) and (subcritical) pressure (p).
$n_l = n_l\ (T, m^s)$	...	amount of monolayer adsorbed on sorbent mass (m^s).
C	...	BET-parameter. It can be related to the monolayer gas pressure (p_l) and also to the inflection point of the AI (7.70) in the (n, p/p_s)-diagram. Statistical interpretations are given in [7.38].
$p_s = p_s(T)$	...	saturation pressure of adsorptive gas.

Gas pressure (p_l) necessary for a monolayer adsorbate ($n = n_l$):

$$\left(\frac{p}{p_s}\right)_{n_l} = \frac{p_l}{p_s} = \frac{1}{1+\sqrt{C}} \qquad 7.72$$

BET-parameter:

$$C = \left(\frac{p_s}{p_l} - 1\right)^2 \qquad 7.73$$

Inflection point of the BET-isotherm [7.3]

$$C > 2: \quad \frac{p_{inf}}{p_s} = \frac{\gamma - 1}{\gamma\left(\sqrt{\gamma} + 1\right)}, \qquad \gamma \equiv (C-1)^{2/3} \qquad 7.74$$

BET-parameter

$$C = 1 + \gamma^{3/2}, \qquad \gamma = \lim_{n\to\infty} \gamma_n, \qquad \gamma_{n+1} = \frac{\gamma_n - 1}{\left(\sqrt{\gamma_n} + 1\right)\left(p_{inf}/p_s\right)} \qquad 7.75$$

The energetic situation of the admolecules in a BET-adsorbate is sketched in Figure 7.11. A graphical representation of the BET-AI, Eq. (7.70) is given in Figure 7.12.

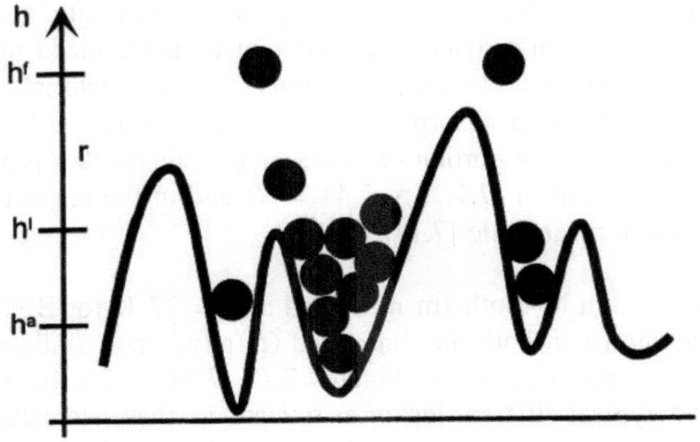

Figure 7.11. Energetic scheme of a BET adsorbate. Adsorption sites can take not only one but several molecules. Interactions between admolecules are not taken into account [7.1-7.5].

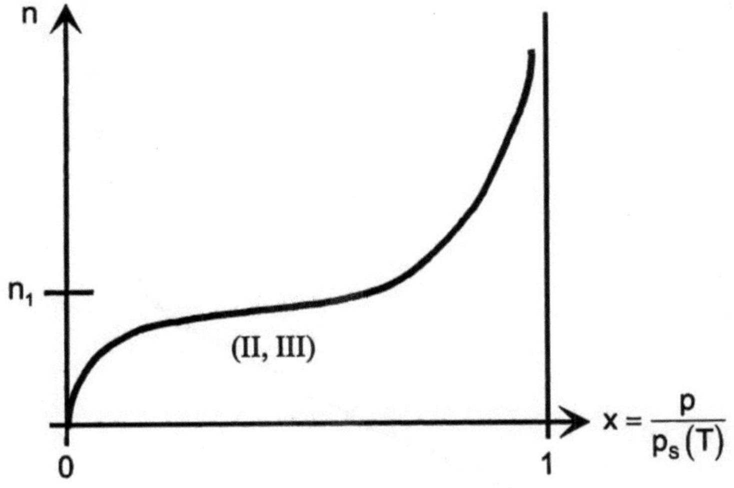

Figure 7.12. The BET-adsorption isotherm (BET AI), Eq. (7.70) showing for C>2 an inflection point and at the sorptive gas saturation pressure $p = p_s(T)$ a singularity $n \rightarrow \infty$ indicating pore condensation and the appearance of a bulk liquid phase [7.1-7.5].

The BET-isotherm has been derived by statistical mechanical methods by Hill and later again by W. Steele (1974) [7.5], [7.43]. It can be shown that the infinitive number of adsorbate layers assumption used in deriving (7.70), is a reasonable approximation for multilayer adsorbates with more than 4 layers. Though admolecular interactions are not explicitly taken into account in (7.70), today this isotherm provides the basis for a standardized method to determine the mesoporous surface of porous sorbent materials by N_2-adsorption at T = 77 K or Ar-adsorption at T = 87 K, i. e. $p_s = 0.1$ MPa. For this normally only the initial portion of a measured isotherm ($0 < p/p_s \leq 0.35$) is used. Details are given in [7.1, 7.5, 7.44-7.47] and in the respective ISO-documents published meanwhile [7.14].

A typical example of a N_2-isotherm measured at T = 77 K for BET-surface determination is shown in both the linearized $(p/(n\,(p_s - p)), p)$-diagram and the traditional $(n, p/p_s)$ –diagram in Figures 7.13, 7.14, [7.25] below. They demonstrate two typical difficulties often related to these so-called BET-measurements: the hysteresis between the adsorption- and the desorption branch indicating existence of mesopores and – in Fig. 7.13 – the deviation from the linear plot for pressures ($p/p_s > 0.4$), which is mainly due to energetic heterogeneities of the surface of the sorbent and to admolecular interactions.

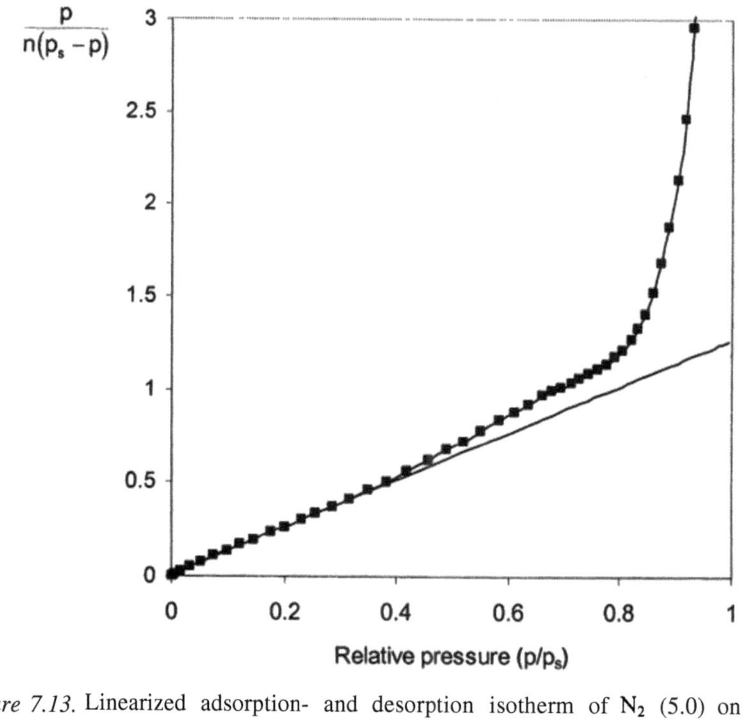

Figure 7.13. Linearized adsorption- and desorption isotherm of N_2 (5.0) on standardized material CRMBAM-PM-104 at 77 K, $p_s = 1$ atm, [7.25]. Correlation: (7.71).

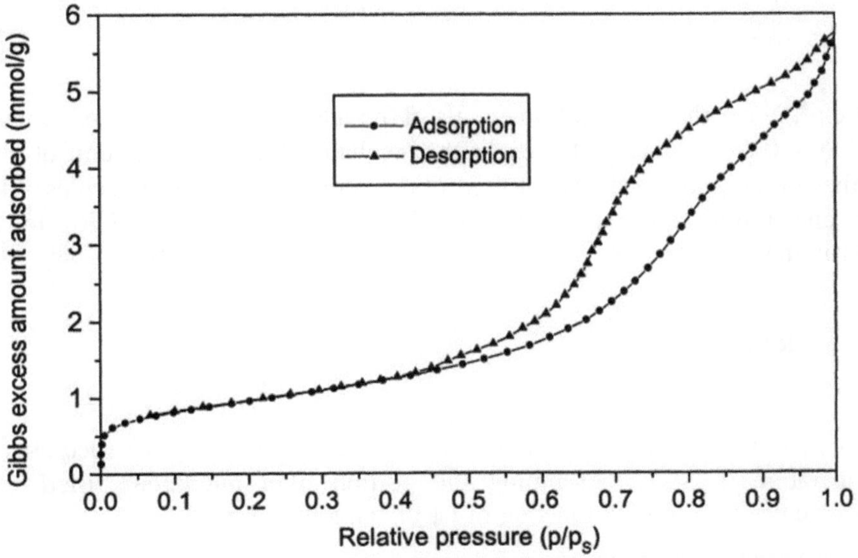

Figure 7.14. Adsorption- and desorption isotherm of N_2 (5.0) on standardized material CRMBAM-PM-104 at T = 77 K, p_s = 1 atm, [7.25].

A thermodynamic consistent extension of the BET-isotherm to multicomponent systems (N > 1) and to real gas adsorptives will be presented in Sect. 4 of this Chapter.

3.5 Dubinin-Polanyi Theory

Dubinin, Polanyi, and Radushkevich proposed about 1947 a simple but very useful empirical theory allowing one to calculate the amount of gas adsorbed in a microporous sorbent. The theory was based on a pore filling model. Today it is used for both characterization of porous solids and also for engineering purposes. It has been extended by several authors among them predominantly Astakhov (1970). The theory is still the subject of further investigations, mainly by statistical mechanics and computational methods (DFT) [7.1-7.3, 7.48-7.55].

The adsorbate is considered as a fluid phase in the sense of thermodynamics, cp. Sect. 7.4, which is exerted to the external forces of the atoms and molecules located on the surface of a sorbent material. The general condition for thermodynamic equilibrium of such a phase (a) against a fluid or gaseous sorptive phase (f) is [7.56, 7.57]:

$$\mu^a - \hat{\phi} = \mu^f - 0 = \text{const} \tag{7.76}$$

Here μ^a, μ^f are the chemical potentials of the adsorbate and the sorptive phase and $\hat{\phi} > 0$ (for attractive forces) is the mechanical potential per unit of mass of the forces acting near the surface of the sorbent on the admolecules. Ideal gas and liquid approximations [7.17] for μ^f, μ^a lead via (7.76) to the approximate expression

$$\hat{\phi} = RT \ln\left(\frac{p_s(T)}{p}\right) \tag{7.77}$$

with p_s (T) indicating the saturation pressure of the sorptive mediums at temperature $T < T_c$. Assuming the portion of a micropore filled with admolecules (W) to be a function of ($\hat{\phi}$) having the form

$$W = W_0 \exp(-(\alpha\hat{\phi})^N), \tag{7.78}$$

with (W_0) indicating the total volume of the pore, we get in view of (7.77):

$$W = W_0 \exp\left(-\left(\alpha RT \ln\left(\frac{p_s(T)}{p}\right)\right)^N\right) \tag{7.79}$$

This is a somewhat generalized version of the so-called Dubinin-Radushkevich (RD) AI [7.48]. The mass of gas adsorbed in the pores then is

$$m = \rho_0 W \tag{7.80}$$

Here ρ_0 is the density of the sorptive medium in a reference liquid state which may be chosen as the density of the saturated boiling liquid at the chosen temperature, i. e. $\rho_0 = \rho_s$ (T), [7.5, 7.3]. The parameter α in the "characteristic curve" of the sorbent material (7.78) is the reciprocal of a specific energy; the exponent N normally is limited to $2 \leq N \leq 6$ and for zeolites and activated carbons often has numerical values about $N \cong 3$. Both parameters are characteristic for a sorbent material and the micropore spectrum included in it. Details of practical applications of (7.79) and a variety of generalizations to real gas adsorptives and multicomponent systems can be found in the (still growing) literature in this field [7.53 7.55, 7.58].

3.6 Integral Equation Approach

The total mass (m) adsorbed on the surface of a porous solid of mass (m^s) can be assumed to be the integral sum of all masses (μ) adsorbed on the surfaces of pores of a certain geometrical structure (Γ), i. e. cylindrical – or slit-like etc. pores, as

$$m(p,T,m^s) = \int_{\Gamma} \mu(p,T,\Gamma) G(\Gamma,T,m^s) d\Gamma \qquad 7.81$$

Here $G = G(\Gamma, T, m^s)$ is the pore distribution function of the sorbent material which due to thermal expansion / contraction of the material also (weakly) depends on the temperature (T) of the system. The function $\mu = \mu(p, T, \Gamma)$ is modelled by statistical mechanical methods, for example by the Density Functional Theory (DFT), cp. [7.44, 7.54, 7.59]. A fairly successful example for (7.81) has been developed among others by Horvath and Kawazoe (1983) [7.21, 7.44, 7.60, 7.61]. Indeed the proposed AI conversely can be used to calculate the pore distribution function $G(\Gamma, T, m^s)$ from known adsorbate masses (m) and a model function (μ) (inverse problem) [7.3, 7.21, 7.61]. Using probe molecules (He, Ar, N_2, CO_2) this method is used today to determine micro- and mesoporous distributions in sorbent materials as well [7.25, 7.44, 7.54].

We here restrict ourselves to the presentation of a fairly simple example for the AI (7.81) originally proposed by Jaroniec and Choma [7.21, 7.62]. Assuming the sorbent material to include only micropores of simple cylindrical shape of different diameters which do not interconnect, adsorption on a single pore can be described by the DR-isotherm (7.79) with N = 2. Assuming also the micropores to be statistically distributed according to a Γ-distribution function [7.63] of degree (n), one gets from (7.81) the isotherm

$$\frac{m(p,T,m^s)}{m_s(T,m^s)} = \left[\frac{\kappa}{\kappa + \left(\alpha RT \ln\left(\frac{p_s(T)}{p} \right) \right)} \right]^{n+1} \qquad 7.82$$

Parameters:

$m_s(T, m^s)$ ... mass of adsorbate in the saturation state , i. e. at completely filled pores and gas pressure $p = p_s(T)$.

α, κ, n ... characteristic parameters of the sorbent material to be determined from adsorption experiments with probe or yardstick molecules [7.50].

p_s (T) ... saturation pressure of sorptive gas at $T < T_C$. For $T > T_C$ the so-called Riedel pressure can be used, cp. Sect. 2.1.1, (7.23).

Though the AI (7.82) has proved to be useful for characterizing microporous materials [7.62], it should always be taken into account that it refers to absolute amounts adsorbed (m) which cannot easily be measured today but in practice have to be determined approximately from Gibbs excess adsorption data, cp. Chap. 1. Hence the sorbent parameters (α, κ, n) may vary considerable depending on the type of probe molecules actually used and also the reference density of the adsorbed phase introduced, cp. Eq. (7.80).

In concluding this Section we want to emphasize that literature shows a variety of other empirical adsorption isotherms developed for special purposes and using quite different physical pictures and concepts [7.65, 7.66] and also new mathematical techniques like neural networks and/or genetic codes [7.67]. The interested reader is referred to the Proceedings of the most important Int. Conferences in the field of adsorption like COPS, FoA, PBAC, etc., cp. Chap. 1.

4. THERMODYNAMIC ISOTHERMS

4.1 Gibbs's Approach

Josiah Willard Gibbs was the first to apply thermodynamic methods to describe physisorption equilibria of gases on porous solids [7.68]. After him many other authors extended his work in several directions, especially to multicomponent systems and to its statistical mechanical foundations. Also a variety of adsorption isotherms for both single- and multicomponent systems were derived from the respective thermal equations of state for the (two-dimensional) adsorbate phase. We here mention only the classical books and papers by Young [7.69], Myers and Prausnitz [7.70], van Ness [7.71], Sircar [7.72], and Talu [7.73]. Today there are many other representations of Gibbsian adsorption thermodynamics available in literature. All of them are nearly equivalent and differ only slightly in their aims and goals. Hence, we here are not going to discuss the Gibbsian approach again but mention only some references which also are only a selection of the existing literature and in no way claim completeness: [7.2, 7.5, 7.22, 7.33, 7.40, 7.75, 7.76], [7.7,

7.53, 7.65]. The interested reader is well advised if he tries to get and read the original papers of the fathers of adsorption thermodynamics, namely J. W. Gibbs [7.74], I. Langmuir [7.64], and W. Ostwald [7.34].

4.2 Internal Variable Approach

In this Section we will present another approach to adsorption thermodynamics, the main statement being a mathematical representation theorem for adsorption isotherms of both single and multicomponent systems with real gas adsorptives on either (energetically) homogenous or inhomogeneous sorbent materials [7.15, 7.77]. From this a variety of thermodynamically consistent isotherms for multicomponent systems can easily be derived, among them generalizations of the BET- and the Aranovich-isotherms which proved to be very useful in describing coadsorption isotherms of gas mixtures including subcritical components.

Let us consider an adsorption system consisting of $N \geq 1$ many gaseous components and a chemically inert sorbent phase. In addition we have at any temperature and any set of partial gas pressures a sorbate phase which is assumed to be a thermodynamic system of fractal dimension ($1 \leq D \leq 3$) in the sense of W. Schottky and J. Meixner [7.78, 7.57, 7.56]. The parameters of the sorptive gas, the adsorbate, and the sorbent are:

Sorptive gas (f)

p_i, $i=1 \ldots N$ ... partial pressure of component i

T ... temperature

Sorbate phase (a)

n_i, $i=1 \ldots N$ ... total amount of component i adsorbed.
This is defined experimentally by combined dielectric-calorimetric and manometric measurements, i. e. without introducing the concept of a dividing surface or that of the (accessible) volume of a porous solid, cp. Chap. 1 and [7.79].

T ... temperature,

$\pi = \gamma^s - \gamma$ ... spreading pressure of sorbate, i. e. the difference between the surface tensions between the (f, s)- and the (f, a, s)-phases.

Inert sorbent phase (s)

m^s	... mass
T	... temperature,
$A = A\,(m^s, T)$	... extensive parameter of the fractal surface of the sorbent material.

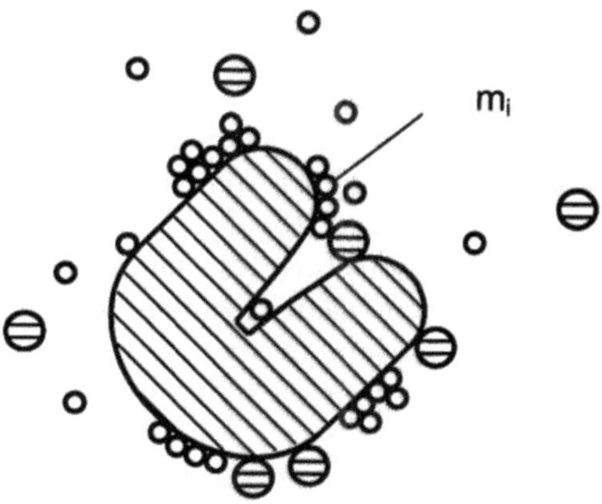

Figure 7.15. Model of an adsorption system with N =2 components on the (possibly fractal) surface of an inert porous solid.

The Gibbs equation corresponding to the (N+1) external operations possible on the adsorption system, namely, changing of partial pressures (p_i, i = 1...N) and temperature (T), is [7.15, 7.17, 7.56, 7.57]

$$dG = -SdT + Ad\pi + \sum_{i=1}^{N} \mu_i dn_i \qquad 7.90$$

Here the spreading pressure (π) is considered as an internal variable in the sense of Bridgeman, Kestin, and Meixner [7.56, 7.57], i. e. it is not an independent variable, but a function of the external parameters $\pi = \pi\,(T, p_1 \ldots p_N)$ of the system. From the Gibbs-Duhem-equation or, likewise, the Euler equation for $G = G(T, n_1 \ldots n_N, m^s)$ being a homogeneous function of degree 1 in the mole numbers ($n_1 \ldots n_N$,) of the sorbate phase, we get the relations

$$A\left(\frac{\partial \pi}{\partial T}\right)_p = S + \sum_{i=1}^{N}\left(\frac{\partial \mu_i}{\partial T}\right)_p n_i \qquad 7.91$$

$$p_k: \quad A\left(\frac{\partial\pi}{\partial p_k}\right)_T = \sum_{i=1}^{N}\left(\frac{\partial\mu_i}{\partial p_k}\right)n_i, \qquad k=1...N \qquad 7.92$$

These can be substituted in the total differential of the spreading pressure

$$d\pi(T,p_1...p_N)=\left(\frac{\partial\pi}{\partial T}\right)_p dT+\sum_{k=1}^{N}\left(\frac{\partial\pi}{\partial p_k}\right)_T dp_k \qquad 7.93$$

which can then be inserted into (7.90) to give

$$dG=\left(-S+A\left(\frac{\partial\pi}{\partial T}\right)p+\sum_i^N\mu_i\left(\frac{\partial n_i}{\partial T}\right)_p\right)dT$$
$$+\sum_k^N\left(A\left(\frac{\partial\pi}{\partial p_k}\right)_T+\sum_i^N\mu_i\left(\frac{\partial n_i}{\partial p_k}\right)_T\right)dp_k \qquad 7.94$$

with $p=\sum^N p_i$ being the total gas pressure in the system. The Maxwell relation for partial pressures (p_k, p_l) is

$$\sum_i^N\frac{1}{f_i}\left(\frac{\partial f_i}{\partial p_k}\right)_T\left(\frac{\partial n_i}{\partial p_l}\right)_T=\sum_i^N\frac{1}{f_i}\left(\frac{\partial f_i}{\partial p_l}\right)_T\left(\frac{\partial n_i}{\partial p_k}\right)_T, \qquad k\neq l \qquad 7.95$$
$$k,l=1...N$$

Here the thermodynamic equilibria conditions

$$\mu_i=\mu_i^f, \qquad i=1...N \qquad 7.96$$

for the isothermal phases (a, f) have been used, μ_i^f indicating the chemical potential of component (i = 1…N) in the sorptive phase. This quantity can be written as

$$\mu_i^f=\mu_{i0}^+(p^+,T)+RT\ln\left(\frac{f_i}{p^+}\right), \quad i=1...N \qquad 7.97$$

with (μ_{i0}^+) being the chemical potential of pure component (i) at a certain reference pressure (p^+) and

$$f_i=f_i(p_1...p_N,T), \qquad i=1...N \qquad 7.98$$

indicating the fugacity of component (i) in the gaseous mixture [7.17].

The Maxwell relations (7.95) provide a set of N(N-1)/2 linear partial differential equations for the adsorption isotherms

$$n_i = n_i(p_1,...,p_N,T,m^s) \ ,i = 1...N \qquad 7.99$$

A class of solutions of (7.95) is given by [7.15]*)

$$n_i(p_1,...,p_N,T,m^s) = n_\infty(T,m^s)\left[\alpha_i c_i^{\alpha_i} + \frac{f_i}{RT}\sum_{k=1}^{N}\alpha_k c_k^{\alpha_k}\left(\frac{\partial q_k}{\partial f_i}\right)\right]\Phi(c^*) \qquad 7.100$$

with

$$c_i = \frac{f_i}{p_{i0}} e^{\frac{q_i}{RT}} \quad , \quad i = 1...N, \qquad 7.101$$

$$c^* = \sum_{i}^{N} c_i^{\alpha_i} \qquad 7.102$$

and

$$q_i = q_i(f_1,...,f_N,T) \quad , \quad i = 1...N \qquad 7.103$$

being the pressure (or – equivalently – the fugacity) dependent molar adsorption energies. Assuming for the fugacities the limiting behavior

$$\lim_{p_i \to \infty} f_i(p_1...p_N,T) \to \infty, \quad i = 1...N \qquad 7.104$$

and restricting ourselves to "characteristic functions" Φ obeying the limiting condition $\lim_{x\to\infty}(x\Phi(x)) = 1$, one can derive from (7.100) that there exist limiting loadings for $p_i \to \infty$, p_k = const, k = 1...i – 1, i+1...N:

*) More solutions can be found provided one can spend a sabbatical term in a stimulating environment being far away from any kind of administration.

$$\lim_{p_i \to \infty} n_i = \alpha_i n_\infty, \quad i = 1 \ldots N \tag{7.105}$$

For spherical or nearly spherical molecules the exponents ($\alpha_i > 0$) are related to their molecular radii as

$$\frac{\alpha_i}{\alpha_0} = \left(\frac{r_i}{r_0}\right)^{-D}, \quad i = 1 \ldots N \tag{7.106}$$

Here (D) is the fractal dimension of the surface of the sorbent material [7.35-7.37, 7.5, p. 183]. The index "0" refers to a reference molecule, which on principle can be chosen arbitrarily. In practice Argon has turned out to be the most useful probe sorptive gas.

As an example some sets of (α, r, D)-data which have been obtained from pure supercritical gas adsorption data are presented in Tab. 7.2 below [7.76].

	α			2r / Å			D
AC/Z	N_2	Ar	CH_4	N_2	Ar	CH_4	
WS IV	0.8404	0.8662	0.7758	3.7	3.65	3.81	2.7
Norit	0.8264	0.8706	0.8028	3.7	3.62	3.75	2.4
AKF	0.670	-	0.540	3.7	-	4.0	2.85
DAY	0.8293	-	0.7838	3.7	-	3.8	2.3

Table 7.2. Numerical values of exponents (α_i) and molecular radii (r_i) for adsorption of gases ($i = N_2$, Ar, CH_4) on zeolites WS IV and DAX and activated carbons Norit R1 and AKF, [7.76].

The difference of enthalpies in the sorptive gas and the sorbate state corresponding to the isotherms (7.100) can be calculated from the general relation

$$H^f - H^a = -AT\left(\frac{\partial \pi}{\partial T}\right)_p \tag{7.107}$$

and Eq. (7.91). The result is

$$H^f - H^a = \sum_i^N \left[\left(q_i - \frac{RT^2}{f_i}\left(\frac{\partial f_i}{\partial T}\right)\right) n_i - n_\infty \Phi(c^*) \frac{q_i f_i}{RT} \sum_k^N \alpha_k c_l^{\alpha_k} \left(\frac{\partial q_k}{\partial f_i}\right)_T\right] \tag{7.108}$$

The function Φ in representations (7.100), (7.107) is a so-called characteristic function of the adsorption system. It is due to several restrictions which are based on thermodynamic stability conditions, i. e. the Second Law [7.17, 7.56]:

$$\Phi = \Phi(x) \geq 0 \quad \text{all } x \geq 0 \tag{7.109}$$

$$\frac{d}{dx}(x\Phi(x)) \geq 0 \tag{7.110}$$

Examples of characteristic functions obeying these conditions are:

Langmuir

$$\Phi(x) = \frac{1}{1+x}, \tag{7.111}$$

BET

$$\Phi(x) = \frac{C}{(1-x)1+(c-1)x)}, \qquad C>0 \tag{7.112}$$

Aranovich

$$\Phi(x) = \frac{D}{(1-x)^{1/2}(1+Dx)}, \qquad D>0 \tag{7.113}$$

Many more functions obeying inequalities (7.109, 7.110) exist but are still subject to investigation.

Inserting the functions (7.111-7.113) into (7.100), (7.108) one gets thermodynamic consistent generalizations of the Langmuir, BET, and Aranovich AIs to adsorption systems with

a) an arbitrary number of components ($N \geq 1$),
b) real gas sorptive media described by their fugacities ($f_1 \ldots f_N$),
c) loading or pressure dependent adsorption energies ($q_1 \ldots q_N$), cp. (7.103), and

d) a sorbent characteristic fractal exponent (D) relating the limiting loadings of sorptive gases according to the size of their molecules, cp. Eqs. (7.105, 7.106).

As an example we here only mention the generalized form of the Langmuir AI resulting from Eqs. (7.100, 7.111) with an sorptive ideal gas mixture, i. e. $f_i = p_i$, i = 1...N, but loading dependent adsorption energies an example of which is given by [7.15]:

$$q_i = \frac{1+\sum_m a_m p_i}{1+\sum_m b_m p_i} q_{io} \qquad 7.114$$

with empirical parameters (a_{im}, b_{im}, $i=1\ldots N$) and initial values $q_i\,(p_k = 0) = q_{io}$, $i=1\ldots N$).

Adsorption isotherms:

$$n_i(p_1 \ldots p_N, T) = n_\infty(T)\left[\alpha_i c_i^{\alpha_i} + \frac{p_i}{RT}\sum_m^N \alpha_m c_m^{\alpha_m}\left(\frac{\partial q_m}{\partial p_i}\right)_T\right]\frac{1}{1+\sum_k^N c_k^{\alpha_k}} \qquad 7.115$$

Adsorption enthalpy:

$$H^f - H^a = \sum_i^N\left[q_i n_i - n_\infty \frac{p_i q_i}{\left(1+\sum_k c_k^{\alpha_k}\right)RT}\sum_m^N \alpha_m c_m^{\alpha_m}\left(\frac{\partial q_m}{\partial p_i}\right)_T\right] \qquad 7.116$$

with

$$c_i = \frac{p_i}{p_{io}} e^{q_i/RT}, \quad i=1\ldots N \qquad 7.117$$

Equations (7.115) and (7.116) have very similar structures: The first terms on their r.h.s always are due to the components (i), whereas the second terms describe mixture effects mainly caused by the loading (or pressure) dependence of the adsorption energies $q_i = q_i\,(p_i \ldots p_N,\, T)$, i = 1...N, cp.

example (7.114), i. e. these terms vanish if q_i = const, i = 1...N, can be assumed.

Finally we want to emphasize that the many parameter isotherms represented in Eqs. (7.100) (and – as an example – in Eq. (7.115)) together with the respective enthalpy functions (7.108) and (7.116) only should be used if there are enough experimental adsorption data available. This means, if the number of (reliable) data points well exceeds the square of the number of parameters included in these equations. The numerical values of these parameters have to be determined by appropriate data correlation procedures, i. e. those which take uncertainties of experimental data well into account.

5. CONCLUSIONS

As has been mentioned in the Introduction to this Chapter, there is no single isotherm which can describe all physisorption phenomena occurring when pure or mixed gases contact the surface of a porous solid. This mainly is due to the complexity of the admolecule – sorbent atom/molecule interactions and, at higher loadings, also the admolecule – admolecule interaction. This complexity is a consequence of the energetic heterogeneity of the adsorption sites, the great variety of pores in shape, size, and connectivity, i. e. the pore spectrum of the sorbent, and also of the various properties of the adsorptive molecules, for example their electrical moments (dipole, quadrupole etc.) and their topography. In case of gas mixtures special mixture effects may occur in addition, for example pore blocking by the larger molecules thus preventing the smaller to enter.

Nevertheless, for practical applications very rough recommendations about adsorption isotherms to be expected in experiments or to be used in industrial processes can be given. They are summarized in Table 7.3 below where sorbent materials simply are classified according to their pore spectrum as "narrow" (zeolites, molecular sieves, etc.) or "wide" (activated carbons, silica, etc.). In the first case we mainly expect microporous sorbent materials, whereas in the second case micro-, meso, and macropores as well can be included in the sorbent.[*)] The adsorptive molecules are classified by their polarity as either non-polar like (He, Ar, N_2, O_2, CH_4) or polar like (CO, NO,

[*)] We do not consider so-called "periodic mesoporous materials", for example certain organosilicas, which do have a very narrow mesopore spectrum and presently are subject to intensive investigations.

H_2O, SO_2, $CCl_iF_jH_{4-i-j}$, etc.). For sake of simplicity, no structure or size arguments are taken into account.

For all isotherms mentioned in Table 7.3, extensions to multicomponent systems exist and have been mentioned in Sects. 2-4. However, it must be emphasized that in mixture adsorption it is often very difficult to predict or calculate the amounts of the lesser (or weakly) adsorbed components. Hence reliable measurements of coadsorption equilibria are recommended again.

Generally speaking for low gas pressures and high temperatures conditions ($p \rightarrow 0$, $T \rightarrow \infty$) physisorption of gas molecules decreases and isotherms are normally of either linear, i. e. Henry-type or of Freundlich-type. The linear region of the isotherm strongly depends on the nature of the sorptive gas. For H_2 at 298 K (supercritical) AIs on ACs are often linear up to 8 MPa, whereas for CO_2 at 298 K (subcritical) deviations from the linear region at about 0.5 MPa can be observed. For high pressures and high temperatures ($p \rightarrow \infty$, $T \rightarrow \infty$) physisorption of gas molecules increases and isotherms of Langmuir – type can be expected, cp. Sect. 2.1.1. For high pressures and low temperatures ($p \rightarrow \infty$, $T \rightarrow 0$) physisorption normally will lead to pore condensation and isotherms of BET- or Freundlich-type should be expected.

Table 7.3. Adsorption isotherms recommended to describe gas adsorption equilibria in sorbent materials with
a) narrow pore spectra:
Zeolites, molecular sieves, carbon molecular sieves etc.
b) wide pore spectra:
Activated carbons, silica, activated alumina etc.

		Sorbent Material Pore Spectrum	
	s \ f	Narrow	Wide
Sorptive Gases	Nonpolar He, Ar, N_2 O_2, CH_4 CO2	Langmuir Virial Expansion Dubinin-Polanyi	BET Integral Equation Ideal Adsorbed Solution Theory
Sorptive Gases	Polar H_2O, H_2S SO_2 VOCs	Langmuir Freundlich et al. Toth	BET Integral Equations Toth
		Zeolite, Molecular Sieve, Carbon Sieves	Activated Carbons, Silica, Act. Alumina

As far as industrial adsorption processes are concerned it always should be taken into account that isotherms which are favorable for adsorption normally are unfavorable for desorption processes. Also, for column performance, for example packed bed dynamics, the velocity of the mass break through front is inverse proportional to the steepness of the adsorption isotherm. Hence it can be decisive to have accurate equilibria data at hand to get reasonably accurate values of the respective differential quotients [7.2, 7.4, 7.40]. For mixture adsorption this argument becomes even more important. S. D. G.

6. LIST OF SYMBOLS

A	m^D	areal extension parameter of a sorbate considered as thermodynamic phase of fractal dimension $(1 \leq D \leq 3)$
$b = (p(T, m = m_\infty/2, m^s))^{-1}$	Pa^{-1}	Langmuir parameter, also reciprocal of half-loading pressure
$b_i = (p_i(T, m_i = m_{i\infty}/2, m^s))^{-1}$	Pa^{-1}	Langmuir parameter of component i = 1…N, also reciprocal of half loading pressure of pure component (i)
$f_i(p, T)$	Pa	fugacity of component i = 1…N in a sorptive gas mixture, cp. Eq. (7.97)
$H = \left(\frac{\partial m}{\partial p}\right)_{0,T}$	gPa^{-1}	Henry's constant of an adsorbate in the low pressure limit $(p \rightarrow 0)$
$H_i = \left(\frac{\partial m_i}{\partial p_i}\right)_{0,T,p_k}$	gPa^{-1}	Henry's constant of component i = 1…N of a coadsorbate in the low pressure limit $(p_i \rightarrow 0)$
H^a	J	enthalpy of mass (m) of adsorbate
H^f	J	enthalpy of mass (m) of sorptive gas
m	kg	mass of an adsorbate on a certain mass (m^s) of sorbent
M	g/mol	molar mass of sorptive gas
m_∞	kg	limiting value of mass adsorbed on sorbent of mass (m^s) for high pressures $(p \rightarrow \infty)$

m_{eq}	kg	equilibrium value of the mass of an adsorbed phase during an ad- or desorption process
m_i	kg	mass of component (i = 1…N) of a coadsorbate on a certain mass (m^s) of sorbent
$m_{i\infty}$	kg	limiting mass of pure component i = 1…N of a gas mixture adsorbed for $p_i \rightarrow \infty$; $p_k = const, k \neq i$
m^s	kg	mass of sorbent material
n	mol	amount of gas adsorbed on a sorbent of mass (m^s)
n_∞	mol	maximum number of adsorption sites of a sorbent of mass (m^s)
n_i	mol	amount of component i = 1…N of a gas mixture adsorbed on sorbent material of mass (m^s)
p	Pa	pressure of sorptive gas
p_0	Pa	reference pressure, often chosen as saturation pressure of sorptive gas at temperature T, i. e. $p_0 = p_s(T)$
p_c	Pa	critical pressure of a sorptive gas
$p_i = y_i p$	Pa	partial pressure of component (i = 1…N) in a sorptive gas mixture
q>0	J/g	specific enthalpy of desorption
q(r, T)	J/g	specific enthalpy of desorption of molecules being adsorbed in pores of radius ($r < r_{pore} < r+dr$)
$\mathbf{R} = 8.314$	J/(molK)	universal gas constant
$R = \mathbf{R}/M$	J/gK	specific gas constant
$S_{ik} = H_i/H_k$	1	separation factor or selectivity of component i relative to component k in a multicomponent adsorbate, cp. Eq. (7.21)
T	K	absolute temperature
T_0	K	temperature of a reference state

T_c	K	critical temperature of a sorptive gas
$V(r, T, m^s)$	m^3	integral distribution function of the pore spectrum of the sorbent mass (m^s), i. e. volume of all pores in (m^s) with radii ($0 < r_{pore} < r$)
$W < W_0$	m^3	portion of volume of micropores filled with sorbate molecules (admolecules)
W_0	m^3	volume of micropores in a sorbent material of mass (m^s)
x_i	1	molar fraction of component i = 1 . . .N in a multicomponent adsorbate
y_i	1	molar fraction of component i = 1 . . .N in a sorptive gas mixture
α_0	1	characteristic exponent of a reference component adsorbate (Ar) considered as thermodynamic phase of fractal dimension (D)
α_i	1	characteristic exponent of a component (i) in a coadsorbate considered as thermodynamic phase of fractal dimension (D)
$\alpha = \alpha_0 \left(\frac{r}{r_0}\right)^{-D}$	1	characteristic exponent of an adsorbed phase whose molecules have the diameter (2r) and are located on a surface of a fractal dimension ($1 \leq D \leq 3$)
π	Jm^{-D}	spreading pressure of an adsorbed phase considered as thermodynamic system of fractal dimension ($1 \leq D \leq 3$)
ρ_c	kg/m^3	critical density of a sorptive gas
$\theta = \frac{m}{m_\infty} = \frac{n}{n_\infty}$	1	fractional loading of an adsorbate with limiting loading (m_∞) or (n_∞) respectively
$\theta_i = \frac{m_i}{m_{i\infty}} = \frac{n_i}{n_{i\infty}}$	1	fractional loading of component i = 1 . . .N of a coadsorbate with (pure component) limiting loading ($m_{i\infty}$) or ($n_{i\infty}$) respectively

REFERENCES

[7.1] **Gregg S. J., Sing K. S. W.**
Adsorption, Surface Area and Porosity,
Academic Press, London etc., 1982.

[7.2] **Ruthven D. M.**
Principles of Adsorption and Adsorption Processes,
J. Wiley & Sons, New York etc., 1984.

[7.3] **Do D. D.**
Adsorption Analysis: Equilibria and Kinetics, Imperial College Press, London, 1998.

[7.4] **Kast W.**
Adsorption aus der Gasphase, Ingenieurwissenschaftliche Grundlagen und technische Verfahren, Verlag Chemie, Wertheim, Germany, 1988.

[7.5] **Rouquerol F., Rouquerol J., Sing K.S.W.**
Adsorption by Powders and Porous Solids,
Academic Press, San Diego, USA, 1999.

[7.6] **Steele W.**
The Interaction of Gases with Solid Surfaces,
Pergamon, New York, 1974.

[7.7] **Adamson A. W., Gast A. P.**
Physical Chemistry of Surfaces,
J. Wiley & Sons, 6th Ed., New York, 1997.

[7.8] **Neimark A. V., Ravikovitch P. I.**
Calibration of Adsorption Theories,
Fundamentals of Adsorption 6, p. 159-164, F. Meunier, Ed., Elsevier, Paris, 1998.

[7.9] **Keller J. U., Zimmermann W., Schein A.**
Determination of Absolute Gas Adsorption Isotherms by Combined Calorimetric and Dielectric Measurements,
Adsorption, 9 (2003), p. 177-188.

[7.10] **Yang R.**
Gas Separation by Adsorption Processes, Imperial College Press, London, 1997.

[7.11] **Mersmann A. B., von Gemmingen U.**
Equilibria in Pressure Swing Adsorption Processes –
Experimental and Theoretical Needs
Fundamental of Adsorption 6, p. 1001-1008, M. D. Le Van, Ed., FoA5 Conference, Asilomar, 1995, Kluwer Academic Publishers Group, Dordrecht, Netherlands, 1996.

[7.12] **Sing K.S.W. et al.**
Reporting Physisorption Data for Gas/Solid Systems with Special Reference to the Determination of Surface Area and Porosity, IUPAC Recommendations 1984, Pure & Appl. Chem., 57 (1985), 603-619.

[7.13] **Rouquerol J., Sing K.S.W. et al.**
Guidelines for the Characterization of Porous Solids, article in: Characterization of Porous Solids III, J. Rouquerol et al. (Eds.), Elsevier, Amsterdam, 1994.

[7.14] **International Standardization Organization (ISO), Geneva Working Group ISO/TC24/SC4,**
"Determination of the specific surface area of solids by gas adsorption using the BET method", ISO 9277: 1995.

[7.15] **Rudzinski W., Panczyk T.**
Phenomenological Kinetics of Real Gas-Adsorption-Systems: Isothermal Adsorption, J. Non.-Equilib. Thermodyn., 27 (2002), p. 149-204.

[7.16] **Langmuir I.**
The Adsorption of Gases on Plane Surfaces of Glass, Mica and Platinium, J. American Chem. Soc., 40 (1918), p. 1361-1403.

[7.17] **Sandler S. I.**
Chemical and Engineering Thermodynamics, J. Wiley & Sons, Inc., New York etc., 3rd Ed., 1999.

[7.18] **Valenzuela D. P., Myers A. L.**
Adsorption Equilibrium Data Handbook, Prentice Hall, New Jersey, 1989.

[7.19] **Kärger J., Ruthven D. M.**
Diffusion in Zeolites and Other Microporous Solids, J. Wiley & Sons, New York, 1992.

[7.20] **Panczyk T., Rudzinski W.**
Phenomenological Kinetics of Real Gas Adsorption Systems: Isothermal Kinetics and Kinetics of Thermodesorption, J. Non-Equilib. Thermodynamics, 28 (2003), 341-397.

[7.21] **Yang R. T.**
Adsorbents, Fundamentals and Applications, Wiley-Interscience, Hoboken, New Jersey, 2003.

[7.22] **Rudzinski W., Everett D. H.**
Adsorption of gases on heterogeneous surfaces, Academic Press, London etc., 1992.

[7.23] **Staudt R.**
„Analytische und experimentelle Untersuchungen von Adsorptionsgleichgewichten von reinen Gasen und Gasgemischen an Aktivkohlen und Zeolithen.“, PHD-Thesis, IFT, University of Siegen, Siegen, 1994.

[7.24] **Dreisbach F.**
„Untersuchung von Adsorptionsgleichgewichten methanhaltiger Gasgemische an Aktivkohle als Grundlage zur Auslegung technischer Adsorptionsanlagen"
PHD-Thesis, IFT University of Siegen, Siegen, 1998, cp. also
Fortschritt-Berichte VDI, Reihe 3, Verfahrenstechnik, No. 547, VDI-Verlag, Düsseldorf, 1998.

[7.25] **He R., Zimmermann W., Keller J. U., Roehl-Kuhn B., Jakob J., Heil V., Kümmel R.**
Characterization of Impregnated Activated Carbon SC-44,
AIChE Annual Meeting 2003, San Francisco, Poster 320 C,
cp. also AIF-Research Project 46 Z, Final Report, "Imprägnierung mikroporöser Substanzen aus der überkritischen Phase", Technische Zentralbibliothek Hannover, Hannover 2003.

[7.26] **Kaneko K., Kanoh H, Hanzawa Y., Eds.**
Fundamentals of Adsorption 7
Proceedings of FoA 7 Conference, Nagasaki, May 2001, IK International, Chiba, Japan 2002.

[7.27] **Fowler R. H., Guggenheim E. A.**
Statistical Thermodynamics,
Cambridge University Press, London, 1939.

[7.28] **Keller J. U., Popernack J. D., Staudt R.**
A Generalization of Langmuir's Adsorption Isotherm to Admolecules with Interaction,
Adsorption Science and Technology, Proceedings of 2nd Pacific Basin Conference on Adsorption, Brisbane, May 2000,
D. D. Do, Ed., p. 336, World Scientific, Singapore, 2000.

[7.29] **Staudt R., Dreisbach F., Keller J. U.**
Generalized Isotherms for Mono- and Multicomponent Adsorption,
Fundamentals of Adsorption, Proceedings of 5th FoA Conference, May 1995, Asilomar, p. 865-872, M. D. Le Van, Ed.,
Kluwer Academic Publishers, Boston, MA, 1996.

[7.30] **Staudt R.**
Gasadsorption, Ingenieurwissenschaftliche Grundlagen, Messmethoden zur Bestimmung und Modelle zur Beschreibung von Adsorptionsgleichgewichten reiner Gase und Gasgemische an porösen Stoffen,
Habilitationsschrift, IFT, University of Siegen, Siegen, 2000.

[7.31] **Keller J. U.**
Equations of State of Adsorbates with Fractal Dimension,
Physica A 166 (1990), p. 180-192.

[7.32] **Stencl J.**
Moisture Sorption Isotherms of Whey Powder Spray in the Temperature Range 10 °C– 40 °C,
Adsorption Science and Technology, Paper Number 128 r, (2002).

[7.33] **Toth J., Editor**
Adsorption, Theory, Modeling, and Analysis, Surfactant Science Series Vol. 107, M. Dekker, New York etc., 2002.

[7.34] **Freundlich H., Boedecker A., Ostwald W.**
Zeitschrift Physikalische Chemie, Leipzig, 57 (1906), p. 385.

[7.35] **Mandelbrot B.**
The Fractal Geometry of Nature,
Freeman, New York, 1983.

[7.36] **Bunde A., Havlin Sh. (Eds.)**
Fractals and Disordered Systems,
Springer, Berlin etc., 1991.

[7.37] **Farin D., Avnir D.**
The fractal nature of molecule-surface activities and physical interactions in porous materials, article in: Characterization of Porous Solids,
p. 421, K. K. Unger et al., Eds., Elsevier, Amsterdam, 1988.

[7.38] **Bottani E., Steele W. A.**
A New Approach to the Theory for Adsorption Isotherms on Heterogeneous Surfaces,
Adsorption, 5 (1999), p. 81-89.

[7.39] **Toth J.**
State equations of the solid-gas interface Layers,
Acta Chim. Acad. Sci. Hung., 69 (1971), p. 311.

[7.40] **Bathen D., Breitbach M.**
Adsorptionstechnik, VDI-Buch, Springer, Berlin, New York etc., 2001.

[7.41] **Wang X., Zimmermann W., Ng K. Ch., Chakraboty A., Keller J. U.**
Investigation of the isotherm of silica gel + water systems by using TG and volumetric/manometric methods,
J. of Thermal Analysis and Calorimetry, in print, 2004.

[7.42] **Brunauer S., Emmet P. H., Teller E.**
J. American Chemical Society, 60 (1938), 309.

[7.43] **Hill T. L.**
Thermodynamics for Chemists and Biologists, Addison-Wesley, Reading, MA, USA, 1968.

[7.44] **Webb P. A., Orr C.**
Analytical Methods in Fine Particle Technology,
Micromeritics Inc., Norcross, GA, 1997.

[7.45] **Sing K.S.W. et al.**
Reporting Physisorption Data for Gas/Solid Systems with Special Reference to the Determination of Surface Area and Porosity, IUPAC Recommendations 1984, Pure & Appl. Chem., 57 (1985), 603-619.

[7.46] **Sing K.S.W. et al. (Eds.)**
IUPAC Recommendations 1994 for Reporting Physisorption Data,
Pure & Appl. Chem., 66 (1994), 1739.

[7.47] **Cheng L. S., Yang R. T.**
Predicting Isotherms in Micropores for Different Molecules and Temperatures from a Known Isotherm by Improved Horvath-Kawazoe Equations, Adsorption 1 (1995), p. 187-196.

[7.48] **Jaroniec M.**
Fifty Years of the Theory of the Volume Filling of Micropores,
Adsorption, 3 (1997), p. 187.

[7.49] **Stoeckli F., Lavanchy A. Hugi-Cleari D.**
Dubinin's Theory, A Versatile Tool in Adsorption Science, Fundamentals of Adsorption 6, p. 75-80,
F. Meunier, E., Elsevier, Paris, 1998.

[7.50] **Hutson N. D., Yang R. T.**
Theoretical Basis for the Dubinin-Radushkevich (D-R) Adsorption Isotherm Equation,
Adsorption 3 (1997), p. 189-195.

[7.51] **Kaneko K., Murata K.**
An Analytical Method of Micropore Filling of a Supercritical Gas,
Adsorption, 3 (1997), p. 197-208.

[7.52] **Kruk M., Jaroniec M., Choma J.**
Critical Discussion of Simple Adsorption Methods Used to Evaluate the Micropore Size Distribution,
Adsorption, 3 (1997), p. 209-219.

[7.53] **Stoeckli F., Couderc G., Sobota R., Lavanchy A.**
The Myers-Prausnitz-Dubinin Theory and Non-ideal Adsorption in Microporous Solids,
Adsorption Science & Technology, 20 (2002), p. 189-198.

[7.54] **Dabrowski A., Robens E., Klobes P., Meyer K., Podkoscielny P.**
Standardization of Methods for Characterizing the Surface Geometry of Solids,
Particle & Particle Systems Characterization, 20 (2003) 5, p. 311-322.

[7.55] **Nieszporek K.**
On the correct use of the Dubinin-Astakhov equation to study the mixed-gas adsorption equilibria,
Adsorption, 8 (2002), p. 45-57.

[7.56] **Keller J. U.**
Thermodynamik der irreversiblen Prozesse mit Aufgaben, Rechenweg und Lösungen,
Teil I: Thermodynamik und Grundbegriffe, Kap. A 14,
W. de Gruyter, Berlin – New York, 1977.

[7.57] **Kestin J.**
A Course in Thermodynamics, Vols. I, II, Blaisdell Publ. Comp., London, 1968.

[7.58] **Marczewski A. W.**
A Practical Guide to Isotherms of Adsorption on Heterogeneous Surfaces, An Overview, electronic document: http://adsorption.org/awm/ads/Ads.htm, 2004.

[7.59] **Neimark A. V., Ravikovich P. I., Vishnyakov A.**
Sorption and Phase Transitions in Nanopores, Fundamental of Adsorption 7, Proceedings of FoA Conference 7, Nagasaki, May 2001, p. 319-326, K. Kaneko et al., Edts., Int. Adsorption Society, IK International, Chiba, Japan, 2002.

[7.60] **Horvath G., Kavazoe K.**
Journal of Chemical Engineering, Japan, 16 (1983), p. 470.

[7.61] **Cheng L. S., Yang R. T.**
Predicting Isotherms in Micropores for Different Molecules and Temperatures From a Known Isotherm by Improved Horvath-Kawazoe Equations, Adsorption, 1 (1995), p. 187.

[7.62] **Basmadjian D.**
The Little Adsorption Book
CRC Press, Boca Raton, 1996.

[7.63] **Madelung E.**
Die mathematischen Hilfsmittel des Physikers, Springer, Berlin etc., 4th Ed., 1957.
Cp. also math programs like MATHCAD, or MATHLAB etc.

[7.64] **Atkins P. W.**
Physical Chemistry, Oxford University Press, 5th Ed., Oxford UK, 1994.

[7.65] **Myers A. L.**
Equation of State for Adsorption of Gases and Their Mixtures in Porous Materials, Adsorption, 9 (2003), p. 9-16.

[7.66] **Zhou L., Bai S. Zhou Y., Yang B.**
Adsorption of nitrogen on silica gel over a large range of temperatures, Adsorption, 8 (2002), p. 79-87.

[7.67] **Carsky M., Do D. D.**
Neural network modelling of adsorption of binary vapour mixtures, Adsorption, 5 (1999), p. 183-192.

[7.68] **Gibbs J. W.**
On the Equilibrium of Heterogeneous Substances, American Journal of Sciences and Arts, 16 (1878), p. 441-458.

[7.69] **Young D. M., Crowell A. D.**
Physical Adsorption of Gases, London, Butterworth, 1962.

[7.70] **Myers A. L., Prausnitz J. M.**
Thermodynamics of Mixed Gas Adsorption
AIChE Journal, 11 (1965), p. 121-127.

[7.71] **van Ness H. C.**
Adsorption of Gases on Solids,
Industrial & Engineering Chemistry Fundamentals, 8 (1969), p. 464-473.

[7.72] **Sircar S.**
Excess Properties and Thermodynamics of Multicomponent Gas Adsorption,
J. Chem. Soc., Faraday Trans. 81 (1985), p. 1527-1540.

[7.73] **Talu O., Myers A. L.**
Rigorous Thermodynamic Treatment of Gas Adsorption, AIChE Journal,
34 (1988), p. 1887-1893.

[7.74] **Gibbs J. W.**
The Collected Work of J. Willard Gibbs, Vol. 1, Thermodynamics,
Yale University Press, New Haven, 1948.

[7.75] **Markham E. C., Benton A. F.**
The Adsorption of Gas Mixtures by Silica,
J. Am. Chem. Soc., 53 (1931), p. 497-507.

[7.76] **Keller J. U.**
Thermodynamics of Non-Isothermal Coadsorption Processes,
Ber. Bunsenges. Phys, Chem., 91 (1987), p. 528-536.

[7.77] **Keller J. U.**
Interrelations Between Thermodynamic Equations of State of Single- and Multi-Component Adsorbates,
Ber. Bunsenges. Phys. Chem. 92 (1988), p. 1510-1516.

[7.78] **Schottky W., Ulrich H., Wagner C.**
Thermodynamik (in German),
Reprint of the 1929 Edition, Springer, Berlin (West), 1973.

[7.79] **Keller J. U., Zimmermann W., Schein E.**
Determination of Absolute Gas Adsorption Isotherms by Combined Calorimetric and Dielectric Measurements, Adsorption, 9 (2003) p. 177-188.

[7.80] **Herbst A.**
Exzessadsorption reiner Gase im Druckbereich bis 50 MPa, PH-D Thesis, University of Leipzig, 2003, Fortschrittberichte VDI, Reihe 3, Verfahrenstechnik, in preparation, VDI-Verlag Düsseldorf, 2003.

[7.81] **Sengers J. V., Kayser R. F., Peters C. J., White Jr. H. F., Eds.**
Equations of State for Fluids and Fluid Mixtures, Part I
IUPAC Series in Experimental Thermodynamics, Vol. V,
Elsevier, New York etc., 2000.

This page intentionally left blank

Subject Index

Author Index

MIX
Papier aus verantwortungsvollen Quellen
Paper from responsible sources
FSC® C105338

If you have any concerns about our products,
you can contact us on
ProductSafety@springernature.com

In case Publisher is established outside the EU,
the EU authorized representative is:
Springer Nature Customer Service Center GmbH
Europaplatz 3, 69115 Heidelberg, Germany

Printed by Libri Plureos GmbH
in Hamburg, Germany